Lecture Notes in Artificial Intelligence   5834

Edited by R. Goebel, J. Siekmann, and W. Wahlster

Subseries of Lecture Notes in Computer Science

T0223387

Xiangdong He   John Horty   Eric Pacuit (Eds.)

Logic, Rationality, and Interaction

Second International Workshop, LORI 2009
Chongqing, China, October 8-11, 2009
Proceedings

 Springer

Series Editors

Randy Goebel, University of Alberta, Edmonton, Canada
Jörg Siekmann, University of Saarland, Saarbrücken, Germany
Wolfgang Wahlster, DFKI and University of Saarland, Saarbrücken, Germany

Volume Editors

Xiangdong He
Southwest University, Institute of Logic and Intelligence
2, Tiansheng Road, 400715 Chongqing, China
E-mail: hexd@swu.edu.cn

John Horty
University of Maryland
Institute for Advanced Computer Studies, Philosophy Department
College Park, MD 20742, USA
E-mail: horty@umiacs.umd.edu

Eric Pacuit
Tilburg University, Center for Logic and Philosophy of Science
Dante Building, 5000 LE Tilburg, The Netherlands
E-mail: E.J.Pacuit@uvt.nl

Library of Congress Control Number: 2009935707

CR Subject Classification (1998): F.4, G.2, I.2.6, K.3

LNCS Sublibrary: SL 7 – Artificial Intelligence

ISSN	0302-9743
ISBN-10	3-642-04892-7 Springer Berlin Heidelberg New York
ISBN-13	978-3-642-04892-0 Springer Berlin Heidelberg New York

springer.com

© Springer-Verlag Berlin Heidelberg 2009
Printed in Germany

Typesetting: Camera-ready by author, data conversion by Scientific Publishing Services, Chennai, India
Printed on acid-free paper   SPIN: 12773466   06/3180   5 4 3 2 1 0

Preface

The First International Workshop on Logic, Rationality and Interaction (LORI-I) took place in Beijing in August 2007, with participation by researchers from the fields of artificial intelligence, game theory, linguistics, logic, philosophy, and cognitive science. The workshop led to great advances in mutual understanding, both academically and culturally, between Chinese and foreign logicians.

Due to the success of LORI-I, it was decided that the series should be continued, at various places in China and possibly other Asian countries, under the overall guidance of a LORI Standing Committee consisting of Johan van Benthem, Shier Ju, Frank Veltman, and Jialong Zhang.

This volume contains the proceedings of the next installment in the series, The Second International Workshop on Logic, Rationality and Interaction (LORI-II), which took place in Chongqing in October 2009. From a flood of submissions, we have, with great difficulty, selected 24 papers for presentation at the conference, and eight more for presentation during a poster session. These contributed presentations were supplemented with invited addresses by Hans van Ditmarsch, Fangzhen Lin, Rohit Parikh, Henry Prakken, Jeremy Seligman, Leon van der Torre, and Ming Xu.

For assistance in constructing the program, we owe a great debt to Johan van Benthem, the Conference Chair, and to all the members of the Program Committee for their hard work under the pressure of a very tight reviewing deadline. We especially wish to thank Fenrong Liu for her many-sided help, not only with the program, but in every possible way.

August 2009

John Horty
Eric Pacuit

Organization

LORI-II was organized at the Institute of Logic and Intelligence, Southwest University in Chongqing, China during October 8 - 11, 2009.

Executive Committee

Conference Chair	Johan van Benthem (Amsterdam and Stanford)
Program Chairs	John Horty (Univeristy of Maryland)
	Eric Pacuit (Univeristy of Tilburg)
Organizing Chair	Xiangdong He (Southwest University)
Local Organizing Committee	Xiaojia Tang (Southwest University)
	Huiwen Deng (Southwest University)
	Ziqiang Peng (Southwest University)
	Jing Wang (Southwest University)
	Meiyun Guo (Southwest University)

Program Committee

Horacio Arlo-Costa	Carnigie Mellon University
Alexandru Baltag	Oxford University
James Delgrande	Simon Fraser University
Hans van Ditmarsch	University of Sevilla and University of Otago
Phan Minh Dung	Asian Institute of Technology
Wiebe van der Hoek	Liverpool University
Shier Ju	Sun Yat-sen University
Barteld Kooi	University of Groningen
Fenrong Liu	Tsinghua University
Larry Moss	Indiana University
Gabriella Pigozzi	University of Luxembourg
Rohit Parikh	City University of New York
Henry Prakken	Utrecht University
Ram Ramnujam	Institute of Mathematical Sciences, Chennai
Hans Rott	University of Regensburg
Olivier Roy	University of Groningen
Sonja Smets	Free University of Brussels
Katie Steele	London School of Economics
Leon van der Torre	University of Luxembourg
Frank Veltman	University of Amsterdam
Minghui Xiong	Sun Yat-sen University

Ming Xu   Wuhan University
Tomoyuki Yamada   Hokkaido University
Audrey Yap   University of Victoria
Jialong Zhang   Yanshan University and Chinese Academy of
                           Social Sciences
Beihai Zhou   Peking University

Table of Contents

Contributed Papers

Posters

Expressing Properties of Coalitional Ability under Resource Bounds*

Natasha Alechina, Brian Logan, Nguyen Hoang Nga, and Abdur Rakib

School of Computer Science
University of Nottingham
Nottingham, UK
{nza,bsl,hnn,rza}@cs.nott.ac.uk

Abstract. We introduce Coalition Logic for Resource Games (CLRG) which extends Coalition Logic by allowing explicit reasoning about resource endowments of coalitions of agents and resource bounds on strategies. We show how to express interesting properties of coalitional ability under resource bounds in this logic, including properties of Coalitional Resource Games introduced by Wooldridge and Dunne in [1]. We also give an efficient model-checking algorithm for CLRG which makes it possible to verify the properties automatically.

1 Introduction

There are many problems in multi-agent systems which are not reducible to the abilities of individual agents in the system, and which can only be usefully analysed in terms of the combined abilities of groups of agents. For example, it may be that no single agent has a strategy to reach a particular state on its own, but two agents cooperating with each other are capable of achieving this outcome. It is often natural to formulate reasoning about the abilities of coalitions of agents in terms of games, and there is a considerable amount of work on coalitional games and logics in the literature, e.g., [2,3,4,5,6,7,8,9]. For example, *Coalition Logic* can be used to reason about the coalitional ability of agents in strategic games. Coalition Logic generalises the notion of a strategic game, in that it is interpreted over state transition systems where each state has an associated strategic game. It can be seen either as a way of reasoning about a sequence of strategic games or as reasoning about a single complex game where each move corresponds to a transition to a new state [2]. *Alternating-Time Temporal Logic* (ATL) was originally developed to reason about computational processes in adversarial environments, and has been shown to generalise Coalition Logic [4]. Instead of talking about the outcome of a strategic game in the next state, ATL can express properties holding in arbitrary future states, or maintained during a series of moves. Using these logics we can express properties such as 'coalition C has a strategy to bring about a state satisfying ϕ' (no matter what the other

---

* An earlier version of this work was presented at the workshop on Logics for Agents and Mobility 2009.

X. He, J. Horty, and E. Pacuit (Eds.): LORI 2009, LNAI 5834, pp. 1–14, 2009.

agents in the system do) where ϕ characterises, e.g., the solution to a problem or the successful execution of a protocol.

However, none of the existing coalition logics can express properties of coalitional abilities under resource restrictions, that is, directly express properties such as 'coalition C has a strategy to bring about ϕ under resource bound b_1, but not under a tighter resource bound b_2'. Examples of situations where such properties are critical are numerous; e.g., whether a team of agents can achieve a task under a given allocation of resources [1], or whether a group of distributed reasoning agents can answer a query given specified memory, communication and time resources [10].

In this paper we present *Coalition Logic for (Strategic) Resource Games* (CLRG). CLRG allows us to express properties such as 'coalition C can enforce a state satisfying ϕ under a resource bound b', that is, the cost of their joint action is at most b (where b is a tuple of resource value pairs $\langle r_1{:}b_1, \ldots, r_j{:}b_j \rangle$ from a fixed set of resources R). We extend strategic games to *strategic resource games* by adding resources, endowments and costs, and define Coalition Logic for (Strategic) Resource Games in much the same way as Coalition Logic is defined with respect to strategic games. As illustrations of how CLRG can be used to express the abilities of coalitions under resource bounds, we show how to express some of the decision problems for Coalitional Resource Games (CRGs) given in [1], and how to express properties of agents in a simple multi-shot game. Finally, we show how to automatically verify properties expressed in CLRG, in particular properties of CRGs, using a standard model-checker, and give a more efficient model-checking algorithm specifically for CLRG formulas.

2   Coalition Logic

In this section we briefly describe Coalition Logic (CL) introduced by Pauly in [3] and state the main results on the complexity of model-checking CL.

Coalition Logic is used for reasoning about coalitional ability in strategic games. The language of coalition logic contains modalities, $[C]$, for each possible set of agents (coalition) C. The meaning of $[C]\phi$ is that the coalition C can choose a tuple of actions so that whatever actions are chosen by the other agents in the system, the outcome state satisfies ϕ (C can enforce an outcome state satisfying ϕ).

Definition 1. *A* strategic game form *is a tuple* $(A, \{Act_i \mid i \in A\}, S, o)$ *where*

- $A = \{1, \ldots, n\}$ *is a set of agents,*
- $\{Act_i \mid i \in A\}$ *is a set of sets of actions (or strategies) for each agent* $i \in A$,
- S *is a non-empty set of states,*
- $o : \Pi_{i \in A} Act_i \to S$ *is an outcome function which associates with every tuple of actions, by all agents in parallel, an outcome state in* S.

The set of all game forms for a set of players A and a set of states S will be denoted $\Gamma(A, S)$. For a set of agents $C \subseteq A$, we will denote a tuple of actions by the agents in C by a_C, and a tuple of actions by all agents where agents in C execute actions a_C and agents $\bar{C} = A \setminus C$ execute actions $a_{\bar{C}}$ as $(a_C, a_{A \setminus C})$.

Definition 2. *A game model for a set of players A over a set of propositions Prop is a triple (S, γ, V), where*

- *S is a non-empty set of states,*
- *$\gamma : S \to \Gamma(A, S)$ is a mapping associating a strategic game form with each state in S (we will use an extra argument s to distinguish components of $\gamma(s)$, for example $o(s, a_1, \ldots, a_n)$ for the outcome function in state s),*
- *$V : S \to 2^{Prop}$ is a valuation function which labels every state in S with a set of propositions that are true in that state.*

The language L of CL (parameterised by a set of propositional variables *Prop* and a set of agents A) is as follows:

$$p \mid \neg \phi \mid \phi_1 \wedge \phi_2 \mid [C]\phi$$

where $p \in Prop$ and $C \subseteq A$.

Formulas of L are evaluated with respect to game models as follows:

- $M, s \models p$ iff $p \in V(s)$
- $M, s \models \neg\phi$ iff $M, s \not\models \phi$
- $M, s \models \phi \wedge \psi$ iff $M, s \models \phi$ and $M, s \models \psi$
- $M, s \models [C]\phi$ iff there exists a_C such that for every $a_{A\setminus C}$, $M, o(s, a_C, a_{A\setminus C})$ $\models \phi$.

The model-checking problem for CL (given a formula ϕ and a model M, return the set $[\phi]_M$ of states satisfying ϕ) can be reduced to the model-checking problem for ATL using the embedding into ATL given in [4], and therefore can be done in time $O(|\phi| \times m)$ where m is the number of transitions in the model.

Coalition logic allows us to express interesting properties both of strategic games and of multi-shot games, using nested $[C]$ operators. For example, we can express that a group of agents C has a three-step winning strategy in a game by saying $[C][C][C]$ win. However CL does not allow us to express properties of coalitional abilities under resource restrictions.

3   Strategic Resource Game Forms

In this section we extend strategic games to strategic resource games (or rather game forms, since we do not have a notion of preference over outcomes) by adding resources to strategic games. Strategic resource game forms will be used to define models of coalition logic for resource games in the next section.

Definition 3. *A strategic resource game form is a tuple $(A, \{Act_i \mid i \in A\}, S, o, R, en, c)$ where*

- *$A = \{1, \ldots, n\}$ is a set of agents,*
- *$\{Act_i \mid i \in A\}$ is a set of sets of actions for each agent $i \in A$,*
- *S is a non-empty set of states,*
- *$o : \Pi_{i \in A} Act_i \to S$ is an outcome function,*

- $R = \{r_1, \ldots, r_t\}$ *is a set of resources,*
- $en : A \times R \to \mathbb{N}$ *is a resource endowment function,*
- $c : \cup_{i \in A} Act_i \times R \to \mathbb{N}$ *is an action cost function.*

We extend strategic games with resources, endowments and costs. Each action an agent can perform requires the consumption of zero or more units of each of a set of resources R (the cost of the action), and each agent is given an endowment of each resource $en(i, r)$ (which may be zero). An action can only be executed by an agent if the endowment of the agent is greater than or equal to the cost of the action: $a \in Act_i$ iff $c(a, r) \le en(i, r)$ for every resource $r \in R$. The set of all resource game forms for a set of players A, a set of resources R and a set of states S will be denoted $\Gamma(A, S, R)$.

We will use $en(C, r)$ for the sum of $en(i, r)$ for all $i \in C$ and $c(a_C, r)$ for the sum of costs of all actions in a_C: $\Sigma_{i \in C} c(a_i, r)$. It is sometimes convenient to talk about resource endowment of an agent as a single value, namely a vector $en(i) = \langle r_1 : n_1, \ldots, r_t : n_t \rangle$, where $en(i, r_j) = n_j$. The sum of vectors is defined in the usual way (pointwise), so $en(C) = \langle r_1 : n_1, \ldots, r_t : n_t \rangle$ where $n_j = \Sigma_{i \in C} en(i, r_j)$. Similarly, we can talk about a cost of an action $c(a_i)$ as a vector of values, one for each resource, and a cost of a joint action by a coalition C, $c(a_C)$, as a vector corresponding to a pointwise sum of vectors $c(a_i)$ where $i \in C$.

Similarly, it is sometimes convenient to talk about tuples of resource values b of the form $\langle r_1 : n_1, \ldots, r_j : n_j \rangle$ (where the indices r_j do not necessarily cover all the values in R). For two tuples of resource values over the same resource indices (referring to the same resources) $b = \langle r_1 : b_1, \ldots, r_j : b_j \rangle$ and $v = \langle r_1 : v_1, \ldots, r_j : v_j \rangle$ we will also use the usual pointwise comparisons and operations, for example $b \le v$ iff $b_1 \le v_1, \ldots, b_j \le v_j$ and $b + v = \langle r_1 : b_1 + v_1, \ldots, r_j : b_j + v_j \rangle$.

4   Resource Game Models

We can now define models of the logic corresponding to strategic resource game forms:

Definition 4. A resource game model *for a set of players A and resources R over a set of propositions Prop is a triple (S, γ, V), where*

- *S is a non-empty set of states,*
- *$\gamma : S \to \Gamma(A, S, R)$ is a mapping associating a strategic resource game form with each state in S,*
- *$V : S \to 2^{Prop}$ is a valuation function which labels every state in S with a set of propositions that are true in that state.*

Endowments in reachable states reflect the resources which were required to obtain that state: if $s' = o(s, a_1, \ldots, a_i)$, then for every resource r, $en(s', i, r) = en(s, i, r) - c(a_i, r)$. Actions are only executable by an agent in a state if the agent's endowment in that state is greater than or equal to the cost of the action. We assume one of the actions is *noop* such that for all resources r,

$c(noop, r) = 0$ (this simply allows other agents to execute actions when some agents have run out of resources). A *noop* action by all agents does not change the state: $o(s, noop, \ldots, noop) = s$. We assume that all other actions cost some non-zero amount for at least one resource.

5   Coalition Logic for Resource Games

In this section we extend Coalition Logic to a logic for reasoning about resource games. We want to be able to express and verify properties such as 'coalition C can enforce ϕ under a resource bound b', that is, the cost of their joint action is less than b (where b is a tuple of resource value pairs $\langle r_1 : b_1, \ldots, r_j : b_j \rangle$, from a fixed set of resources R). While endowments constrain the abilities of individual agents to perform actions, resource bounds constrain the abilities of a coalition of agents to perform a joint action. With this in mind, we introduce quantifiers $[C^b]\phi$ where b is a tuple of resource values. It is also useful to be able to refer to the endowments of agents in each state; for this purpose we introduce nullary modalities (endowment counters) $e^=(C, b)$ where C is a coalition and b is a tuple of resource values, which means that the coalitions's endowment for the given resources in a given state is equal to b.

The language $L^{b,e}$ of CLRG (parameterised by a set of propositional variables *Prop*, a set of agents A and a set of resources R) is as follows:

$$p \mid \neg\phi \mid \phi_1 \wedge \phi_2 \mid [C]\phi \mid [C^b]\phi \mid e^=(C, b)$$

where $p \in Prop$, $C \subseteq A$ and b is a tuple of resource values. Note that $e^\leq(C, b)$ can be defined as $\bigvee_{b' \leq b} e^=(C, b')$ and $e^\geq(C, b)$ as $\neg e^\leq(C, b) \vee e^=(C, b)$. We will write $e^=(i, b)$ for $e^=(\{i\}, b)$.

We will refer to the language obtained from $L^{b,e}$ by omitting modalities $[C^b]$ by L^e, the language obtained from $L^{b,e}$ by omitting endowment counters $e^=(C, b)$ as L^b, and the language of coalition logic as L.

Formulas of $L^{b,e}$ are evaluated with respect to resource game models as follows:

- $M, s \models p$ iff $p \in V(s)$
- $M, s \models \neg\phi$ iff $M, s \not\models \phi$
- $M, s \models \phi \wedge \psi$ iff $M, s \models \phi$ and $M, s \models \psi$
- $M, s \models [C]\phi$ iff there exists a_C such that for every $a_{A\backslash C}$, $M, o(s, a_C, a_{A\backslash C}) \models \phi$.
- $M, s \models [C^b]\phi$ iff there exists a_C with $c(a_C) \leq b$ such that for every $a_{\bar{C}}$, the outcome of the resulting tuple of actions executed in s satisfies ϕ: $M, o (s, a_C, a_{A\backslash C}) \models \phi$
- $M, s \models e^=(C, b)$ iff $en(s, C) = b$.

6   Expressing Properties in CLRG

In this section we show how to express some properties of coalitional ability under resource restrictions in the language of CLRG.

6.1   Properties of CRGs

We first show how to express properties of Coalitional Resource Games (CRGs) introduced by Wooldridge and Dunne in [1]. A coalitional resource game Γ is defined as a tuple $(A, G, R, G_1, \ldots, G_n, en, req)$ where

- $A = \{1, \ldots, n\}$ is a set of agents,
- $G = \{g_1, \ldots, g_m\}$ is a set of goals,
- $R = \{r_1, \ldots, r_t\}$ is a set of resources,
- $G_i \subseteq G$ is the set of goals for agent i,
- $en : A \times R \to \mathbb{N}$ is the resource endowment function (how many units of a given resource is allocated to an agent),
- $req : G \times R \to \mathbb{N}$ is the resource requirement function (how many units of a particular resource is required to achieve a goal). It is assumed that each goal requires a non-zero amount for at least one resource.

In CRGs, the endowment of a coalition is equal to the sum of the endowments of its members: $en(C, r) = \Sigma_{i \in C} en(i, r)$. Similarly, the resource requirement for a set of goals is the sum of the requirements for each of the goals in the set: $req(X, r) = \Sigma_{g \in X} req(g, r)$. A set of goals satisfies a coalition if the intersection with each of the members' goal sets is non empty: X satisfies C if for every $i \in C$, $X \cap G_i \neq \emptyset$. The set of such sets for a coalition C is denoted by $sat(C)$. A set of goals is feasible for a coalition if the coalition has sufficient resources to achieve it: X is feasible for C if for every resource r, $req(X, r) \leq en(C, r)$. The set containing all feasible sets of goals for C is denoted by $feas(C)$. The effectivity function sf returns, for each coalition, the set of sets of goals which both satisfy a coalition and are feasible for it, namely $sat(C) \cap feas(C)$.

Coalitional resource games are defined in terms of goals whereas CLRG is defined in terms of actions. However we show that we can encode a CRG without loss of information in the initial state of a CLRG model, i.e., in a strategic resource game form.

For each CRG Γ, we define a corresponding CLRG model M_Γ. Let $Prop = G \cup \{sat_i \mid i \in A\}$. Intuitively, $g \in G$ holds in a state if this goal has been achieved in this state and sat_i holds in a state if one of agent i's goals has been achieved in this state. Given a CRG $\Gamma = (A, G, R, G_1, \ldots, G_n, en, req)$, we define $M_\Gamma = (S, \gamma, V)$ as follows:

- $S = 2^G$. Intuitively, the initial state is $s_0 = \emptyset$, where no goals have been achieved.
- $V : S \to 2^{Prop}$ assigns to a state $s = Q \subseteq G$ the set of goals which 'hold' in that state, namely $g \in V(s)$ iff $g \in s$; $sat_i \in V(s)$ if for some $g \in G_i$, $g \in V(s)$.
- for s_0, $\gamma(s_0)$ is as follows:
  - $Act_i(s_0)$: the actions of each agent are vectors of the form $\langle g_1 : x_1, \ldots, g_m : x_m \rangle$ where $g_j : x_j$ means that the agent contributes a vector of resources x_j to the goal g_j. $\langle g_1 : x_1, \ldots, g_m : x_m \rangle \in Act_i(s_0)$ iff $\Sigma_j x_j \leq en(i)$ and $x_j \leq req(g_j)$. In other words, the agents don't contribute more than their endowment or more than the requirement for the goal.

- $o(s_0, \langle g_1 : x_1^1, \ldots, g_m : x_m^1 \rangle, \ldots, \langle g_1 : x_1^n, \ldots, g_m : x_m^n \rangle) = s$ where $g_j \in s$ iff $\Sigma_i x_j^i \geq req(g_j)$, that is, g_j is achieved in s iff the agents together contributed sufficient resources to achieve g_j, and otherwise g_j is false in s.
- $en(s_0, i) = en(i)$.
- $c(\langle g_1 : x_1, \ldots, g_m : x_m \rangle) = \Sigma_j x_j$, that is the cost of an action is the (vector) sum of resources committed to all the goals in this action.
- for $s \neq s_0$, $Act_i(s) = \{noop\}$ for all i.

Proposition 1. *Given M_Γ as defined above, there is a unique (up to isomorphism) CRG Γ corresponding to M_Γ.*

To show that we can recover Γ from M_Γ, observe that s_0 is the state with the maximal endowment. The set of goals Γ is given, but we can also reconstruct it from the fact that $S = 2^G$. Once we know s_0, the outcome function in s_0, and Γ, we can compute the requirements for each goal: the minimal cost of a joint action which achieves (exactly) that goal. Finally, sat_i allows us to compute G_i for each i.

We now show how to express some CRG decision problems introduced in [1] in CLRG. Below for a set $G' \subseteq G$, we define $G'^\vee$ and $G'^\wedge$ as the disjunction and conjunction respectively of goals in G'.

Successful coalition (Γ, C): in a CRG Γ, C is a successful coalition. An agent is successful if it achieves at least one of its goals. A coalition is successful if each agent in the coalition is successful. A coalition C is successful in a CRG Γ iff the following formula is true in M_Γ, s_0:

$$[C] \bigwedge_{i \in C} G_i^\vee$$

Maximal coalition (Γ, C): in a CRG Γ, C is a maximal coalition if any larger coalition is not successful. A coalition C is maximal in a CRG Γ iff the following formula is true in M_Γ, s_0:

$$\bigwedge_{C':C \subseteq C'} \neg [C'] \bigwedge_{i \in C'} G_i^\vee$$

Maximally successful coalition (Γ, C): in a CRG Γ, C is a maximally successful coalition if it is maximal and successful. A coalition C is maximally successful in Γ iff the conjunction of the two previous properties holds in M_Γ.

Coalition successful under resource bound b (Γ, C, b): in a CRG Γ, C is a coalition successful under resource bound b if it can be successful while staying within a resource bound b. A coalition C is successful under resource bound b in Γ iff the following formula is true in M_Γ, s_0:

$$[C^b] \bigwedge_{i \in C} G_i^\vee$$

Necessary resource (Γ, C, r): in a CRG Γ, for a coalition C, r is a necessary resource if for every $G' \in sf(C)$, $req(G', r) > 0$. For readability assume that $C = \{1, \ldots, k\}$. This property holds iff the following formula is true in M_Γ, s_0:

$$\bigwedge_{G' \in sat(C)} ([C]G' \rightarrow \neg[C^{\langle r:0 \rangle}]G')$$

where $sat(C) = \{Q_1^\wedge \wedge \ldots \wedge Q_k^\wedge \mid i \in C, \; Q_i \subseteq G_i\}$ (intuitively, a set of formulas saying that at least one of the goals of each agent in C is satisfied).

Strictly necessary resource (Γ, C, r): r is a strictly necessary resource for C in Γ if $sf(C) \neq \emptyset$ and for every $G' \in sf(C)$, $req(G', r) > 0$. This can be expressed as a conjunction of the successful coalition formula and the necessary resource formula.

G' is optimal for C (Γ, C, G', r): a set of goals G' is optimal for C in Γ if $G' \in sf(C)$ and for every $G'' \in sf(C)$, $req(G'', r) \geq req(G', r)$. This property holds iff the following formula is true in M_Γ, s_0:

$$[C]G'^\wedge \wedge \bigwedge_{G'' \in sat(C)} ([C^b]G'' \rightarrow [C^b]G'^\wedge)$$

where $sat(C)$ is as above and $G'^\wedge \in sat(C)$.

G' is Pareto-efficient for C (Γ, C, G'): G' is Pareto-efficient for C in Γ if for every $G'' \in sf(C)$, if for some r_1, $req(G'', r_1) < req(G', r_1)$, then there exists r_2 such that $req(G'', r_2) > req(G', r_2)$. This property holds iff the following formula is true in M_Γ, s_0:

$$\bigwedge_{G'' \in sat(C)} ([C^{r_1:v}]G'' \wedge \neg[C^{r_1:v}]G'^\wedge \rightarrow \bigvee_{r \neq r_1} \bigvee_{v \leq en(C,r)} ([C^{r:v}]G'^\wedge \wedge \neg[C^{r:v}]G''))$$

where $sat(C)$ is as above.

Conflicting coalitions (Γ, C_1, C_2, b): C_1 and C_2 are conflicting coalitions if for every $G_1 \in sf(C_1)$, $G_2 \in sf(C_2)$, G_1 and G_2 are achievable under the resource bound b, but $G_1 \cup G_2$ is not achievable under b. This property holds iff the following formula is true in M_Γ, s_0:

$$\bigwedge_{G_1 \in sat(C_1), \; G_2 \in sat(C_2)} ([C_1]G_1 \wedge [C_2]G_2 \rightarrow ([A^b]G_1 \wedge [A^b]G_2 \wedge \neg[A^b](G_1 \wedge G_2)))$$

Positive goal set(Γ, G'): G' is a positive goal set in Γ if there exists a coalition C such that $G' \in sf(C)$. This holds iff the following formula is true in M_Γ, s_0:

$$\bigvee_{C \subseteq A: \; G'^\wedge \in sat(C)} [C]G'^\wedge$$

6.2   Properties of Multi-shot Games

In addition to expressing properties of CRGs, CLRG can express properties of multi-shot resource games. Consider a simple example. Three agents 1, 2 and 3 are playing two consecutive games in which there is only one resource which is money m. In the first game, they decide whether to get into town (which costs 1 dollar) or stay at home (*noop* action). Once an agent is in town, it can decide to eat in a restaurant X, where dinner costs 25 dollars, or in a restaurant Y, where dinner costs 50 dollars (or skip dinner). The goal of the agents is that two or more agents have dinner together. Assume that their endowments are above 26 dollars. Then although no single agent can enforce the goal, each two-agent coalition can, and so can the grand coalition: $[1,2][1,2]$ 12dinner and $[1,2,3][1,2,3]$ 123dinner (where 12dinner stands for agents 1 and 2 having a joint dinner, similarly for 123dinner). However if the dinner costs at most 50 dollars (for all participants), the agents can claim the money as project meeting expenses. This can be expressed as achieving a joint dinner under the resource bound of 50 in the second game. Now $[1,2][1, 2^{\langle m:50\rangle}]$ 12dinner but not $[1,2,3][1,2,3^{\langle m:50\rangle}]$ 123dinner.

7   Automated Verification of CLRG Formulas

While CLRG allows us to express properties of resource games, there are no existing automated verification tools which accept $L^{b,e}$ formulas as specifications. However, in this section we show that it is possible to use a standard Alternating-Time Temporal Logic (ATL) model-checker (such as MOCHA [11]) to automatically verify a property ψ stated in the language $L^{b,e}$ of CLRG by translating it into a formula of L^e which is equivalent to ψ with respect to a fixed model.

Note that under our assumption that each action apart from *noop* has a non-zero cost, each model has a state with the maximal endowment value for each agent (intuitively, the initial state of the system), and all other states have a lower endowment value on at least one resource for one of the agents. This has several implications. First of all, it implies that each model has a finite number of non-identical states. Second, if we know the endowment in the initial state and the cost of each action, we can calculate the set of all possible endowments for each agent for each resource. Let us denote the set of resource endowment tuples (one value for each resource) which agent i can have in a model M by $Q(M,i)$. As a trivial example, if the set of resources $R = \{r_1, r_2\}$, the highest endowment of agent i is $\langle r_1 : 4, r_2 : 3\rangle$ and in addition to *noop* agent i can execute a single action a with cost $\langle r_1 : 2, r_2 : 2\rangle$, then the possible endowments of i are $\langle r_1 : 4, r_2 : 3\rangle$ and $\langle r_1 : 2, r_2 : 1\rangle$.

Given a finite set $Q(M,i)$ for each i, we can eliminate $[C^b]$ modalities from any formula of CLRG:

Theorem 1. *Given a model M such that $Q(M,i)$ is finite for every i, and a formula ϕ in the language $L^{b,e}$, there exists a formula ϕ' in the language L^e such that for every state s in M, $M, s \models \phi$ iff $M, s \models \phi'$.*

Proof. We define a translation function t which takes a formula $\phi \in L^{b,e}$ and returns $t(\phi) \in L^e$:

- $t(p) = p$, $t(e^=(C,r)) = e^=(C,r)$,
- t commutes with the booleans and $[C]$,
- $t([C^b]\psi) = \bigwedge_{v_1,\ldots,v_k \in Q(M,i)} (e^=(1,v_1) \wedge \ldots \wedge e^=(k,v_k) \to [C](t(\psi) \wedge \bigvee_{b_1+\cdots+b_k \leq b} e^=(1,v_1-b_1) \wedge \cdots \wedge e^=(k,v_k-b_k)))$ where, for readability, we assume that $C = \{1,\ldots,k\}$

It is straightforward to translate formulas of L^e into ATL. The endowment modalities can be encoded as propositional variables (which are true in a state s if and only if the value of the resource counter in the state satisfies the corresponding condition), and formulas of the form $[C]\phi$ are translated as $\langle\langle C \rangle\rangle X \phi$ (see, for example, [4]).

As an illustration, we show how to translate the following example CRG from [1] into ATL.

$A = \{1,2,3\}$; $G = \{g_1,g_2\}$; $R = \{r_1,r_2\}$; $G_1 = \{g_1\}$, $G_2 = \{g_2\}$, $G_3 = \{g_1,g_2\}$; $en(1,r_1) = 2, en(1,r_2) = 0, en(2,r_1) = 0, en(2,r_2) = 1, en(3,r_1) = 1, en(3,r_2) = 2$; $req(g_1,r_1) = 3, req(g_1,r_2) = 2, req(g_2,r_1) = 2$, and $req(g_2,r_2) = 1$.

The corresponding model $M_\Gamma = (S, \gamma, V)$ is as follows:

$S = \{s_0 = \emptyset, s_1 = \{g_1\}, s_2 = \{g_2\}, s_3 = \{g_1,g_2\}\}$
$V(s_0) = \emptyset$, $V(s_1) = \{g_1, sat_1, sat_3\}$, $V(s_2) = \{g_2, sat_2, sat_3\}$, $V(s_3) = \{g_1, g_2, sat_1, sat_2, sat_3\}$;
$Act_1(s_0) = \langle g_1 : \langle 0,0 \rangle, g_2 : \langle 0,0 \rangle \rangle$, $\langle g_1 : \langle 1,0 \rangle, g_2 : \langle 0,0 \rangle \rangle$, $\langle g_1 : \langle 2,0 \rangle, g_2 : \langle 0,0 \rangle \rangle$, $\langle g_1 : \langle 0,0 \rangle, g_2 : \langle 1,0 \rangle \rangle$, $\langle g_1 : \langle 0,0 \rangle, g_2 : \langle 2,0 \rangle \rangle$, $\langle g_1 : \langle 1,0 \rangle, g_2 : \langle 1,0 \rangle \rangle$ (the actions available to agents 2 and 3 can be enumerated in a similar way);
$o(s_0, \langle g_1 : \langle 0,0 \rangle, g_2 : \langle 0,0 \rangle \rangle, \langle g_1 : \langle 0,0 \rangle, g_2 : \langle 0,0 \rangle \rangle, \langle g_1 : \langle 0,0 \rangle, g_2 : \langle 0,0 \rangle \rangle) = \emptyset$ (other action transitions can be enumerated similarly);
$c(\langle g_1 : \langle 0,0 \rangle, g_2 : \langle 0,0 \rangle \rangle) = \langle 0,0 \rangle$, for agent 1; the cost of actions 2 and 4 are $\langle 1,0 \rangle$ and the costs of actions 3, 5 and 6 are $\langle 2,0 \rangle$ (costs of actions by other agents can be computed in a similar way).

The endowment in s_0 is as in Γ.

Using this translation, we can state properties such as the coalition of agents 1 and 3 can achieve g_1 under the resource bound corresponding to the sum of their endowments: $[1, 3^{\langle r_1:3, r_2:2 \rangle}]g_1$. In ATL, to simplify the translation, we just give the conjunct for the endowment values which evaluate to true:

$$e^=(1, \langle r_1:2, r_2:0 \rangle) \wedge e^=(3, \langle r_1:1, r_2:2 \rangle) \to$$

$$\langle\langle 1,3 \rangle\rangle X(g_1 \wedge \bigvee_{b_1+b_2 \leq \langle r_1:3, r_2:2 \rangle} e^=(1, \langle r_1:2, r_2:0 \rangle - b_1) \wedge e^=(3, \langle r_1:1, r_2:2 \rangle - b_2))$$

The possible values of b_1 and b_2 above are all b_1 and b_2 such that $b_1 + b_2 \leq \langle r_1 : 3, r_2 : 2 \rangle$, but the values which make the disjunction in the consequent true are $b_1 = \langle r_1:2, r_2:0 \rangle$ and $b_2 = \langle r_1:1, r_2:2 \rangle$.

It is straightforward to encode a CLRG model for an ATL model checker. As an example, we sketch an encoding for the *reactive modules* system description language used by MOCHA model checker.

States of the CLRG models correspond to an assignment of values to state variables in the model checker. The agent's resource endowments are encoded as vectors of state variables (of range type) and the goal propositions are encoded as a set of boolean state variables. We also define a set of vectors of 'contribution variables' for each agent and each goal, which represent the amount of each resource the agent has contributed to achieving the goal. The actions available to the agents are encoded as MOCHA atoms which describe the initial condition and transition relation for a set of related state variables. An action is enabled if the agent's remaining endowment of a resource is greater than or equal to the cost of the action on that resource, i.e., if $en(i, r) \geq c(a_i, r)$. Performing an action decrements the endowment variables for the agent by the cost of the action on each resource, and increments the contribution variables for the proposition(s) affected by the action by the same amount.

Each agent is encoded as a MOCHA module. A module is a collection of atoms and a specification of which of the state variables updated by those atoms are visible from outside the module. For technical reasons, it is convenient to associate the state variables encoding the goal propositions with a separate 'valuation' module. The valuation module aggregates the effects of each individual agent's actions as encoded in the contribution variables and determines the truth value of the goal propositions affected by the actions. A particular game is then simply a parallel composition of the appropriate agent modules and the valuation module.

Using the translation above, CRG problems can be translated from CLRG into ATL with resource counters with only a polynomial increase in the size of the formula (since the CLRG formulas contain no nested modalities). However the translation may lead to an exponential increase in the size of the ATL formula if the CLRG formula contains nested $[C^b]$ modalities. This naturally raises the question of whether there is a special purpose model-checking algorithm for $L^{b,e}$ formulas which avoids the exponential blowup.

8 Model-Checking Problem for CLRG

In this section we describe a model-checking algorithm for CLRG, which is very similar to the model-checking algorithm for ATL [12] except that it only considers 'next state' formulas. We show that bounds on coalition modalities and endowment counters do not increase the complexity of the algorithm which has the same complexity as the ATL algorithm (namely linear).

Theorem 2. *The model-checking problem for CLRG is solvable in linear time. Given a model $M = (S, V, \gamma)$ and an $L^{b,e}$ formula ϕ, there is an algorithm which returns the set of states $[\phi]_M$ satisfying ϕ: $[\phi]_M = \{s \mid M, s \models \phi\}$, which runs in time $O(|\phi| \times m)$ where m is the number of transitions in M.*

Proof. Consider the following symbolic model-checking algorithm:

for every ϕ' in the set of subformulas of ϕ:
case $\phi' == p$: $[\phi']_M = \{s \mid p \in V(s)\}$
case $\phi' == e^=(C, b)$: $[\phi']_M = \{s \mid \Sigma_{i \in C} en(i) = b\}$
case $\phi' == \neg\psi$: $[\phi']_M = S \setminus [\psi]_M$
case $\phi' == \psi_1 \wedge \psi_2$: $[\phi']_M = [\psi_1]_M \cap [\psi_2]_M$
case $\phi' == [C]\psi$: $[\phi']_M = Pre(C, [\psi]_M)$
case $\phi' == [C^b]\psi$: $[\phi']_M = Pre^b(C, [\psi]_M)$

where Pre and Pre^b can be computed in $O(m)$ as follows:

Computation of $Pre(C, [\psi]_M)$:

1. bucket sort the set of transitions possible from a state. In other words, iterate through the set of all possible transitions $(a_C, a_{A \setminus C})$ from a state s collecting all transitions with the same a_C (i.e., the same set of actions by the agents in C) in the same 'bucket' (note that given the mapping γ from states to strategic game forms, the number of buckets is known in advance);[1]
2. iterate through the set of buckets and put s in $Pre(C, [\psi]_M)$ if there is a bucket where for all the transitions (joint actions) in the bucket, $o(s, a_C, a_{A \setminus C}) \in [\psi]_M$.

Computation of $Pre^b(C, [\psi]_M)$:

1. bucket sort the set of transitions possible from a state: iterate through the set of all possible transitions $(a_C, a_{A \setminus C})$ from a state s collecting all transitions with the same a_C and where $c(a_C) \leq b$ in the same bucket;
2. iterate through the set of buckets and put s in $Pre^b(C, [\psi]_M)$ if there is a bucket where for all the transitions (joint actions) in the bucket, $o(s, a_C, a_{A \setminus C}) \in [\psi]_M$.

We might hope that the resource bounds in CLRG might reduce the complexity of the model checking problem for formulas with bounds, but this is not the case. In ATL, for an explicitly enumerated model with m states, the maximum length of the pre-image computation for computing $\langle\langle C \rangle\rangle \Box \phi$ is bounded by m. In our case, if $b/d < m$, where d is the smallest resource cost for any action possible by agents in C and any resource, then the maximum length of the pre-image computation may be shorter than for ATL. However the saving cannot be greater than m for each subformula of the property to be checked, and so the overall saving can be no better than linear in the size of the property, which doesn't change the complexity class.

9   Conclusion and Further Work

We propose a logic CLRG for reasoning about resource limitations on coalitional ability, and show how to express some example problems—including CRG decision problems—in CLRG. As far as we are aware, there are no other automatic

---

[1] The buckets serve essentially the same role as the C-move states in [12].

tools for solving CRG problems. We show how to use a standard ATL model-checker for verifying CLRG properties. However for nested modalities with resource bounds the translation into ATL may cause an exponential blow-up. We show that this can be avoided with a special purpose model-checking algorithm for CLRG which is (as expected) linear in the size of the formula and the number of transitions in the model. This gives an automatic verification procedure for CRG properties which is EXPTIME (since the model is exponential in the size of a CRG). For NP- and co-NP complete properties this is the best that can be expected. In other work [13] we provide a complete and sound axiomatisation for a related logic (without endowment counters and unbounded [C] operators). Axiomatisation of CLRG is a subject of future work.

Acknowledgements

This work was supported by the Engineering and Physical Sciences Research Council [grant number EP/E031226/1.

References

1. Wooldridge, M., Dunne, P.E.: On the computational complexity of coalitional resource games. Artif. Intell. 170(10), 835–871 (2006)
2. Pauly, M.: Logic for Social Software. Ph.D. thesis, ILLC, University of Amsterdam (2001)
3. Pauly, M.: A modal logic for coalitional power in games. J. Log. Comput. 12(1), 149–166 (2002)
4. Goranko, V.: Coalition games and alternating temporal logics. In: Proceeding of the Eighth Conference on Theoretical Aspects of Rationality and Knowledge (TARK VIII), pp. 259–272. Morgan Kaufmann, San Francisco (2001)
5. Wooldridge, M., Dunne, P.E.: On the computational complexity of qualitative coalitional games. Artif. Intell. 158(1), 27–73 (2004)
6. Ågotnes, T., van der Hoek, W., Wooldridge, M.: On the logic of coalitional games. In: Nakashima, H., Wellman, M.P., Weiss, G., Stone, P. (eds.) 5th International Joint Conference on Autonomous Agents and Multiagent Systems (AAMAS 2006), Hakodate, Japan, May 8-12, pp. 153–160. ACM, New York (2006)
7. Ågotnes, T., van der Hoek, W., Wooldridge, M.: Temporal qualitative coalitional games. In: Nakashima, H., Wellman, M.P., Weiss, G., Stone, P. (eds.) 5th International Joint Conference on Autonomous Agents and Multiagent Systems (AAMAS 2006), Hakodate, Japan, May 8-12, pp. 177–184. ACM, New York (2006)
8. Wooldridge, M., Ågotnes, T., Dunne, P.E., van der Hoek, W.: Logic for automated mechanism design - a progress report. In: Proceedings of the Twenty-Second AAAI Conference on Artificial Intelligence, Vancouver, British Columbia, Canada, July 22-26, pp. 9–16. AAAI Press, Menlo Park (2007)
9. Dunne, P.E., van der Hoek, W., Kraus, S., Wooldridge, M.: Cooperative boolean games. In: Padgham, L., Parkes, D.C., Müller, J., Parsons, S. (eds.) 7th International Joint Conference on Autonomous Agents and Multiagent Systems (AAMAS 2008), Estoril, Portugal, May 12-16, vol. 2, pp. 1015–1022. IFAAMAS (2008)

10. Alechina, N., Logan, B., Nguyen, H.N., Rakib, A.: Verifying time, memory and communication bounds in systems of reasoning agents. In: Padgham, L., Parkes, D., Müller, J., Parsons, S. (eds.) Proceedings of the Seventh International Conference on Autonomous Agents and Multiagent Systems (AAMAS 2008), Estoril, Portugal, May 2008, vol. 2, pp. 736–743. IFAAMAS (2008)
11. Alur, R., Henzinger, T.A., Mang, F.Y.C., Qadeer, S., Rajamani, S.K., Tasiran, S.: MOCHA: Modularity in model checking. In: Vardi, M.Y. (ed.) CAV 1998. LNCS, vol. 1427, pp. 521–525. Springer, Heidelberg (1998)
12. Alur, R., Henzinger, T., Kupferman, O.: Alternating-time temporal logic. Journal of the ACM 49(5), 672–713 (2002)
13. Alechina, N., Logan, B., Nguyen, H.N., Rakib, A.: A logic for coalitions with bounded resources. In: Boutilier, C. (ed.) Proceedings of the Twenty First International Joint Conference on Artificial Intelligence, IJCAI, Pasadena CA, USA, vol. 2, pp. 659–664. AAAI Press, Menlo Park (2009)

Dynamic Context Logic

Guillaume Aucher[1], Davide Grossi[2], Andreas Herzig[3], and Emiliano Lorini[3]

[1] University of Luxembourg, ICR
`guillaume.aucher@uni.lu`
[2] University of Amsterdam, ILLC
`d.grossi@uva.nl`
[3] University of Toulouse, IRIT
`{herzig,lorini}@irit.fr`

Abstract. Building on a simple modal logic of context, the paper presents a dynamic logic characterizing operations of contraction and expansion on theories. We investigate the mathematical properties of the logic, and show how it can capture some aspects of the dynamics of normative systems once they are viewed as logical theories.

1 Introduction

In artificial intelligence as well as in philosophy, contexts—when addressed formally—are often thought of as sets of models of given theories [7] or, more simply, as sets of possible worlds [17]. Once a context is viewed as a set of models/possible worlds, its content is nothing but the set of logical formulae it validates or, otherwise, its theory or intension.

This perspective has been used, in [9], to develop a simple modal logic for representing and reasoning about contexts. This logic is based on a set of modal operators $[X]$ where X is a label denoting the context of a theory. A formula $[X]\varphi$ reads 'in the context of X it is the case that φ'. The present paper develops a 'dynamification' of such logic by studying the two following operations:

- *Context expansion* (in symbols $X{+}\psi$), by means of which a context is restricted and hence, its intension—its logical theory—strengthened. Such operation is similar to the operation for announcement studied in DEL [21]. Its function is to restrict the space of possible worlds accepted by the context X to the worlds where ψ is true.
- *Context contraction* (in symbols $X{-}\psi$), by means of which a context is enlarged giving rise to a weaker logical theory. The function of this operation is to add to the space of possible worlds accepted by context X some worlds in which ψ is false.

The resulting dynamic logic is studied from the point of view of its mathematical properties and illustrated through a running example. Just like the context logic introduced in [9] was developed to cope with some problems in the analysis of normative systems, its dynamic version will be illustrated by resorting to examples taken from the normative domain. In particular, we will link context expansion to some kind of "promulgation" of norms and context contraction to some kind of "derogation" of norms.[1]

---

[1] The terms promulgation and derogation are borrowed from the first paper addressing norm dynamics by formal means, that is, [2].

X. He, J. Horty, and E. Pacuit (Eds.): LORI 2009, LNAI 5834, pp. 15–26, 2009.

This illustration will highlight the features that norm dynamics shares with theory dynamics once norms are viewed as logical statements of the type: "a certain fact φ implies a violation". Such view of norms builds on those approaches to deontic logic which stemmed from the work of Anderson [4] and Kanger [12]. Although rather abstract and, under many respects, simplistic such view of norms has received considerable attention by recent developments in the logical analysis of normative systems within the artificial intelligence community (e.g., [1, 13, 18]).

The paper is organized as follows. In Section 2 we will briefly present the modal logic of context of [9]. Section 3 is devoted to extend this logic with the two events $X{+}\psi$ and $X{-}\psi$ which allow to model context dynamics. Finally, in Section 4, we will apply our logical framework to norm change, *i.e.* norm promulgation and norm derogation. The results of this application are compared with related work in Section 5. Conclusions follow.

2   A Modal Logic of Context

The logic presented in this section is a simple modal logic designed to represent and reason about a localized notion of validity, that is, of validity with respect to all models in a given set. Such a given set is what is here called a *context*.

Let $\Phi = \{p, q, \ldots\}$ be a countable non-empty set of propositional letters, and let $\mathcal{C} = \{X, Y, \ldots\}$ be a countable set of contexts. $\mathcal{L}_{Prop}$ is the propositional language.

2.1   Models

Definition 1. *A context model (**Cxt**-model) $\mathcal{M} = (W, R, \mathcal{I})$ is a tuple such that:*

 – *W is a nonempty set of possible worlds;*
 – *$R : \mathcal{C} \longrightarrow 2^W$ maps each context X to a subset of W;*
 – *$\mathcal{I} : \Phi \longrightarrow 2^W$ is a valuation.*

We write R_X for $R(X)$ and $w \in \mathcal{M}$ for $w \in W$. For $w \in \mathcal{M}$, the couple $(\mathcal{M}, w)$ is a pointed context model.

A **Cxt**-model represents a logical space together with some of its possible restrictions, i.e., the contexts. In our case, contexts are used to represent the restrictions to those sets of propositional models satisfying the rules stated by a given normative system [9]. Let us illustrate how they can be used to model normative systems.

Example 1 (A toy normative system). Consider a normative system according to which: motorized vehicles must have a numberplate ; motorized vehicles must have an insurance; bikes should not have an insurance; bikes are classified as not being a motorized vehicle. Once a designated atom V is introduced in the language, which represents a notion of "violation" [4], the statements above obtain a simple representation:

> **Rule 1:** $(mt \wedge \neg pl) \rightarrow$ V **Rule 2:** $(mt \wedge \neg in) \rightarrow$ V
> **Rule 3:** $(bk \wedge in) \rightarrow$ V **Rule 4:** $bk \rightarrow \neg mt$

A **Cxt**-model $\mathcal{M} = (W, R, \mathcal{I})$, where $\mathcal{I}$ maps atoms mt, pl, in, bk and V to subsets of W, models the normative system above as a context X if R_X coincides with the subset of W where Rules 1-4 are true according to propositional logic.

2.2 Logic

The logic **Cxt** is now presented which captures the notion of validity with respect to a context. To talk about **Cxt**-models we use a modal language $\mathcal{L}_{\mathbf{Cxt}}$ containing modal operators $[X]$ for every $X \in \mathcal{C}$, plus the universal modal operator $[\mathsf{U}]$. The set of well-formed formulae of $\mathcal{L}_{\mathbf{Cxt}}$ is defined by the following BNF:

$$\mathcal{L}_{\mathbf{Cxt}} : \varphi ::= p \mid \neg\varphi \mid \varphi \wedge \varphi \mid [\mathsf{U}]\varphi \mid [X]\varphi$$

where p ranges over Φ and X over $\mathcal{C}$. The connectives $\top, \vee, \rightarrow, \leftrightarrow$ and the dual operators $\langle X \rangle$ are defined as usual within $\mathcal{L}_{\mathbf{Cxt}}$ as: $\langle X \rangle \varphi = \neg[X]\neg\varphi$, for $X \in \mathcal{C} \cup \{\mathsf{U}\}$.

We interpret formulas of $\mathcal{L}_{\mathbf{Cxt}}$ in a **Cxt**-models as follows: the $[\mathsf{U}]$ operator is interpreted as the universal modality [5], and the $[X]$ operators model restricted validity.

Definition 2. *Let $\mathcal{M}$ be a **Cxt**-model, and let $w \in \mathcal{M}$.*

$\mathcal{M}, w \models p$ *iff* $w \in \mathcal{I}(p)$;
$\mathcal{M}, w \models [X]\varphi$ *iff for all* $w' \in R_X$, $\mathcal{M}, w' \models \varphi$;
$\mathcal{M}, w \models [\mathsf{U}]\varphi$ *iff for all* $w' \in W$, $\mathcal{M}, w' \models \varphi$.

*and as usual for the Boolean operators. Formula φ is valid in $\mathcal{M}$, noted $\mathcal{M} \models \varphi$, iff $\mathcal{M}, w \models \varphi$ for all $w \in \mathcal{M}$. φ is **Cxt**-valid, noted $\models_{\mathbf{Cxt}} \varphi$, iff $\mathcal{M} \models \varphi$ for all **Cxt**-models $\mathcal{M}$.*

Cxt-validity is axiomatized by the following schemas:

$$(\mathrm{P}) \quad \text{all propositional axiom schemas and rules}$$
$$(4^{XY}) \quad [X]\varphi \rightarrow [Y][X]\varphi$$
$$(5^{XY}) \quad \langle X \rangle \varphi \rightarrow [Y]\langle X \rangle \varphi$$
$$(\mathrm{T}^{\mathsf{U}}) \quad [\mathsf{U}]\varphi \rightarrow \varphi$$
$$(\mathrm{K}^X) \quad [X](\varphi \rightarrow \varphi') \rightarrow ([X]\varphi \rightarrow [X]\varphi')$$
$$(\mathrm{N}^X) \quad \text{IF} \vdash \varphi \text{ THEN} \vdash [X]\varphi$$

where $X, Y \in \mathcal{C} \cup \{\mathsf{U}\}$. The $[X]$ and $[Y]$ operators are **K45** modalities strengthened with the two inter-contextual interaction axioms 4^{XY} and 5^{XY}. $[\mathsf{U}]$ is an **S5** modality. Provability of a formula φ, noted $\vdash_{\mathbf{Cxt}} \varphi$, is defined as usual.

Logic **Cxt** is well-behaved for both axiomatizability and complexity.

Theorem 1 ([9]). $\models_{\mathbf{Cxt}} \varphi$ *iff* $\vdash_{\mathbf{Cxt}} \varphi$.

Theorem 2. *Deciding **Cxt**-validity is coNP-complete.*

Proof (Sketch of proof). Satisfiability of **S5** formulas is decidable in nondeterministic polynomial time [5]. Let $\mathcal{L}^{[\mathsf{U}]}$ be the language built from the set of atoms $\Phi \cup \mathcal{C}$ (supposing Φ and $\mathcal{C}$ are disjoint) and containing only one modal operator $[\mathsf{U}]$. That is:

$$\mathcal{L}^{[\mathsf{U}]} : \varphi ::= p \mid \neg\varphi \mid \varphi \wedge \varphi \mid [\mathsf{U}]\varphi$$

where p ranges over $\Phi \cup \mathcal{C}$. It gets a natural interpretation on context models where $[\mathsf{U}]$ is the global modality. Then one can show that the following is a satisfiability-preserving polytime reduction f of $\mathcal{L}_{\mathbf{Cxt}}$ to $\mathcal{L}^{[\mathsf{U}]}$: $f(p) = p$; $f(\neg\varphi) = \neg f(\varphi)$; $f(\varphi \wedge \varphi') = f(\varphi) \wedge f(\varphi')$; $f([\mathsf{U}]\varphi) = [\mathsf{U}]f(\varphi)$; $f([X]\varphi) = [\mathsf{U}](X \rightarrow f(\varphi))$.

The same argument proves linear time complexity if the alphabet Φ is finite.

Another interesting property of **Cxt** is that every formula of $\mathcal{L}_{\mathbf{Cxt}}$ is provably equivalent to a formula without nested modalities, as the following proposition shows. We first formally define the language without nested modalities:

$$\mathcal{L}^1_{\mathbf{Cxt}} : \varphi ::= \alpha \mid [X]\alpha \mid [\mathsf{U}]\alpha \mid \neg\varphi \mid \varphi \wedge \varphi$$

where α ranges over $\mathcal{L}_{Prop}$ and X over $\mathcal{C}$. This result is of use in Proposition 3.

Proposition 1. *For all $\varphi \in \mathcal{L}_{\mathbf{Cxt}}$ there is $\varphi^1 \in \mathcal{L}^1_{\mathbf{Cxt}}$ such that $\vdash_{\mathbf{Cxt}} \varphi \leftrightarrow \varphi^1$.*

Proof. By induction on φ. The Boolean cases clearly work. If φ is of the form $[X]\psi$ with $X \in \mathcal{C} \cup \{\mathsf{U}\}$ then by IH there are $\alpha_k, \alpha^i_j, \beta^i \in \mathcal{L}_{Prop}$ such that

$$\varphi \leftrightarrow [X] \bigwedge_{k\in\mathbb{N}_l} (\alpha_k \vee \bigvee_{i\in\mathbb{N}_{n_k}} ([X_i]\alpha^i_1 \vee \ldots \vee [X_i]\alpha^i_{n_i} \vee \langle X_i\rangle\beta^i))).$$

However, using (4^{XY}) and (5^{XY}), one can easily show that

$$\vdash_{\mathbf{Cxt}} [X](\alpha_k \vee \bigvee_{i\in\mathbb{N}_{n_k}} ([X_i]\alpha^i_1 \vee \ldots \vee [X_i]\alpha^i_{n_i} \vee \langle X_i\rangle\beta^i))) \leftrightarrow$$

$$([X]\alpha_k \vee \bigvee_{i\in\mathbb{N}_{n_k}} ([X_i]\alpha^i_1 \vee \ldots \vee [X_i]\alpha^i_{n_i} \vee \langle X_i\rangle\beta^i))).$$

This completes the proof.

2.3   Normative Systems in Cxt

We are ready to provide an object-level representation of Example 1. The contextual operators $[X]$ and the universal operator $[\mathsf{U}]$ can be used to define the concepts of *classificatory rule*, *obligation* and *permission* which are needed to model normative systems. Classificatory rules are of the form "φ counts as ψ in the normative system X" and their function in a normative systems is to specify classifications between different concepts [15]. For example, according to the classificatory rule "in the context of Europe, a piece of paper with a certain shape, color, *etc.* counts as a 5 Euro bill", in Europe a piece of paper with a certain shape, color, etc. should be classified as a 5 Euro bill. The concept of classificatory rule is expressed by the following abbreviation:

$$\varphi \Rightarrow_X \psi \overset{def}{=} [X](\varphi \to \psi)$$

where $\varphi \Rightarrow_X \psi$ reads 'φ counts as ψ in normative system X'. As done already in Example 1, by introducing the violation atom V we can obtain a reduction of deontic logic to logic **Cxt** along the lines first explored by Anderson [4]. As far as obligations are concerned, we introduce operators of the form $\mathbf{O}_X$ which are used to specify what is obligatory in the context of a certain normative system X:

$$\mathbf{O}_X\varphi \overset{def}{=} \neg\varphi \Rightarrow_X \mathsf{V}$$

According to this definition, 'φ is obligatory within context X' is identified with '$\neg\varphi$ counts as a violation in normative system X'. Note that we have the following **Cxt**-theorem:

$$\vdash_{\mathbf{Cxt}} ((\varphi \Rightarrow_X \psi) \wedge (\varphi \Rightarrow_X \neg\psi)) \rightarrow \mathbf{O}_X \neg\varphi \tag{1}$$

Every $\mathbf{O}_X$ obeys axiom K and necessitation, and is therefore a normal modal operator.

$$\vdash_{\mathbf{Cxt}} \quad \mathbf{O}_X(\varphi \rightarrow \psi) \rightarrow (\mathbf{O}_X\varphi \rightarrow \mathbf{O}_X\psi) \tag{2}$$

$$\text{IF } \vdash_{\mathbf{Cxt}} \varphi \text{ THEN } \vdash_{\mathbf{Cxt}} \mathbf{O}_X\varphi \tag{3}$$

Note that the formula $\mathbf{O}_X\bot$ is consistent, hence our deontic operator does not satisfy the D axiom.

We define the permission operator in the standard way as the dual of the obligation operator: "φ is permitted within context X", noted $\mathbf{P}_X\varphi$. Formally:

$$\mathbf{P}_X\varphi \overset{def}{=} \neg\mathbf{O}_X\neg\varphi$$

Formula $\mathbf{P}_U\varphi$ should be read "φ is deontically possible".

Example 2 (Talking about a toy normative system). Consider again the normative system of Example 1. We can now express in $\mathbf{Cxt}$ that Rules 1-4 explicitly belong to context X:

Rule 1: $\mathbf{O}_X(mt \rightarrow pl)$ **Rule 2:** $\mathbf{O}_X(mt \rightarrow in)$
Rule 3: $\mathbf{O}_X(bk \rightarrow \neg in)$ **Rule 4:** $bk \Rightarrow_X \neg mt$

Rules 1'-4' explicitly localize the validity of Rules 1-4 of Example 1 to context X. Logic $\mathbf{Cxt}$ is therefore enough expressive to represent several (possibly inconsistent) normative systems at the same time.

3 Dynamic Context Logic

In the present section we 'dynamify' logic $\mathbf{Cxt}$.

3.1 Two Relations on Models

We first define the relations $\overset{X+\psi}{\longrightarrow}$ and $\overset{X-\psi}{\longrightarrow}$ on the set of pointed $\mathbf{Cxt}$-models.

Definition 3. *Let* $(\mathcal{M}, w) = (W, R, \mathcal{I}, w)$ *and* $(\mathcal{M}', w') = (W', R', \mathcal{I}', w')$ *be two pointed* $\mathbf{Cxt}$*-models, and let* $\varphi \in \mathcal{L}_{\mathbf{Cxt}}$ *and* $X \in \mathcal{C}$.
We set $(\mathcal{M}, w) \overset{X+\psi}{\longrightarrow} (\mathcal{M}', w')$ *iff* $W = W', w = w', \mathcal{I} = \mathcal{I}'$*, and*

- $R'_Y = R_Y$ *if* $Y \neq X$;
- $R'_X = R_X \cap ||\psi||_{\mathcal{M}}$.

We set $(\mathcal{M}, w) \overset{X-\psi}{\longrightarrow} (\mathcal{M}', w')$ *iff* $W = W', w = w', \mathcal{I} = \mathcal{I}'$*, and*

- $R'_Y = R_Y$ *if* $Y \neq X$;
- $R'_X = \begin{cases} R_X \text{ if } \mathcal{M}, w \models \neg[X]\psi \vee [U]\psi \\ R_X \cup S \text{ otherwise, for some } \emptyset \neq S \subseteq ||\neg\psi||_{\mathcal{M}} \end{cases}$

In case $(\mathcal{M}, w) \overset{X+\psi}{\longrightarrow} (\mathcal{M}', w')$ *(resp.* $(\mathcal{M}, w) \overset{X-\psi}{\longrightarrow} (\mathcal{M}', w')$*), we say that* $\mathcal{M}'$ *is a (context) expansion (resp. contraction) of* $\mathcal{M}$.

In the above definition, $||\psi||_{\mathcal{M}} = \{w \in \mathcal{M} : \mathcal{M}, w \models \psi\}$. So in both cases, it is only the context X which changes from $\mathcal{M}$ to $\mathcal{M}'$. In the first case, it is restricted to the worlds that satisfy ψ, and in the second case, it is enlarged with some worlds which satisfy $\neg\psi$, except if such worlds do not exist in the model ($[U]\psi$) or if $\neg\varphi$ is already consistent with the context ($\neg[X]\psi$). Note that there might be several contractions of a given Cxt-model but there is always a unique expansion. The relation $\xrightarrow{X-\psi}$ thus defines implicitly a *family* of contraction operations. The following proposition shows that $\xrightarrow{X-\psi}$ is essentially the converse relation of $\xrightarrow{X+\psi}$.

Proposition 2. *Let $(\mathcal{M}, w)$ and $(\mathcal{M}', w')$ be two pointed Cxt-models and $\psi \in \mathcal{L}_{Cxt}$. Then $(\mathcal{M}, w) \xrightarrow{X+\psi} (\mathcal{M}', w')$ iff $(\mathcal{M}', w') \xrightarrow{X-\psi} (\mathcal{M}, w)$ and $\mathcal{M}', w' \models [X]\psi$.*

Proof. The left to right direction is clear. Assume that $(\mathcal{M}', w') \xrightarrow{X-\psi} (\mathcal{M}, w)$ and $\mathcal{M}', w' \models [X]\psi$. Then $R'_Y = R_Y$ if $Y \neq X$ by definition. If $\mathcal{M}', w' \models \neg[U]\psi$ then $R'_X = R_X \cup S$ for some $\emptyset \neq S \subseteq ||\neg\psi||_{\mathcal{M}}$ because $\mathcal{M}', w' \models [X]\psi \wedge \neg[U]\psi$. So $R'_X = R_X \cap ||\psi||_{\mathcal{M}}$. Otherwise, if $\mathcal{M}', w' \models [U]\psi$ then $R'_X = R_X$ by definition. So $R'_X = R_X \cap ||\psi||_{\mathcal{M}}$ because $\mathcal{M}', w' \models [X]\psi$. In both cases $R'_X = R_X \cap ||\psi||_{\mathcal{M}}$. Therefore $(\mathcal{M}, w) \xrightarrow{X+\psi} (\mathcal{M}', w')$.

3.2 Logic

The language of the logic **DCxt** is obtained by adding the dynamic operators $[X+\psi]$ and $[X-\psi]$ to the language $\mathcal{L}_{Cxt}$:

$$\mathcal{L}_{DCxt} : \varphi ::= p \mid \neg\varphi \mid \varphi \wedge \varphi \mid [X]\varphi \mid [U]\varphi \mid [X+\psi]\varphi \mid [X-\psi]\varphi$$

where p ranges over Φ, X over $\mathcal{C}$ and ψ over $\mathcal{L}_{Cxt}$. $[X+\psi]\varphi$ reads 'after the expansion of the context X by ψ, φ is true', and $[X-\psi]\varphi$ reads 'after *any* contraction of the context X by ψ, φ is true'.

Definition 4. *Let $\mathcal{M}$ be a Cxt-model. The truth conditions for $\mathcal{L}_{DCxt}$ in $\mathcal{M}$ are those of Definition 2, plus:*

$\mathcal{M}, w \models [X+\psi]\varphi$ iff $\mathcal{M}', w' \models \varphi$ for all **Cxt**-models $(\mathcal{M}', w')$
    such that $(\mathcal{M}, w) \xrightarrow{X+\psi} (\mathcal{M}', w')$;
$\mathcal{M}, w \models [X-\psi]\varphi$ iff $\mathcal{M}', w' \models \varphi$ for all **Cxt**-models $(\mathcal{M}', w')$
    such that $(\mathcal{M}, w) \xrightarrow{X-\psi} (\mathcal{M}', w')$.

*As before, $\mathcal{M} \models \varphi$ iff $\mathcal{M}, w \models \varphi$ for all $w \in \mathcal{M}$, and φ is **DCxt**-valid ($\models_{DCxt} \varphi$) iff $\mathcal{M} \models \varphi$ for all **Cxt**-models $\mathcal{M}$.*

The operator $[X-\psi]$ is thus useful if we want to have general properties about our family of contractions or about a situation; for example, given some formulas $\psi_1, \ldots, \psi_n$, what would be true after any sequence of contractions and expansions by these formulas? Can we get an inconsistency with a specific choice of contractions?

In order to axiomatize the **DCxt**-validities we define for every $X \in \mathcal{C}$ two auxiliary languages $\mathcal{L}_{\neq X}$ and $\mathcal{L}_{=X}$:

$$\mathcal{L}_{=X} : \varphi ::= [X]\alpha \mid \neg\varphi \mid \varphi \wedge \varphi$$
$$\mathcal{L}_{\neq X} : \varphi ::= \alpha \mid [Y]\alpha \mid \neg\varphi \mid \varphi \wedge \varphi$$

where α ranges over $\mathcal{L}_{Prop}$ and Y over $(\mathcal{C} \cup \{U\}) - \{X\}$.

Logic **DCxt** is axiomatized by the following schemata:

(Cxt)   All axiom schemas and inference rules of **Cxt**

(R+1) $[X+\psi]\varphi_{\neq X} \leftrightarrow \varphi_{\neq X}$

(R+2) $[X+\psi][X]\alpha \leftrightarrow [X](\psi \rightarrow \alpha)$

(R+3) $[X+\psi]\neg\varphi \leftrightarrow \neg[X+\psi]\varphi$

(R−1) $[X-\psi](\varphi_{\neq X} \vee \varphi_{=X}) \leftrightarrow (\varphi_{\neq X} \vee [X-\psi]\varphi_X)$

(R−2) $\neg[X-\psi]\bot$

(R−3) $[X-\psi]([X]\alpha_1 \vee \ldots \vee [X]\alpha_n \vee \langle X\rangle\alpha) \leftrightarrow$
$$((\neg[X]\psi \vee [U]\psi) \wedge ([X]\alpha_1 \vee \ldots \vee [X]\alpha_n \vee \langle X\rangle\alpha))$$
$$\vee \, (([X]\psi \wedge \neg[U]\psi) \wedge$$
$$((\bigvee_i ([X]\alpha_i \wedge [U](\psi \vee \alpha_i))) \vee \langle X\rangle\alpha \vee [U](\psi \vee \alpha)))$$

(K$^+$) $[X+\psi](\varphi \rightarrow \varphi') \rightarrow ([X+\psi]\varphi \rightarrow [X+\psi]\varphi')$

(K$^-$) $[X-\psi](\varphi \rightarrow \varphi') \rightarrow ([X-\psi]\varphi \rightarrow [X-\psi]\varphi')$

(RRE)   Rule of replacement of proved equivalence

where $X \in \mathcal{C}, \varphi, \varphi' \in \mathcal{L}_{\mathbf{DCxt}}, \psi \in \mathcal{L}_{\mathbf{Cxt}}, \varphi_{=X} \in \mathcal{L}_{=X}, \varphi_{\neq X} \in \mathcal{L}_{\neq X}$, and $\alpha, \alpha_i \ldots \in \mathcal{L}_{Prop}$.

Note that from (R−1) and (R−2) one can deduce $[X-\psi]\varphi_{\neq X} \leftrightarrow \varphi_{\neq X}$. The formulae above are reduction axioms:

Proposition 3. *For all* $\varphi_{\mathbf{DCxt}} \in \mathcal{L}_{\mathbf{DCxt}}$ *there is* $\varphi_{\mathbf{Cxt}} \in \mathcal{L}_{\mathbf{Cxt}}$ *such that* $\vdash_{\mathbf{DCxt}} \varphi_{\mathbf{DCxt}} \leftrightarrow \varphi_{\mathbf{Cxt}}$.

Proof (Sketch of proof). (By induction on the number of occurrences of dynamic operators.) Let $\varphi_{\mathbf{DCxt}} \in \mathcal{L}_{\mathbf{DCxt}}$ and $\varphi'_{\mathbf{DCxt}}$ be one of its sub-formulas of the form $[X+\psi]\varphi_{\mathbf{Cxt}}$ or $[X-\psi]\varphi_{\mathbf{Cxt}}$, with $\varphi_{\mathbf{Cxt}} \in \mathcal{L}_{\mathbf{Cxt}}$. By Proposition 1, there is $\varphi^1_{\mathbf{Cxt}} \in \mathcal{L}^1_{\mathbf{Cxt}}$ such that $\vdash_{\mathbf{Cxt}} \varphi_{\mathbf{Cxt}} \leftrightarrow \varphi^1_{\mathbf{Cxt}}$. So $\vdash_{\mathbf{DCxt}} [X+\psi]\varphi_{\mathbf{Cxt}} \leftrightarrow [X+\psi]\varphi^1_{\mathbf{Cxt}}$ by (REE) and (K$^+$). Now, thanks to axioms (R+1), (R+2) and (R+3) and because $\varphi^1_{\mathbf{Cxt}} \in \mathcal{L}^1_{\mathbf{Cxt}}$, one can easily show that there is $\psi_{\mathbf{Cxt}} \in \mathcal{L}_{\mathbf{Cxt}}$ such that $\vdash_{\mathbf{DCxt}} [X+\psi]\varphi^1_{\mathbf{Cxt}} \leftrightarrow \psi_{\mathbf{Cxt}}$. For the case $[X-\psi]\varphi_{\mathbf{Cxt}}$ we apply the same method using (R−1), (R−2) and (R−3). So $\vdash_{\mathbf{DCxt}} \varphi'_{\mathbf{DCxt}} \leftrightarrow \psi_{\mathbf{Cxt}}$. Now we replace $\varphi'_{\mathbf{DCxt}}$ by $\psi_{\mathbf{Cxt}}$ in $\varphi_{\mathbf{DCxt}}$. This yields an equivalent formula (thanks to (RRE)) with one dynamic operator less. We then apply to this formula the same process we applied to $\varphi_{\mathbf{Cxt}}$ until we get rid of all the dynamic operators.

So, if we want to check that a given formula of the form $[X \pm \psi_1] \ldots [X \pm \psi_n]\varphi$ holds in a **Cxt**-model, instead of computing all the corresponding sequences of contractions and expansions $\psi_1, \ldots, \psi_n$ of the **Cxt**-model, we can also reduce the formula to one of $\mathcal{L}_{\mathbf{Cxt}}$ and check it on the original **Cxt**-model. This way to proceed might be computationally less costly. For example, $\vdash_{\mathbf{DCxt}} [X-\alpha]\neg[X]\alpha \leftrightarrow \langle U \rangle \neg \alpha$. As in DEL, soundness, completeness and decidability follow from Proposition 3:

Theorem 3. $\models_{\mathbf{DCxt}} \varphi$ *iff* $\vdash_{\mathbf{DCxt}} \varphi$. *Deciding* **DCxt**-*validity is decidable.*

Finally, it should be noted that we could easily enrich this formalism with specific contraction operators. For example we could add to $\mathcal{L}_{\mathbf{DCxt}}$ the contraction operator $[X \stackrel{\circ}{=} \psi]\varphi$ whose semantics would be defined as follows: for $\mathcal{M} = (W, R, \mathcal{I})$, $\mathcal{M}, w \models [X \stackrel{\circ}{=} \psi]\varphi$ iff $\mathcal{M}', w \models \varphi$, where $\mathcal{M}' = (W, R', \mathcal{I})$ with $R'_Y = R_Y$ for $Y \neq X$ and $R'_X = R_X \cup \{w \in W \mid \mathcal{M}, w \models \neg\psi\}$. To get a complete axiomatization, we just have to add to **DCxt** the following axiom schemas: (1) $[X \stackrel{\circ}{=} \psi]\varphi_{\neq X} \leftrightarrow \varphi_{\neq X}$; (2) $[X \stackrel{\circ}{=} \psi]\neg\varphi \leftrightarrow \neg[X \stackrel{\circ}{=} \psi]\varphi$; (3) $[X \stackrel{\circ}{=} \psi][X]\alpha \leftrightarrow [X]\alpha \wedge [U](\neg\psi \rightarrow \alpha)$; and the distribution axiom ($K^{\stackrel{\circ}{=}}$). In fact this contraction $\stackrel{\circ}{=}$ belongs to the family of contractions defined in Definition 3, and so we get $\vdash_{\mathbf{DCxt}} [X-\psi]\varphi \rightarrow [X \stackrel{\circ}{=} \psi]\varphi$.

4   A Logical Account of Norm Change

Just as we defined the static notions of obligation and classificatory rules on the basis of **Cxt**, we can in the same spirit define the dynamic notions of promulgation and derogation of obligation and classificatory rules on the basis of **DCxt**:

$$+(\varphi \Rightarrow_X \psi) \stackrel{def}{=} X+(\varphi \rightarrow \psi) \quad +\mathbf{O}_X\psi \stackrel{def}{=} X+(\neg\psi \rightarrow V)$$
$$-(\varphi \Rightarrow_X \psi) \stackrel{def}{=} X-(\varphi \rightarrow \psi) \quad -\mathbf{O}_X\psi \stackrel{def}{=} X-(\neg\psi \rightarrow V)$$

Operator $[+(\varphi \Rightarrow_X \psi)]\chi$ (resp. $[-(\varphi \Rightarrow_X \psi)]\chi$) should be read 'after the promulgation (resp. after *any* derogation) of the classificatory rule $\varphi \Rightarrow_X \psi$, χ is true'. Likewise, $[+\mathbf{O}_X\psi]\varphi$ (resp. $[-\mathbf{O}_X\psi]\varphi$) should be read 'after the promulgation (resp. after *any* derogation) within context X of the obligation ψ, χ is true'.

Example 3 (Changing a toy normative system). In Example 2, after the legislator's proclamation that motorized vehicles having more than 50cc (mf) are obliged to have a numberplate (event $+\mathbf{O}_X((mt \wedge mf) \rightarrow pl)$ and that motorized vehicles having less than 50cc ($\neg mf$) are not obliged to have a numberplate (event $-\mathbf{O}_X((mt \wedge \neg mf) \rightarrow pl)$ we should expect that motorbikes having more than 50cc have the obligation to have a numberplate and motorbikes having less than 50cc have the permission not to have a numberplate. This is indeed the case as the following formula is a theorem:

$$\mathbf{P}_U(mt \wedge \neg mf \wedge \neg pl) \rightarrow ([+\mathbf{O}_X((mt \wedge mf) \rightarrow pl)][-\mathbf{O}_X((mt \wedge \neg mf) \rightarrow pl)]$$
$$\mathbf{O}_X((mt \wedge mf) \rightarrow pl) \wedge \mathbf{P}_X(mt \wedge \neg mf \wedge \neg pl)).$$

More generally, we have the following proposition.

Proposition 4. *The following formulae are* **DCxt**-*theorems:*

$$[\vdash(\varphi \Rightarrow_X \psi)]\varphi \Rightarrow_X \psi \tag{4}$$

$$[\vdash\mathbf{O}_X\psi]\mathbf{O}_X\psi \tag{5}$$

$$\mathbf{P}_\mathsf{U}\neg\psi \rightarrow [\vdash\mathbf{O}_X\psi]\mathbf{P}_X\neg\psi \tag{6}$$

$$((\varphi \Rightarrow_X \psi) \wedge \langle\mathsf{U}\rangle\neg(\varphi \rightarrow \psi)) \rightarrow \tag{7}$$
$$[\vdash(\varphi \Rightarrow_X \psi)][\vdash(\varphi \Rightarrow_X \neg\psi)]\neg((\varphi \Rightarrow_X \neg\psi) \wedge (\varphi \Rightarrow_X \psi))$$

$$(\mathbf{O}_X(\varphi \rightarrow \psi) \wedge \mathbf{P}_\mathsf{U}\neg(\varphi \rightarrow \psi)) \rightarrow \tag{8}$$
$$[\vdash\mathbf{O}_X(\varphi \rightarrow \psi)][\vdash\mathbf{O}_X(\varphi \rightarrow \neg\psi)]\neg(\mathbf{O}_X(\varphi \rightarrow \neg\psi) \wedge \mathbf{O}_X(\varphi \rightarrow \psi))$$

$$(\mathbf{P}_\mathsf{U}\neg(\psi \rightarrow \varphi) \wedge \mathbf{O}_X\varphi) \rightarrow [\vdash\mathbf{O}_X(\psi \rightarrow \varphi)]\neg\mathbf{O}_X\varphi \tag{9}$$

$$(\varphi \Rightarrow_X \psi) \rightarrow ([\vdash\mathbf{O}_X\varphi]\xi \rightarrow [\vdash\mathbf{O}_X\psi]\xi) \tag{10}$$

$$((\langle\mathsf{U}\rangle\neg(\varphi \rightarrow \psi) \wedge (\varphi \Rightarrow_X \psi) \wedge (\psi \Rightarrow_X \xi)) \rightarrow \langle\dashv\varphi \Rightarrow_X \psi\rangle\neg(\varphi \Rightarrow_X \xi) \tag{11}$$

$$\neg[X]\psi \rightarrow (\varphi \leftrightarrow [X\dashv\psi]\varphi) \tag{12}$$

$$\alpha \rightarrow [X\dashv\psi][X\dashv\psi]\alpha \quad for\ \alpha \in \mathcal{L}_{Prop} \tag{13}$$

$$[Y]\alpha \rightarrow [X\dashv\psi][X\dashv\psi][Y]\alpha \quad for\ \alpha \in \mathcal{L}_{Prop} \tag{14}$$

Proofs are omitted for space reasons but the theorems can easily be checked semantically. Let us spell out the intuitive readings of these formulae. Formulae 4 and 5 simply state the obvious consequences of the expansion of a context with a classificatory rule and with an obligation. Formula 6 states that if a state of affairs can possibly be permitted, then derogating the obligation for that state of affairs gives rise to a permission for that state of affairs. It is worth noticing that this captures a notion of "strong permission", as it is often called in the literature on deontic logic (see, for instance, [11]), that is, a permission which is obtained as the effect of an explicit derogation to norms in force. Formulae 8 and 9 describe recipes for appropriately updating contexts. For instance, Formula 9 roughly says that if I want to make $\neg\psi$ obligatory in φ-situations starting from a context where ψ is instead obligatory, I have to first derogate this latter obligation and then promulgate the desired one if I do not want to end up in situations where both ψ and $\neg\psi$ are obligatory. Formula 9 states that if φ is obligatory, then by derogating that φ is obligatory in ψ-situations, an exception is introduced so that φ is not obligatory in an unconditional way any more. Formula 10 says that, in the presence of a classificatory rule, by derogating the obligatoriness of the antecedent of the rule, we obtain a derogation of the obligatoriness of its consequent too. Finally, Formula 11 states that if I have two interpolated classificatory rules, by derogating one of them I undercut the conclusion I could draw by transitivity before the derogation. Formulae 12-14 are reminiscent of AGM postulates. Formula 12 expresses a form of minimality criterion, while Formulae 13 and 14 state two recovery principles for formulae belonging to a restricted language.

5   Related Work on Norm Change

Formal models of norm change have been drawing attention since the seminal work of Alchourrón and Makinson on the logical structure of derogation in legal codes [3] which expanded into a more general investigation of the logic of theory change (alias belief change) [2]. In this section we position our work with respect to AGM and related approaches to norm change available in the literature.

The first thing to notice about AGM is that its models are about the contraction of $\mathcal{L}_{Prop}$-theories, and focus on minimal change. In contrast, we consider here a modal language $\mathcal{L}_{Cxt}$. Our contraction operator "$-$" allows to express properties about a *family* of contractions, which actually do not necessarily satisfy the AGM criteria of minimal change. However, as shown in Proposition 4, our operator enjoys a minimality criterion (Formula 12) and two forms of recovery (Formulae 13 and 14). With respect to recovery it should be noticed, on the other hand, that formula $\neg[X]p \rightarrow [X-p][X+p]\neg[X]p$ is instead invalid, and hence that Formulae 13 and 14 do not generalize to all formulae in $\mathcal{L}_{Cxt}$.

Recently, norm change has gained quite some attention in the multi-agent systems community. As it is often the case, two main methodological approaches are recognizable: on the one hand syntactic approaches—inspired by legal practice—where norm change is considered as an operation performed directly on the explicit provisions contained in the "code" of the normative system [6, 8], and on the other hand semantic approaches, which are inspired by the dynamic logic paradigm [21] and which look at norm change as some form of model-update. Our contribution clearly falls in the second group and for this very reason our logic can be used for the formal specification and verification of computational models of norm-based interaction. Our approach is in fact close in spirit to Segerberg's [16], who argued for an integration of AGM belief revision with Hintikka-like static logics of belief: we here do the same for 'Andersonian' deontic logic.

From the proposals belonging to this latter group, it is worth comparing our work in particular with the approach proposed in [14]. There, an extension of the dynamic logic of permission (DLP) of [20] with operations of granting or revoking a permission was proposed. They call DLP_{dyn} this DLP extension. Their operations are similar to our operations of norm promulgation and norm derogation. DLP is itself an extension of PDL (propositional dynamic logic) [10] where actions are used to label transitions from one state to another state in a model. The DLP_{dyn} operation of granting a permission just augments the number of *permitted* transitions in a model, whereas the operation of revoking a permission reduces the number of *permitted* transitions. However there are important differences between our approach and Pucella & Weissman's. For us, normative systems are more basic than obligations and permissions, and the latter are defined from (and grounded on) the former. Moreover, dynamics of obligations and permissions are particular cases of normative system change (normative system expansion and contraction). Thus, we can safely argue that our approach is more general than Pucella & Weissman's in which only dynamics of permissions are considered. It is also to be noted that, while in our approach classificatory rules and their dynamics are crucial concepts in normative change, in DLP_{dyn} they are not considered and even not expressible. In future work we will analyze the relationships between DLP_{dyn} and our logic, and possibly a reduction of DLP_{dyn} to our logic **DCxt**.

While Pucella & Weissman's revocation of permissions corresponds to public announcements in DEL, no DEL approaches have proposed the counterpart of their operation of granting permissions, alias contractions (with the exception of [19], but in the framework of a logic of preference). Arguably, the reason for it is that it is difficult to define contraction operations both preserving standard properties of epistemic models such as transitivity and Euclidianity and allowing for reduction axioms. This is made instead possible in **DCxt** by the intercontextual interaction axioms 4^{XY} and 5^{XY}.

6   Conclusions

We have introduced and studied a dynamic logic accounting for context change, and have applied it to analyze several aspects of the dynamics of norms, viz. the dynamics of permissions, obligations and classificatory rules. Although the logic has been applied here only to provide a formal analysis of norm-change, it is clear that its range of application is much broader. Viewed in its generality, the logic is a logic of the dynamics of propositional theories, and as such, can be naturally applied to formal epistemology by studying theory-change, or to non-monotonic reasoning by studying how the context of an argumentation evolves during, for instance, a dialogue game. This kind of applications are future research.

Acknowledgments. Davide Grossi is supported by *Nederlandse Organisatie voor Weten-schappelijk Onderzoek* (VENI grant Nr. 639.021.816). Guillaume Aucher is supported by *Fond National de la Recherche* (Luxembourg). We also thank the anonymous reviewers for their comments.

References

1. Ågotnes, T., van der Hoek, W., Rodriguez-Aguilar, J.A., Sierra, C., Wooldridge, M.: On the logic of normative systems. In: Proc. of IJCAI 2007, pp. 1181–1186. AAAI Press, Menlo Park (2007)
2. Alchourrón, C., Gärdenfors, P., Makinson, D.: On the logic of theory change: Partial meet contraction and revision functions. J. of Symbolic Logic 50, 510–530 (1985)
3. Alchourrón, C., Makinson, D.: Hierarchies of regulations and their logic. In: Hilpinen, R. (ed.) Deontic Logic: Introductory and Systematic Readings. D. Reidel (1981)
4. Anderson, A.: A reduction of deontic logic to alethic modal logic. Mind 22, 100–103 (1958)
5. Blackburn, P., de Rijke, M., Venema, Y.: Modal Logic. Cambridge Univ. Press, Cambridge (2001)
6. Boella, G., Pigozzi, G., van der Torre, L.: Normative framework for normative system change. AAMAS (1), 169–176 (2009)
7. Ghidini, C., Giunchiglia, F.: Local models semantics, or contextual reasoning = locality + compatibility. Artificial Intelligence 127(2), 221–259 (2001)
8. Governatori, G., Rotolo, A.: Changing legal systems: abrogation and annulment (part I: revision of defeasible theories). In: van der Meyden, R., van der Torre, L. (eds.) DEON 2008. LNCS (LNAI), vol. 5076, pp. 3–18. Springer, Heidelberg (2008)
9. Grossi, D., Meyer, J.-J.C., Dignum, F.: The many faces of counts-as: A formal analysis of constitutive-rules. J. of Applied Logic 6(2), 192–217 (2008)
10. Harel, D., Kozen, D., Tiuryn, J.: Dynamic Logic. MIT Press, Cambridge (2000)
11. Hilpinen, R. (ed.): New Studies in Deontic Logic. Synthese Library Series. Reidel (1981)
12. Kanger, S.: New fondations for ethical theory. In: Hilpinen, R. (ed.) Deontic Logic: Introductory and Systematic Readings, pp. 36–58. Reidel Publishing Company (1971)

13. Lomuscio, A., Sergot, M.: Deontic intepreted systems. Studia Logica 75, 63–92 (2003)
14. Pucella, R., Weissman, V.: Reasoning about dynamic policies. In: Walukiewicz, I. (ed.) FOSSACS 2004. LNCS, vol. 2987, pp. 453–467. Springer, Heidelberg (2004)
15. Searle, J.R.: Speech acts: An essay in the philosophy of language. Cambridge Univ. Press, Cambridge (1969)
16. Segerberg, K.: Two traditions in the logic of belief: bringing them together. In: Ohlbach, H.J., Reyle, U. (eds.) Logic, Language and Reasoning: essays in honour of Dov Gabbay. Kluwer, Dordrecht (1999)
17. Stalnaker, R.: On the representation of context. J. of Logic, Language, and Information 7, 3–19 (1998)
18. Ågotnes, T., Wiebe van der Hoek, W., Tennenholtz, M., Wooldridge, M.: Power in normative systems. In: Decker, P., Sichman, J., Sierra, C., Castelfranchi, C. (eds.) Proceedings of the Eighth International Conference on Autonomous Agents and Multiagent Systems (AAMAS 2009), pp. 145–152 (2009)
19. van Benthem, J., Liu, F.: Dynamic logic of preference upgrade. J. of Applied Non-Classical Logics 17(2), 157–182 (2007)
20. Van der Meyden, R.: The dynamic logic of permission. J. of Logic and Computation 6, 465–479 (1996)
21. van Ditmarsch, H., van der Hoek, W., Kooi, B.: Dynamic Epistemic Logic. Synthese Library Series, vol. 337. Springer, Heidelberg (2007)

Toward a Dynamic Logic of Questions

Johan van Benthem[1,2] and Ştefan Minică[1]

[1] Institute for Logic, Language and Computation, University of Amsterdam,
P.O. Box 94242, 1090 GE Amsterdam, The Netherlands
{johan.vanbenthem, s.a.minica}@uva.nl
[2] Department of Philosophy, Stanford University, Stanford, CA, USA

Abstract. Questions are triggers for explicit events of 'issue management'. We give a complete logic in dynamic-epistemic style for events of raising, refining, and resolving an issue, all in the presence of information flow through observation or communication. We explore extensions of the framework to longer-term temporal protocols and multi-agent scenarios. We sketch a comparison with two main alternative accounts: Hintikka's interrogative logic and Groenendijk's inquisitive semantics.

Keywords: question, issue management, logical dynamics.

1 Introduction and Motivation

Questions are different from statements, but they are just as important in driving reasoning, communication, and general processes of investigation. The first logical studies merging questions and propositions seem to have come from the Polish tradition: cf. [12]. A forceful modern defender of this dual perspective is Hintikka, who has long pointed out how any form of inquiry depends on an interplay of inference and answers to questions. Cf. [13] and [14] on the resulting 'interrogative logic', and the epistemological views behind it. These logics are mainly about *general inquiry* and learning about the world. But there is also a related stream of work on the *questions* in natural language, as important speech acts with a systematic linguistic vocabulary. Key names are Groenendijk & Stokhof: cf. [16], [17], and the recent 'inquisitive semantics' of [18] ties this in with a broader information-oriented 'dynamic semantics'. Logic of inquiry and logic of questions are related, but there are also differences in thrust: a logic of 'issue management' that fits our intuitions is not necessarily the same as a logic of speech acts that must make do with what natural language provides.

In this paper, we do not choose between these streams, but we propose a different technical approach. Our starting point is a simple observation. Questions are evidently important informational actions in human agency. Now the latter area is the birth place of *dynamic-epistemic logic* of explicit events that make information flow. But surprisingly, existing dynamic-epistemic systems do not give an explicit account of what questions do! In fact, central examples in the area have questions directing the information flow (say, by the Father in the puzzle of the Muddy Children) – but the usual representations in systems like *PAL* or *DEL* leave them out, and merely treat the answers, as events of

X. He, J. Horty, and E. Pacuit (Eds.): LORI 2009, LNAI 5834, pp. 27–41, 2009.
© Springer-Verlag Berlin Heidelberg 2009

public announcement. Can we make questions themselves first-class citizens in dynamic-epistemic logic, and get closer to the dynamics of inquiry? We will show that we can, following exactly the methodology that has already worked in other areas, and pursuing the same general issues: what are natural acts of inquiry, and how can dynamic logics bring out their structure via suitable recursion axioms? Moreover, by doing so, we at once get an account of non-factual questions, multi-agent aspects, temporal sequences, and other themes that have already been studied in a *DEL* setting.

2   A Toy-System of Asking and Announcing

The methodology of dynamic-epistemic logic starts with a static base logic describing states of the relevant phenomenon, and identifies the key informational state-changing events. Then, dynamic modalities are added to the base language, and their complete logic is determined on top of the given static logic. To work in the same style, we need a convenient static semantics to 'dynamify'. We take such a model from existing semantics of public questions, considering only one agent first, for simplicity. We work in the style of public announcement logic *PAL*, though our logic of questions will also have its differences.

2.1   Epistemic Issue Models

A simple framework for representing questions uses an equivalence relation over some relevant domain of alternatives, that we will call the 'issue relation'. This idea is found in many places, from linguistics (cf. [16]) to learning theory (cf. [19]): the current 'issue' is a partition of the set of options, with partition cells standing for the areas where we would like to be. This partition may be induced by a conversation whose current focus are the issues that have been put on the table, or a game where finding out about certain issues has become important to further play, a learning scenario for the language fed to us by our environment, or even a whole research program with an agenda determining what is currently under investigation. The 'alternatives' or worlds may range here from simple finite settings like deals in a card game to complex infinite histories representing a total life experience. Formally, all this reduces to the following structure:

Definition 1 (Epistemic Issue Model). *An* epistemic issue model *is a structure $M = \langle W, \sim, \approx, V \rangle$ where:*
- *W is a set of possible worlds or states (epistemic alternatives),*
- *$\sim$ is an equivalence relation on W (epistemic indistinguishability),*
- *$\approx$ is an equivalence relation on W (the abstract issue relation),*
- *$V : \mathrm{P} \to \wp(W)$ is a valuation for atomic propositions $p \in \mathrm{P}$.*

We could introduce models with more general structure, but equivalence relations will suffice for the points that we are trying to make in this paper.

2.2   Static Language of Information and Issues

To work with these structures, we need matching modalities in our language. Here we make a minimal choice of modal and epistemic logic for state spaces plus two modalities describing the issue structure. First, $K\varphi$ talks about knowledge or semantic information of an agent, its informal reading is "φ is known", and its explanation is as usual: "φ holds in all epistemically indistinguishable worlds". To describe our models a bit further, we add a universal modality $U\varphi$ saying that "φ is true in all worlds". Next, we use $Q\varphi$ to say that, locally in a given world, the current structure of the issue-relation validates φ: "φ holds in all issue-equivalent worlds". While convenient, this local notion does not express the global assertion that the current issue *is* φ, which will be defined later. Finally, we find a need for a notion that mixes the epistemic and issue relations, talking (roughly) about what would be the case if the issue were resolved given what we already know. Technically, we add an intersection modality $R\varphi$ saying that "φ holds in all epistemically indistinguishable and issue equivalent worlds". While such modalities are frequent in many settings, they complicate axiomatization. We will assume the standard device of adding *nominals* naming single worlds (cf. [23] [24] for recent instances of this technique in the *DEL* setting).[1]

Definition 2 (Static Language). *The* language $\mathcal{L}_{\mathbf{ELQ}}(\mathsf{P}, \mathsf{N})$ *has disjoint countable sets* P *and* N *of propositions and nominals, respectively, with* $p \in \mathsf{P}$, $i \in$ N. *Its formulas are defined by:* $\qquad \perp \mid p \mid i \mid \neg\varphi \mid (\varphi \wedge \psi) \mid K\varphi \mid Q\varphi \mid R\varphi \mid U\varphi$.

Modal formulas of this static language are interpreted in the following way:

Definition 3 (Interpretation). *Formulas are interpreted in models* M *at worlds* w *with the usual Boolean clauses, and the following modal ones:*

$$M \models_w K\varphi \quad \textit{iff for all } v \in W : w \sim v \textit{ implies } M \models_v \varphi,$$
$$M \models_w Q\varphi \quad \textit{iff for all } v \in W : w \approx v \textit{ implies } M \models_v \varphi,$$
$$M \models_w R\varphi \quad \textit{iff for all } v \in W : w \, (\sim \cap \approx) \, v \textit{ implies } M \models_v \varphi,$$
$$M \models_w U\varphi \quad \textit{iff for all } w \in W : M \models_w \varphi,$$

This semantics validates a number of obvious principles reflecting connections between our modalities. In particular, the following are valid: $U\varphi \to K\varphi, U\varphi \to Q\varphi, U\varphi \to R\varphi$, and also $K\varphi \to R\varphi, Q\varphi \to R\varphi$. Corresponding facts hold for existential modalities $\widehat{U}$, etc., defined as usual.

Next, the *intersection modality* $R\varphi$ cannot be defined in terms of others. In particular, $\widehat{R}\varphi$ is not equivalent with $\widehat{K}\varphi \wedge \widehat{Q}\varphi$. However, the use of so-called 'nominals' i from hybrid logic helps us to completeness, by the valid converse:

$$\widehat{K}(i \wedge \varphi) \wedge \widehat{Q}(i \wedge \varphi) \to \widehat{R}\varphi$$

Our modal language can define various basic global statements describing the current structure of inquiry. For instance, here is how it says which propositions φ are 'settled' given the current structure of the issue-relation:

Definition 4 (Settlement). *The current issue settles fact* φ *iff* $U(Q\varphi \vee Q\neg\varphi)$.

---

[1] As one illustration, working with nominals requires a modified valuation function in Definition 1, to a $V : \mathsf{P} \uplus \mathsf{N} \to \wp(W)$ mapping every proposition $p \in \mathsf{P}$ to a set of states $V(p) \subseteq W$, but every nominal $i \in \mathsf{N}$ to a singleton set $V(i)$ of a world $w \in W$.

2.3   Static Base Logic of Information and Issues

As for reasoning with our language, we write $\models \varphi$ if the static formula φ is true in every model at every world. The static epistemic logic $\mathbf{EL_Q}$ of questions in our models is defined as the set of all validities: $\mathbf{EL_Q} = \{\varphi \in \mathcal{L}_{\mathbf{EL_Q}} : \ \models \varphi\}$.

We write $\vdash_s \varphi$ iff φ is provable in the proof system given in Table 1.

Table 1. The proof system $\mathbf{EL_Q}$

All propositional tautologies	Inclusion: $U\varphi \rightarrow K\varphi,\ U\varphi \rightarrow Q\varphi,\ U\varphi \rightarrow R\varphi$
S5 axioms for U,K,Q and R	Intersection: $\widehat{R}i \leftrightarrow \widehat{K}i \wedge \widehat{Q}i,\ K\varphi \rightarrow R\varphi, Q\varphi \rightarrow R\varphi$
Nominals: $\widehat{U}(i \wedge \varphi) \rightarrow U(i \rightarrow \varphi)$	Necessitation and Modus Ponens

These laws of reasoning derive many intuitive principles. For instance, here is how agents have introspection about the current public issue: $U(Qp \vee Q\neg p) \vdash UU(Qp \vee Q\neg p) \vdash KU(Qp \vee Q\neg p)$.

Theorem 1 (Completeness of $\mathbf{EL_Q}$). *For every formula* $\varphi \in \mathcal{L}_{\mathbf{EL_Q}}(\mathsf{P}, \mathsf{N})$:

$$\models \varphi \quad \textit{if and only if} \quad \vdash_s \varphi$$

Proof. By standard techniques for multi-modal hybrid logic.

2.4   Dynamic Actions of Issue Management

Now we look into basic actions that change the issue relation in a given model. We do this first by some pictures where epistemic indistinguishability is represented by links, and the issue relation by partition cells. For simplicity, we start with the initial issue as the universal relation, represented by the bordering frame.

In Figure 1, the first transition illustrates the effect of asking a question: the issue relation is split into p and $\neg p$ cells. The second transition illustrates the effect of asking a second question: the issue partition is further refined.

In Figure 2, the first transition models an announcement: indistinguishability links between p and $\neg p$ worlds are removed. The second transition is the effect of a second announcement, the epistemic partition is further refined. Here we use special events congenial to this setting, viz. the *link-cutting announcements* of van Benthem & Liu [9] that do not throw away worlds.

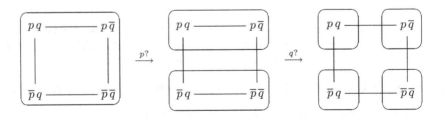

Fig. 1. Effects of Asking Yes/No Questions

There is a certain symmetry between asking a question and making a soft announcement. One refines the issue, the other the information partition:

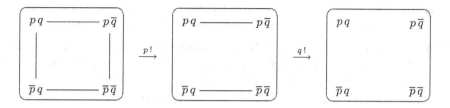

Fig. 2. Almost Symmetrical Effects of 'Soft' Announcing

Definition 5 (Questions & Announcements). *The execution of a φ? action in a given model M results in a changed model $M_{\varphi?} = \langle W_{\varphi?}, \sim_{\varphi?}, \approx_{\varphi?}, V_{\varphi?}\rangle$, with $\overset{\varphi}{\equiv}_M = \{(w,v) \mid \|\varphi\|^M_w = \|\varphi\|^M_v\}$. Likewise, the execution of a φ! action results in $M_{\varphi!} = \langle W_{\varphi!}, \sim_{\varphi!}, \approx_{\varphi!}, V_{\varphi!}\rangle$, and we then have:*

$$W_{\varphi?} = W \qquad\qquad W_{\varphi!} = W$$
$$\sim_{\varphi?} = \sim \qquad\qquad \sim_{\varphi!} = \sim \cap \overset{\varphi}{\equiv}_M$$
$$\approx_{\varphi?} = \approx \cap \overset{\varphi}{\equiv}_M \qquad\qquad \approx_{\varphi!} = \approx$$
$$V_{\varphi?} = V \qquad\qquad V_{\varphi!} = V$$

The symmetry in this mechanism is lost if we let p! be an executable action only if it is *truthful*, while the corresponding question p? is executable in every world in a model, even those not satisfying p.

This attractive setting suggests further operations on information and issues. Figure 3 contains two more issue management actions. In the first example two Yes/No questions p? and q? are asked and a *resolving* action follows on the

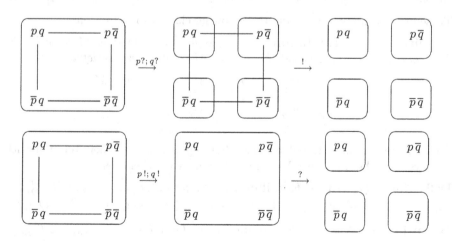

Fig. 3. Resolving and Refining Actions

epistemic relation. In the second, two announcements $p!$ and $q!$ are made, and a *refinement* action follows on the issue relation, adjusting it to what agents already know. These operations are natural generalizations of asking and announcing, that need not have natural language correspondents. They can be formally defined by abstract model operations as indicated in Definition 6.

Definition 6 (Resolution and Refinement). *The execution of the 'resolve' action* !, *and of the 'refine' action* ? *in model* M *results in a changed model* $M_! = \langle W_!, \sim_!, \approx_!, V_! \rangle$, *respectively,* $M_? = \langle W_?, \sim_?, \approx_?, V_? \rangle$ *with:*

$$
\begin{array}{ll}
W_? = W & W_! = W \\
\sim_? = \sim & \sim_! = \sim \cap \approx \\
\approx_? = \approx \cap \sim & \approx_! = \approx \\
V_? = V & V_! = V
\end{array}
$$

Again, these two actions are symmetric - suggesting that we could view, say, the 'issue manager' as an epistemic information agent.

In light of the previous observations we can also note that the 'refine' action behaves analogously to the intersection of indistinguishability relations in the usual treatment of "Distributed group knowledge" in the literature. This notion of group knowledge represents what the agents would know by pooling their information. This observation will become even more relevant in a multi-agent setting and in combination with other group notions for issue managment, defining an issue that is common to a group of agents.

2.5 Dynamic Language of Issue Management

In order to talk about the above changes, dynamic modalities are added to the earlier modal language of static epistemic situations:

Definition 7 (Dynamic Language). *Language* $\mathcal{L}_{\mathbf{DELQ}}(P, N)$ *is defined by adding the following clauses to Definition 2:* $\cdots$ $| \, [\varphi!]\psi \, | \, [\varphi?]\psi \, | \, [?]\varphi \, | \, [!]\varphi$

These are interpreted by adding the following clauses to Definition 3:

Definition 8 (Interpretation). *Formulas are interpreted in* M *at* w *by the following clauses, where models* $M_{\varphi?}$, $M_{\varphi!}$, $M_?$ *and* $M_!$ *are as defined above:*

$$
\begin{array}{lll}
M \models_w [\varphi!]\psi & \text{iff} & M \models_w \varphi \text{ implies } M_{\varphi!} \models_w \psi, \\
M \models_w [\varphi?]\psi & \text{iff} & M_{\varphi?} \models_w \psi, \\
M \models_w [?]\varphi & \text{iff} & M_? \models_w \varphi \\
M \models_w [!]\varphi & \text{iff} & M_! \models_w \varphi
\end{array}
$$

This language defines useful notions about questions and their relation with answers from the literature. We only mention two of them here:

Definition 9 (Question Entailment). *For formulas* $\varphi_0, \ldots, \varphi_n, \psi \in \mathcal{L}_{\mathbf{DELQ}}^{\mathrm{prop}}$
$?\varphi_0, \ldots, ?\varphi_n$ *entail* $?\psi$ *iff* $\models_p [\varphi_0?] \cdots [\varphi_n?] U((\psi \to Q\psi) \wedge (\neg\psi \to Q\neg\psi))$.

Definition 10 (Answer Compliance). *For formulas* $\varphi_0, \ldots, \varphi_n, \psi \in \mathcal{L}_{\mathbf{DELQ}}^{\mathrm{prop}}$
$?\varphi_0; \cdots ; ?\varphi_n$ *license* $!\psi$ *iff* $\models_p [\varphi_0?] \cdots [\varphi_n?] \neg((\neg\psi \wedge \widehat{Q}\psi) \vee (\psi \wedge \widehat{Q}\neg\psi))$.

An interesting property, also observed in a preference change context in [9] and [6], is that, unlike for standard announcements, implemented by world elimination, there is no 'action contraction' principle equating two or more successive questions to a single one with identical effect:

Fact 1 (Proper Iteration). *There is no question composition principle.*

Proof. If one single assertion had the same effect as a sequence $\varphi?; \psi?$, then, starting with the issue as the universal relation, such a sequence will always induce a two, not four, element partition (cf. Figure 3). □

Validities encode reasoning in advance about later epistemic effects of asking questions and answering them. Here is an example:

Fact 2 (Questioning Thrust). *The formula $K[\varphi?][!]U(K\varphi \vee K\neg\varphi)$ is valid.*

This says that agents know that the effect of a question followed by resolution is knowledge. Such results generalize to more complex types of questions. Thus, our logic encodes a formal base theory of question answering, usually investigated in the literature using rather ad-hoc approaches.

Finally, this system brings to light phenomena reminiscent of *DEL*. For instance, asking the same question repeatedly can have different effects on a model, as illustrated by the question: $(\widehat{Q}i \rightarrow (j \vee k)) \wedge ((\widehat{Q}j \wedge p) \rightarrow \widehat{Q}i)$ starting with $\approx$ as the universal relation in a three-world model i, j, k where p is true at k.

2.6   Complete Dynamic Logic of Informational Issues

Our examples show that predicting epistemic effects of asking questions is not always easy, but they also suggest an interesting algebra of operations on models. For both purposes, we axiomatize a complete dynamic epistemic logic of questions. Satisfaction and validity are defined as before. The dynamic epistemic logic of questioning based on a partition modeling (henceforth, **DEL$_Q$**) is defined as:

$$\mathbf{DEL_Q} = \{\varphi \in \mathcal{L}_{\mathbf{DEL_Q}}(\mathsf{P}, \mathsf{N}) : \; \models \varphi\}$$

We introduce a new proof system by adding the reduction axioms in Table 2 to the proof system for the static fragment from Table 1.

To save some space, in this short paper, Table 2 only lists axioms for two of the four dynamic modalities. *Soft announcements* $\langle\varphi!\rangle$ satisfy the usual *PAL*-style axioms given in [9], plus principles for its interaction with the two new base modalities involving questions, such as: $\langle\varphi!\rangle\widehat{Q}\psi \leftrightarrow (\varphi \wedge \widehat{Q}\langle\varphi!\rangle\psi)$. Also, the axioms for the *refinement action* [?] or are those listed below for the 'resolving' action with the modalities K and Q interchanged. We write $\vdash \varphi$ iff φ is provable in the system from Tables 1 and 2.

Theorem 2 (Soundness). *The reduction axioms in Table 2 are sound.*

Proof. We discuss two cases that go beyond mere commutation of operators. The first, (*Asking & Partition*), explains how questions refine a partition:

$$[\varphi?]Q\psi \leftrightarrow (\varphi \wedge Q(\varphi \rightarrow [\varphi?]\psi)) \vee (\neg\varphi \wedge Q(\neg\varphi \rightarrow [\varphi?]\psi))$$

Table 2. Reduction axioms for **DEL_Q**

Asking&Atoms: $[\varphi?]a \leftrightarrow a$	*Asking&Negation:* $[\varphi?]\neg\psi \leftrightarrow \neg[\varphi?]\psi$
Asking&Conj.: $[\varphi?](\psi \wedge \chi) \leftrightarrow [\varphi?]\psi \wedge [\varphi?]\chi$	*Asking&Knowledge:* $[\varphi?]K\psi \leftrightarrow K[\varphi?]\psi$
Asking&Partition: $[\varphi?]Q\psi \leftrightarrow (\varphi \wedge Q(\varphi \to [\varphi?]\psi)) \vee (\neg\varphi \wedge Q(\neg\varphi \to [\varphi?]\psi))$	
Asking&Intersection: $[\varphi?]R\psi \leftrightarrow (\varphi \wedge R(\varphi \to [\varphi?]\psi)) \vee (\neg\varphi \wedge R(\neg\varphi \to [\varphi?]\psi))$	
Asking&Universal: $[\varphi?]U\psi \leftrightarrow U[\varphi?]\psi$	*Resolving&Atoms:* $[!]a \leftrightarrow a$
Resolving&Negation: $[!]\neg\varphi \leftrightarrow \neg[!]\varphi$	*Resolving&Conj.:* $[!](\psi \wedge \chi) \leftrightarrow [!]\psi \wedge [!]\chi$
Resolving&Knowledge: $[!]K\varphi \leftrightarrow R[!]\varphi$	*Resolving&Partition:* $[!]Q\varphi \leftrightarrow Q[!]\varphi$
Resolving&Intersection: $[!]R\varphi \leftrightarrow R[!]\varphi$	*Resolving&Universal:* $[!]U\varphi \leftrightarrow U[!]\varphi$

(from left to right) Assume $M \models_w [\varphi?]Q\psi$ then we also have $M_{\varphi?} \models_w Q\psi$. In case $M \models_w \varphi$, suppose $M \models_w (\varphi \wedge Q(\varphi \to [\varphi?]\psi)) \vee (\neg\varphi \wedge Q(\neg\varphi \to [\varphi?]\psi))$ does not hold, then we can proceed by cases. If $M \models_w \neg Q(\varphi \to [\varphi?]\psi)$ and $M \models_w \varphi$, then we have $\exists v \in W : w \approx v$ and $M \models_v \varphi \wedge \neg[\varphi?]\psi$, therefore, $w \overset{\varphi}{\equiv} v$, and from this we have $w \approx_{\varphi?} v$. But we also have $M_{\varphi?} \models_v \neg\psi$, hence $M_{\varphi?} \models_w \neg Q\psi$, which contradicts our initial assumption. For the remaining interesting case $M \models_w \neg Q(\varphi \to [\varphi?]\psi)$ and $M \models_w \neg Q(\neg\varphi \to [\varphi?]\psi)$ the argument is similar. In case $M \models_w \neg\varphi$ we can reason analogously.

Our second illustration, (*Resolving & Knowledge*), shows how resolution changes knowledge making crucial use of our intersection modality:

$$[!]K\varphi \leftrightarrow R[!]\varphi$$

Let $M \models_w [!]K\varphi$. Then we have equivalently, $M_! \models_w K\varphi$ and from this we get $\forall v \in W_! : w \sim_! v$ implies $M_! \models_v \varphi$. As $\sim_! = \sim \cap \approx$, we can obtain equivalently $\forall v \in W : w (\sim \cap \approx) v$ implies $M_! \models_v \varphi$, and from this, by the semantics of our dynamic modality, we get $M \models_w R[!]\varphi$ as desired.

Theorem 3 (Completeness of DEL_Q). *For every formula* $\varphi \in \mathcal{L}_{\mathbf{DEL_Q}}(\mathtt{P}, \mathtt{N})$:

$$\models \varphi \quad \text{if and only if} \quad \vdash \varphi.$$

Proof. Proceeds by a standard *DEL*-style translation argument. Working inside out, the reduction axioms translate dynamic formulas into corresponding static ones, in the end completeness for the static fragment is invoked.

Remark (Hidden validities). Although DEL_Q is complete, like PAL, it leaves something to be desired. We said already that model operations of issue management have a nice algebraic structure. For instance, resolving is idempotent: $!; ! = !$, while it commutes with refinement: $!; ? = ?; !$. But our logic does not state such facts explicitly, since, by working from innermost occurrences of dynamic modalities, our completeness argument needed no recursion axioms with stacked modalities like $[!][!]$. Still, this is crucial information for a logic of issue management, and schematic validities about operator stacking remain to be investigated.

3   Temporal Protocols with Questions

Single announcements usually only make sense in a longer-term temporal perspective of an informational process: a conversation, experimental protocol, or learning mechanism. To make this procedural information explicit, van Benthem, Gerbrandy, Hoshi & Pacuit 2009 [5] introduced *protocols* into dynamic-epistemic logic. This results in a modified public announcement logic *PAL*, that now encodes procedural as well as factual and epistemic information.

But the same applies to questions: not everything can be asked, because of social convention, resource limitations, etc. Thus, we enrich our dynamic logic with protocols, toward a more realistic theory of inquiry.

Definition 11 (DEL$_Q$ **Protocol**). *Let Σ be an arbitrary set of epistemic events (questioning actions). Let Σ^* be the set of finite strings over Σ (finite histories of questioning events). A questioning protocol is a set $\mathcal{H} \subseteq \Sigma^*$ (containing all non-empty finite histories and all their prefixes, or rooted sub-histories) such that $\mathsf{FinPre}_{-\lambda}(\mathcal{H}) = \{h \mid h \neq \lambda, \exists h' \in \mathcal{H} : h \preceq h'\} \subseteq \mathcal{H}$.*

During the construction in the following definition the only sequences considered are those of the form $w\sigma$, where w is a world in the initial model M, and σ a sequence in the protocol Q, σ_n denotes the sequence σ up to its n-th position and $\sigma_{(n)}$ denotes the n-th element in the sequence.

Definition 12 (**Q-Generated Model**). *Let $M = \langle W, \sim, \approx, V \rangle$ be an arbitrary model and let Q be an arbitrary DEL$_Q$ protocol over model M (a prefix-closed set of finite sequences of questioning events). The Q-Generated Model at level n, $M_Q^n = \langle W_Q^n, \sim_Q^n, \approx_Q^n, V_Q^n \rangle$ is defined by induction on n as follows:*

1 $W_Q^0 = W$, $\quad \sim_Q^0 = \sim$, $\quad \approx_Q^0 = \approx$, $\quad V_Q^0 = V$,

2 $w\sigma \in W_Q^{n+1}$ iff $w \in \mathrm{dom}(M)$, $\sigma \in Q$, $\mathrm{len}(\sigma) = n+1$, and $w\sigma_n \in W_Q^n$,

3 If $\sigma_{(n+1)} = \langle ! \rangle$ then: (a) $(w\sigma, v\sigma') \in \sim_Q^{n+1}$ iff $(w\sigma_n, v\sigma'_n) \in \sim_Q^n$, $(w\sigma_n, v\sigma'_n) \in \approx_Q^n$, and $\sigma_{(n+1)} = \sigma'_{(n+1)}$; and (b) $(w\sigma, v\sigma') \in \approx_Q^{n+1}$ iff $(w\sigma_n, v\sigma'_n) \in \approx_Q^n$, and $\sigma_{(n+1)} = \sigma'_{(n+1)}$,

4 If $\sigma_{(n+1)} = \langle \varphi? \rangle$ then: (a) $(w\sigma, v\sigma') \in \approx_Q^{n+1}$ iff $\sigma_{(n+1)} = \sigma'_{(n+1)}$, $(w\sigma_n, v\sigma'_n) \in \approx_Q^n$, and $(\sigma_{(n+1)}, \sigma'_{(n+1)}) \in \equiv_{M_Q^n}$; and (b) $(w\sigma, v\sigma') \in \sim_Q^{n+1}$ iff $(w\sigma_n, v\sigma'_n) \in \sim_Q^n$, and $\sigma_{(n+1)} = \sigma'_{(n+1)}$.

The class of structures $\mathsf{Forest}(\mathrm{TDEL}_Q)$ consists of all models $\mathsf{Forest}(M, Q)$ for some arbitrary model M and some arbitrary TDEL_Q protocol Q.

Next we give a truth definition for a suitable dynamic language, where we assume that all the dynamic actions involve formulas in the static base language only. Also, in Definition 13 q is used as a variable for issue management actions.

Definition 13 (**Interpretation**). *Formulas are interpreted at state h in model $\mathsf{Forest}(M, \mathcal{Q}) := \mathsf{Fr}(M, \mathcal{Q}) = \langle \mathcal{H}, \sim, \approx, V \rangle$, by the following recursive definition:*

- $\mathsf{Fr}(M, \mathcal{Q}) \models_h K\varphi$ *iff* $\forall h \in \mathcal{H} : h \sim h'$ *implies* $\mathsf{Fr}(M, \mathcal{Q}) \models_{h'} \varphi$
- $\mathsf{Fr}(M, \mathcal{Q}) \models_h Q\varphi$ *iff* $\forall h \in \mathcal{H} : h \approx h'$ *implies* $\mathsf{Fr}(M, \mathcal{Q}) \models_{h'} \varphi$
- $\mathsf{Fr}(M, \mathcal{Q}) \models_h R\varphi$ *iff* $\forall h \in \mathcal{H} : h (\sim \cap \approx) h'$ *implies* $\mathsf{Fr}(M, \mathcal{Q}) \models_{h'} \varphi$
- $\mathsf{Fr}(M, \mathcal{Q}) \models_h \langle q \rangle \varphi$ *iff* $hq \in \mathcal{H}$ *and* $\mathsf{Fr}(M, \mathcal{Q}) \models_{hq} \varphi$

One feature that distinguishes $TDEL_Q$ from our earlier system is this. Even "Yes/No" questions φ? with tautological preconditions $\varphi \vee \neg\varphi$ need not always be available for inquiry. Thus, as in PAL with protocols, the earlier recursion axioms of DEL_Q have to be modified as in these two samples of $TDEL_Q$-axioms:

(Resolving & Knowledge) : $\langle ! \rangle K\varphi \leftrightarrow \langle ! \rangle \top \wedge R \langle ! \rangle \varphi$

(Ask&Part) : $\langle \varphi? \rangle Q\psi \leftrightarrow \langle \varphi? \rangle \top \wedge ((\varphi \wedge Q(\varphi \rightarrow \langle \varphi? \rangle \psi)) \vee (\neg\varphi \wedge Q(\neg\varphi \rightarrow \langle \varphi? \rangle \psi)))$

Such new axioms describe the procedural restrictions that drive conversations or processes of inquiry and discovery and leads to a logical system which is sound and complete (cf. [7] for the full list of axioms and proofs of the results).

In $TDEL_Q$, $\langle \varphi? \rangle \top$ means that the question φ? can be asked. In general, $\langle q \rangle \top$ will mean that the issue management action q is available for execution.[2]

4   Multi-agent Scenarios

Questions typically involve more than one person. Indeed, our system is easily extended, high-lighting aspects lacking in the usual single-agent approaches.

4.1   Multi-agent DEL_Q with Public Issues

It is easy to generalize earlier definitions, marking accessibility relations and modalities with agent subscripts. Complete logics are as before, since as in epistemic logic, we do not expect the logic itself to enforce significant interaction principles tying agents together.

The multi-agent setting is essential with *preconditions* of questions, referring to both questioner and answerer:

> one precondition of $e_1 = $ "b asks φ" is $\neg K_b \varphi \wedge \neg K_b \neg\varphi$.

The questioner must not know the answer to the question she asks. But questions are also asked to be answered, in general, by another agent:

> a complex epistemic event $e_3 = $ "b asks φ to a" also has the
> precondition that the questioner must consider it possible
> that the answerer knows the answer: $\widehat{K}_b(K_a \varphi \vee K_a \neg\varphi)$.

---

[2] We have considered here only uniform protocols restricted to asking φ and resolution questioning actions. Of course, it is possible to add the remaining questioning actions of announcing φ and refinement to this setting in a standard way.

These observations suggest that the following definition might be useful:

$$M \models_w \langle\varphi?\rangle_a^b \psi \text{ iff } M \models_w (\neg K_a\varphi \wedge \neg K_a\neg\varphi) \wedge \widehat{K}_a(K_b\varphi \vee K_b\neg\varphi) \text{ and } M_{\varphi?} \models_w \psi$$

Our logic describes such more realistic questions – and many other events.[3]

There are also interesting further issues here. In the above event, can we *separate* informative preconditions from questions per se? That is, we first announce the precondition, and then perform the issue change? For factual assertions, this works well, but in general, there is a problem, since announcing the precondition may change the correct answer to the question. Thus, we may have to analyze complex question events as one unit, writing their recursion axioms separately.[4]

4.2   Multi-agent DEL_Q with Private Issues

So far, questions were public events. But in many scenarios, there may be private aspects, reflecting partial information or other observational limitations.

One obvious scenario is a public question followed by a private answer, like what happens in many classrooms. This is easily dealt with by attaching our logic of public questions to the logic DEL of private announcements. But there can also be private questions, with either private or public answers. For instance, agent a can ask b if φ, while c does not hear it. Or, c may just have been unable to hear if the question asked was $P?$ or $Q?$. Such scenarios call for events that modify the issue relation in ways that are different for different agents. In the extended version of this paper (van Benthem & Minică 2009 [7]), we give a generalization of the *product update* mechanism of DEL to deal with issue management in the presence of privacy.[5]

Other multi-agent issues concern the formation of *groups* of agents. In particular, when many agents have different views of the issues, they may *merge* their issue relations in one 'common refinement'. This relation is a natural candidate for the 'collective issue' of the whole group, and thus, we now also have group versions of our modalities K, Q, R, linking common issues to common knowledge.

5   Further Directions, Comparisons and Conclusions

Further Agent Attitudes: Beliefs and Preferences. We have studied the interaction of questions with knowledge. But of course, agents' *beliefs* are just as important, and we can also merge the preceding analysis with dynamic logics of belief change. Thus, our question dynamics might be added to the DEL-style belief logics of van Benthem 2007 [4] and Baltag & Smets 2007 [11].

---

[3] Indeed, the logic should be flexible here. Different types of question can have different preconditions: e.g., rhetorical questions have none of the above.

[4] Incidentally, a multi-agent setting may also change our views of the effects of answers. For instance, as observed in van Benthem (One is a Lonely Number) [1], an *answer* like "I do not know" can be highly informative!

[5] Of course, this requires refined epistemic issue models where the structure of the issue for different agents is no longer common knowledge.

Beyond beliefs, questions can also affect other agent attitudes. For instance, a question can give us information about other agents' goals and *preferences*. This would come out concretely by adding question dynamics to the preference logics of Girard 2008 [23] and Liu 2008 [24].[6]

Questioning Games. An interesting further development is an application of our analysis to epistemic games like those developed for public announcements by Ågotnes and van Ditmarsch in [10]. In Public Announcement Games players have to find the optimal announcement to make in order to reach their epistemic goals given their knowledge. Considering games in which the available moves for the players include both announcements and questions is a way in which the value of a question can receive a precise game-theoretical definition.

Update, Inference, and Syntactic Awareness Dynamics. While *DEL* has been largely about observation-based semantic information, some recent proposals have extended it to include more finely grained information produced by inference or introspection. The same sort of move makes sense in our current setting. This would work well in the syntactic approach to inferential and other fine-grained information in van Benthem & Quesada 2009 [8], with questions providing one reason for their acts of 'awareness promotion'. The latter take would also fit well with Hintikka's emphasis on the combination of questions and deductions as driving inquiry.

Multi-agent Behavior over Time. We have already seen that, like assertions, questions make most sense in the context of some longer temporal process. A single question is hard to 'place' without a scenario. Our study of *protocols* was one step in this direction, but obviously, we also need to make our dynamic logics of questions work in analyses of extended conversation, or especially, *games*. Another perspective where this makes eminent sense are *learning* scenarios, where asking successive local questions would seem a very natural addition to the usual input streams of answers (cf. Kelly 1996 [19]) to one unchanging grand question which global hypothesis about the actual history is the correct one.

Structured Issues and Agenda Dynamics. To us, the most striking limitation of our current approach is the lack of structure in our epistemic issue models. Surely, both in conversation and in general investigation, the *agenda* of relevant issues is much more delicate than just some equivalence relation. If we are to have any realistic logical account of, say, the development of research programs, we need to understand this more finely-grained dynamics.

Moreover, there are already models that allow for this sort of dynamics. Girard 2008 [23], Liu 2008 [24] consider, essentially, 'priority graphs' of ordered relevant propositions (first studied in Andreka, Ryan & Schobbens 2001 [21]) that can be used for this purpose. Priority graphs can encode a structured family of issues,

---

[6] Indeed, there are formal analogies between our question update operation and the 'ceteris paribus' preferences of van Benthem, Girard & Roy 2008 [6].

and they allow for a larger repertoire of inserting or deleting questions.[7] For a first system of this more structured sort we refer to ([7]).

5.1 Comparisons with Other Approaches

We have mentioned several other approaches to the logic of questions. There is the tradition of erotetic logic in the sense of Wiśniewski [12], and the slightly later classic Belnap & Steel 1976 [20].

More directly connected to our approach, we have mentioned the still active program of Hintikka for *interrogative logic* [13]. Questions are treated here as requests for new information, which function intertwined with deductive indicative moves in 'interrogative tableaux'. The framework has a number of nice theoretical results, including meta-theorems about the scope of questioning in finding the truth about some given situation. Clearly, several of these results would also be highly relevant to what we are doing here, and a merge of the two approaches might be of interest, bringing out Hintikka's concerns even more explicitly in a dynamic epistemic setting.

The closest comparison to our approach is the inquisitive semantics ([18] [15]). Inquisitive semantics gives propositions an 'interrogative meaning' defined in a universe of information states over propositional valuations, with sets of valuations expressing issues. First, a compositional semantics is given, evaluating complex propositions in sets of worlds, viewed as information states. Based on this semantics, a propositional logic arises that describes valid consequence and other relations between questions, and questions with answers.

At some level of abstraction, the ideas in this system sound very close to ours: there is information dynamics, questions change current partitions, etcetera. But the eventual system turns out to be an intermediate propositional logic in between intuitionistic and classical logic. Comparing the two approaches is an enterprise we leave for another occasion, though Icard 2009 (seminar presentation, Stanford) has suggested that there might be both translations between the two systems, and natural merges. Inquisitive semantics puts the dynamic information about questions in a new account of the meaning of interrogative sentences in a propositional language. By contrast, dynamic-epistemic logic wants to give an explicit account of questions and other actions of issue management, but it does so by means of dynamic modalities on top of a classical logical language. The distinction is similar to one in logic itself (van Benthem 1993, 'Reflections on Epistemic Logic' [3]). Intuitionistic logic studies knowledge and information *implicitly* by changing the meaning of the classical logical constants, and then picking a fight with classical logic in the set of 'validities'. By contrast, epistemic logic analyzes knowledge *explicitly* as an additional operator on top of classical propositional logic: there is no meaning shift, but agenda expansion. In our view, dynamic-epistemic logic of questions stands in exactly the same relationship to inquisitive semantics: it makes the dynamics explicit, and steers away from foundational issues of meaning and logic. Comparisons between the

---

[7] Being good at research seems to imply being able to ask good questions just as much as giving clever answers.

two approaches can be quite delicate (van Benthem 2008, 'The Information in Intuitionistic Logic' [2]), and the same may also be true here.

6 Conclusion

The dynamic calculi of questions in this paper show how dynamic-epistemic logic can incorporate a wide range of what we have dubbed 'issue management' beyond mere information handling. Our contribution is showing how this can be defined precisely, leading to complete dynamic logics that fit naturally with existing systems of DEL, broadly construed. Moreover, we have indicated how these systems can be used to explore properties of issue management beyond what is found in other logics of questions, including complex epistemic assertions, many agents, and explicit dynamics.[8]

Even so, we do feel that our systems are only a first step - still far removed from the complex structures of issues that give direction to rational agency. The insight itself that the latter are crucial comes from other traditions, as we have observed, but we hope to have shown that dynamic-epistemic logic has something of interest to contribute.

Acknowledgments. We first started developing these ideas about half a year ago, inspired by the 'inquisitive semantics' of [18] which, to us, raised the issue how the phenomena covered there (and others) would be dealt with from a dynamic-epistemic perspective. In the meantime, we have profited from comments on various drafts of this paper from Viktoria Denisova, Solomon Feferman, Tomohiro Hoshi, Thomas Icard, Floris Roelofsen, Lena Kurzen, Cedric Degremont, Fernando Velazquez-Quesada, and George Smith. We also thank two anonymous LORI-II referees for interesting comments and improvement suggestions. The junior author aknowledges Nuffic for the Huygens scholarship.

References

1. van Benthem, J.: One is a Lonely Number. In: Chatzidakis, Z., Koepke, P., Pohlers, W. (eds.) Logic Colloquium 2002, pp. 96–129. ASL & A.K. Peters, Wellesley, MA (2006)
2. van Benthem, J.: The Information in Intuitionistic Logic. Synthese 2, 167 (2008)
3. van Benthem, J.: Reflections on Epistemic Logic. Logique et Analyse 34 (1993)
4. van Benthem, J.: Dynamic Logic of Belief Revision. Journal of Applied Non-classical Logics 2, 17 (2007)
5. van Benthem, J., Gerbrandy, J., Hoshi, T., Pacuit, E.: Merging Frameworks of Interaction. Journal of Philosophical Logic (2009)
6. van Benthem, J., Girard, P., Roy, O.: Everything Else Being Equal: A Modal Logic for Ceteris Paribus Preferences. Journal of Philosophical Logic 38, 83–125 (2008)

---

[8] However, we have not arrived at any definite conclusion about the formal relationships between our dynamic logics and existing alternatives. Perhaps all of them are needed to get the full picture of issue management.

7. van Benthem, J., Minică, Ş.: Dynamic Logic of Questions. Institute for Logic, Language and Computation, working paper (2009)
8. van Benthem, J., Velazquez-Quesada, F.R.: Inference, Promotion and the Dynamics of Awareness. ILLC, working paper (2009)
9. van Benthem, J., Liu, F.: Dynamic Logic of Preference Upgrade. Journal of Applied Non-Classical Logic 17, 157–182 (2007)
10. Ågotnes, T., van Ditmarsch, H.: But what will Everyone Say? - Public Announcement Games. In: Proceedings of Logic, Game Theory and Social Choice 6, Tsukuba, August 26-29 (2009)
11. Baltag, A., Smets, S.: From Conditional Probability to the Logic of Doxastic Actions. In: Proceedings of the 11th Conference on Theoretical Aspects of Rationality and Knowledge, New York, pp. 52–61 (2007)
12. Wiśniewski, A.: The Posing of Questions: Logical Foundations of Erotetic Inferences. Kluwer Academic Publishers, Dordrecht (1995)
13. Hintikka, J., Halonen, I., Mutanen, A.: Interogative Logic as a General Theory of Reasoning. In: Gabbay, D., Johnson, R., Ohlbach, H. (eds.) Handbook of the Logic of Argument and Inference. Elsevier, Amsterdam (2002)
14. Hintikka, J.: Socratic Epistemology: Explorations of Knowledge-Seeking by Questioning. Cambridge University Press, Cambridge (2007)
15. Ciardelli, I., Roelofsen, F.: Generalized Inquisitive Logic: Completeness via Intuitionistic Kripke Models. In: Proceedings of the 12th Conference on Theoretical Aspects of Rationality and Knowledge, California, pp. 71–80 (2009)
16. Groenendijk, J., Stokhof, M.: Questions. In: van Benthem, J., ter Meulen, A. (eds.) Handbook of Logic and Language. Elsevier, Amsterdam (1997)
17. Groenendijk, J.: The Logic of Interrogation: Classical Version. In: Matthews, T., Strolovitch, M. (eds.) SALT IX: Semantics and Linguistic Theory (1999)
18. Groenendijk, J.: Inquisitive Semantics: Two Possibilities for Disjunction. In: 7th International Tibilisi Symposium on Language, Logic & Computation, Tibilisi (2008)
19. Kelly, K.: The Logic of Reliable Inquiry. Oxford University Press, Oxford (1996)
20. Belnap, N., Steel, T.: The Logic of Questions and Answers. Yale Univ. Press (1976)
21. Andérka, H., Ryan, M., Schobbens, P.-Y.: Operators and Laws for Combining Preference Relations. Journal of Logic and Computation 1, 12 (2002)
22. van Ditmarsch, H.: Knowledge Games. Ph.D. Thesis, Institute for Logic, Language and Computation, Amsterdam (2000)
23. Girard, P.: Modal Logic for Belief and Preference Change. Ph.D. Thesis. Stanford University, Stanford (2008)
24. Liu, F.: Changing for the Better - Preference Dynamics and Agent Diversity. Ph.D. Thesis, Institute for Logic, Language and Computation, Amsterdam (2008)

A General Family of Preferential Belief Removal Operators

Richard Booth[1], Thomas Meyer[2], and Chattrakul Sombattheera[1]

[1] Mahasarakham University, Faculty of Informatics, Mahasarakham 44150 Thailand
{richard.b, chattrakul.s}@msu.ac.th
[2] Meraka Institute, CSIR and School of Computer Science, University of
Kwazulu-Natal, South Africa
tommie.meyer@meraka.org.za

Abstract. Most belief change operators in the AGM tradition assume
an underlying plausibility ordering over the possible worlds which is transitive and *complete*. A unifying structure for these operators, based on
supplementing the plausibility ordering with a second, guiding, relation
over the worlds was presented in [5]. However it is not always reasonable
to assume completeness of the underlying ordering. In this paper we generalise the structure of [5] to allow incomparabilities between worlds.
We axiomatise the resulting class of belief removal functions, and show
that it includes an important family of removal functions based on *finite
prioritised belief bases*.

1   Introduction

The problem of *belief removal* [1,5,19], i.e., the problem of what an agent, hereafter $\mathcal{A}$, should believe after being directed to remove some sentence from his
stock of beliefs, has been well studied in philosophy and in AI over the last 25
years. During that time many different families of removal functions have been
studied. A great many of them are based on constructions employing *total preor-
ders* over the set of possible worlds which is meant to stand for some notion $\leq$ of
relative *plausibility* [12]. A unifying construction for these families was given in
[5], in which a general construction was proposed which involved supplementing
the relation $\leq$ with a second, guiding, relation $\preceq$ which formed a subset of $\leq$.
By varying the conditions on $\preceq$ and its interaction with $\leq$ many of the different
families can be captured as instances.

The construction in [5] achieves a high level of generality, but one can argue it
fails to be general enough in one important respect: the underlying plausibility
order $\leq$ is *always* assumed to be a total preorder which by definition implies it
is *complete*, i.e., for any two worlds x, y, we have either $x \leq y$ or $y \leq x$. This
implies that agent $\mathcal{A}$ is *always* able to decide which of x, y is more plausible. This
is not always realistic, and so it seems desirable to study belief removal based on
plausibility orderings which allow *incomparabilities*. A little work been done on
this ([3,8,9,12,17], and especially the choice-theoretic approach to belief change
advocated in [18]) but not much. This is in contrast to work in nonmonotonic

X. He, J. Horty, and E. Pacuit (Eds.): LORI 2009, LNAI 5834, pp. 42–54, 2009.

reasoning (NMR), the research area which is so often referred to as the "other side of the coin" to belief change. In NMR, semantic models based on incomplete orderings are the norm, with work dating back to the seminal papers on *preferential models* of [13,20]. Our aim in this paper is to relax the completeness assumption from [5] and to investigate the resulting, even more general class of removal functions.

The plan of the paper is as follows. In Sect. 2 we give our generalised definition of the construction from [5], which we call *(semi-modular) contexts*. We describe their associated removal functions, as well as mention the characterisation from [5]. Then in Sect. 3 we present an axiomatic characterisation of the family of removal functions generated by semi-modular contexts. Then, in Sect. 4 we mention a couple of further restrictions on contexts, leading to two corresponding extra postulates. In Sect. 5 we mention an important subfamily of the general family, i.e., those removals which may be generated by a finite prioritised base of *defaults,* before moving on to AGM style removal in Sect. 6. We conclude in Sect. 7.

Preliminaries: We work in a finitely-generated propositional language L. The set of non-tautologous sentences in L is denoted by L_*. The set of propositional worlds/models is W. For any set of sentences $X \subseteq L$, the set of worlds which satisfy every sentence in X is denoted by $[X]$. Classical logical consequence and equivalence are denoted by $\vdash$ and $\equiv$ respectively. As above, we let $\mathcal{A}$ denote some agent whose beliefs are subject to change. A *belief set* for $\mathcal{A}$ is represented by a single sentence which is meant to stand for all its logical consequences. A *belief removal function* (hereafter just *removal function*) belonging to $\mathcal{A}$ is a unary function $*$ which takes any non-tautologous sentence $\lambda \in L_*$ as input and returns a new belief set $*(\lambda)$ for $\mathcal{A}$ such that $*(\lambda) \nvdash \lambda$. For any removal function $*$ we can always derive an associated belief set. It is just the belief set obtained by removing the contradiction, i.e., $*(\bot)$.

The following definitions about orderings will be useful in what follows. A binary relation R over W is:

- *reflexive* iff $\forall x : xRx$
- *transitive* iff $\forall x, y, z : xRy$ & $yRz \rightarrow xRz$
- *complete* iff $\forall x, y : xRy \lor yRx$
- a *preorder* iff it is reflexive and transitive
- a *total preorder* iff it is a complete preorder

The above notions are used generally when talking of "weak" orderings, where xRy is meant to stand for something like "x is *at least as good as* y". However in this paper, following the lead of [17], we will find it more natural to work under a *strict* reading, where xRy denotes "x is *strictly better than* y". In this setting, the following notions will naturally arise. R is:

- *irreflexive* iff $\forall x : \text{not}(xRx)$
- *modular* iff $\forall x, y, z : xRy \rightarrow (xRz \lor zRy)$
- a *strict partial order (spo)* iff it is both irreflexive and transitive
- the *strict part of* another relation R' iff $\forall x, y : xRy \leftrightarrow (xR'y$ & $\text{not}(yR'x))$
- the *converse complement* of R' iff $\forall x, y : xRy \leftrightarrow \text{not}(yR'x)$

We have that R is a modular spo iff it is the strict part of a total preorder [15]. So in terms of *strict* relations, much of the previous work on belief removal, including [5], assumes an underlying strict order which is a modular spo. It is precisely the modularity condition which we want to relax in this paper.

Given any ordering R and $x \in W$, let $\nabla_R(x) = \{z \in W \mid zRx\}$ be the set of all worlds below x in R. Then we may define a new binary relation $\sqsubseteq^R$ from R by setting $x \sqsubseteq^R y$ iff $\nabla_R(x) \subseteq \nabla_R(y)$. That is, $x \sqsubseteq^R y$ iff every element below x in R is also below y in R. It is easy to check that if R is a modular spo then $x \sqsubseteq^R y$ iff not (yRx), i.e., $\sqsubseteq^R$ is just the converse complement of R.

2   Contexts, Modular Contexts and Removals

In this section we set up our generalised definition of a context, show how each such context yields a removal function and vice versa, and recap the main results from [5].

2.1   Contexts

We assume our agent $\mathcal{A}$ has in his mind *two* binary relations $(<, \prec)$ over the set W. The relation $<$ is a *strict* plausibility relation which forms the basis for $\mathcal{A}$'s actionable beliefs, i.e., $x < y$ means that, to $\mathcal{A}$'s mind, and on the basis of all available evidence, *world x is strictly more plausible than y*. We assume $<$ is a strict partial order. In addition to this there is a second binary relation $\prec$. This relation is open to several different interpretations, but the one we attach is as follows: $x \prec y$ means "*$\mathcal{A}$ has **an explicit reason** to hold x more plausible than y (or to treat x more favourably than y)*". We will use $\preceq$ to denote the converse complement of $<$, i.e.,$x \preceq y$ iff $y \not< x$. Thus $x \preceq y$ iff $\mathcal{A}$ has no reason to treat y more favourably than x. Note $\preceq$ and $\prec$ are interdefinable, and we find it convenient to switch between them freely.

Note the equivalence "$x \prec y$ iff both $x \preceq y$ and $y \not\preceq x$" holds only if $\prec$ is asymmetric, which might not hold in general, since it is perfectly possible for $\mathcal{A}$ to have one explicit reason to hold x more plausible than y, and another to hold y more plausible than x. In this case both these reasons will compete with each other, with at most one of the pairs $\langle x, y \rangle$ or $\langle y, x \rangle$ making it into $\mathcal{A}$'s plausibility relation $<$.

What are the properties of $\prec$? We assume only two things, at least to begin with: *(i)* an agent can never possess a reason to hold a world strictly more plausible than itself, and *(ii)* an agent does not hold a world x to be more plausible than another world y, i.e., $x < y$, *without* being in possession of some reason for doing so. (Note this latter property lends a certain "foundationalist" flavour to our construction.) All this is formalised in the following definition:

Definition 1. *A context $\mathcal{C}$ is a pair of binary relations $(<, \prec)$ over W such that:*
$(\mathcal{C}1)$ $<$ *is a strict partial order*
$(\mathcal{C}2)$ $\prec$ *is irreflexive*
$(\mathcal{C}3)$ $< \subseteq \prec$

If $<$ is modular then we call C a *modular context*. We will later have grounds
for strengthening $(C3)$. How does A use his context C to construct a removal
function $*_C$? In terms of models, the set $[*_C(\lambda)]$ of models of his new belief set,
when removing a sentence λ, *must* include some $\neg\lambda$-worlds. Following the usual
practice in belief revision, he should take the most plausible ones according to $<$,
i.e., the $<$-minimal ones. But which, if any, of the λ-worlds should be included?
The following principle was proposed by Rott and Pagnucco [19]:

Principle of Weak Preference
If one object is held in equal or higher regard than another, the former
should be treated no worse than the latter.

Rott and Pagnucco use this principle to argue that the new set of worlds following
removal should contain all worlds x which are not less plausible than a $<$-minimal
$\neg\lambda$-world y, i.e., $y \not< x$. We propose to apply a tempered version of this principle
using the second ordering $\prec$. We include x if there is *no explicit reason to believe*
that y is more plausible than x, i.e., if $y \not\prec x$.

Definition 2. (* *from* C) *Given a context C we define the removal function*
$*_C$ *by setting, for each* $\lambda \in L_*$, $[*_C(\lambda)] = \bigcup\{\nabla_{\preceq}(y) \mid y \in \min_{<}([\neg\lambda])\}$.

It can be shown that different contexts give rise to different removal functions,
i.e., the mapping $C \mapsto *_C$ is injective. The case of modular contexts was the
one which was studied in detail in [5], where it was shown how, by placing
various restrictions on the interaction between $<$ and $\prec$, this family captures
a wide range of removal operations which have been previously studied, for
example both AGM contraction *and* AGM revision [1][1], severe withdrawal [19],
systematic withdrawal [16] and belief liberation [4]. For the general family in
that paper the following representation result was proved.

Theorem 1. *[5,6] Let C be a modular context. Then $*_C$ satisfies the following*
rules:

$(*1)$ $*(\lambda) \not\vdash \lambda$
$(*2)$ *If* $\lambda_1 \equiv \lambda_2$ *then* $*(\lambda_1) \equiv *(\lambda_2)$
$(*3)$ *If* $*(\lambda \wedge \chi) \vdash \chi$ *then* $*(\lambda \wedge \chi \wedge \psi) \vdash \chi$
$(*4)$ *If* $*(\lambda \wedge \chi) \vdash \chi$ *then* $*(\lambda \wedge \chi) \vdash *(\lambda)$
$(*5)$ $*(\lambda \wedge \chi) \vdash *(\lambda) \vee *(\chi)$
$(*6)$ *If* $*(\lambda \wedge \chi) \not\vdash \lambda$ *then* $*(\lambda) \vdash *(\lambda \wedge \chi)$

Furthermore if $*$ is any removal function satisfying the above 6 rules, there exists
a unique modular context C such that $* = *_C$.

All these rules are familiar from the belief removal literature. $(*1)$ is the Success
postulate while $(*2)$ is a syntax-irrelevance property. $(*3)$ is sometimes known
as Conjunctive Trisection [11,17]. It says if χ is believed after removing the
conjunction $\lambda \wedge \chi$, then it should also be believed when removing the longer

---

[1] The fact that basic removal also covers AGM revision is what motivated our choice
of the contraction-revision "hybrid" symbol $*$ to denote removal functions.

conjunction $\lambda \wedge \chi \wedge \psi$. Rule ($*4$) is closely-related to the rule Cut from non-monotonic reasoning [13], while ($*5$) and ($*6$) are the two AGM supplementary postulates for contraction [1].

Note the non-appearance in this list of the AGM contraction postulates Vacuity ($* (\bot) \not\vdash \lambda$ implies $* (\lambda) \equiv * (\bot)$),Inclusion ($* (\bot) \vdash * (\lambda)$) and Recovery ($* (\lambda) \wedge \lambda \vdash * (\bot)$), none of which are valid in general for removal functions generated from modular contexts. Vacuity has been argued against as a general principle of belief removal in [5,6]. Inclusion has been questioned in [4], while Recovery has long been regarded as controversial (see, e.g., [10]). Nevertheless we will see in Sect. 6 how each of these three rules may be captured within our general framework.

The second part of Theorem 1 was proved using the following construction.

Definition 3. *(C **from** $*$) Given any removal function $*$ we define the context* $\mathcal{C}(*) = (<, \prec)$ *as follows:* $x < y$ *iff* $y \notin [*(\neg x \wedge \neg y)]$ *and* $x \prec y$ *iff* $y \notin [* (\neg x)]$.[2]

[5] showed that if $*$ satisfies ($*1$)-($*6$) then $\mathcal{C}(*)$ is a modular context and $* = *_{\mathcal{C}(*)}$.

3   Characterising the General Family

Now we want to drop the assumption that $<$ is modular and assume only it is a strict partial order. How can we characterise the resulting class of removal functions? We focus first on establishing which of the postulates from Theorem 1 are sound for the general family, modifying our initial construction as and when necessary. Clearly we cannot expect that all the rules remain sound. In particular rule ($*6$) is known to depend on the modularity of $<$ and so might be expected to be the first to go. However we might hope to retain weaker versions of it, for instance:

($*6a$) If $*(\lambda \wedge \chi) \vdash \chi$ then $*(\lambda) \vdash *(\lambda \wedge \chi)$
($*6b$) $*(\lambda) \wedge *(\chi) \vdash *(\lambda \wedge \chi)$

These two rules appear respectively as (-8c) and (-8r) in [18] (see also [9]). ($*6b$) follows from ($*6$) given ($*1$).

Proposition 1. *If C is a general context then $*_C$ satisfies ($*1$), ($*2$), ($*4$), ($*5$) and ($*6a$) but not ($*6b$) (hence also ($*6$)) in general.*

Surprisingly, we lose ($*3$), as the following counterexample shows:

Example 1. Assume $L = \{p, q\}$ and let the 4 valuations of L be $W = \{00, 11, 01, 10\}$, where the first and second numbers denote the truth-values of p, q respectively. Let $<= \{(00, 10)\}$ and $\preceq= \{(10, 01)\}$ (strictly speaking the reflexive closure of this). We have $[*_C(p \wedge q)] = \{00, 10, 01\}$ and $[*_C(q)] = \{00\}$. Hence $10 \in [\neg q \wedge *_C(p \wedge q)]$ but $10 \notin [*_C(q)]$.

---

[2] When a world appears in the scope of a propositional connective, it should be understood as denoting any sentence which has that world as its only model.

This leaves us with a problem, since whereas ($*$**6**) is to be considered somewhat dispensible, ($*$**3**) is a very reasonable property for removal functions. Is there some way we can capture it? It turns out we can capture it if we strengthen the basic property (C**3**) to:

(C**3a**) $\preceq \subseteq \sqsubseteq^<$

In other words if $z < x$ and $x \preceq y$ then $z < y$. (C**3a**) is a *coherence* condition between $\prec$ and $<$. It is saying that if there is a world z which $\mathcal{A}$ judges to be more plausible than x but not to y then $\mathcal{A}$ has a reason to treat y more favourably than x. Note that for modular contexts (C**3**) and (C**3a**) are equivalent, but in general they are not.

Proposition 2. *If C satisfies (C**3a**) then $*_C$ satisfies ($*$**3**).*

Thus (C**3a**) seems necessary. Note rule (C**3a**) may also be interpreted as a restricted form of modularity for $<$, since it may be re-written as $\forall x, y, z \, (z < x \rightarrow (y \prec x \vee z < y))$. For this reason we make the following definition:

Definition 4. *A semi-modular context is any context C satisfying (C**3a**).*

In the rest of the paper we will work only with semi-modular contexts. It can be shown that $*_C$ still fails in general to satisfy ($*$**6b**) even for semi-modular contexts.

So far we have a list of sound properties for the removal functions defined from semi-modular contexts. They are the same as the rules which characterise modular removal, but with ($*$**6**) replaced by the weaker ($*$**6a**). It might be hoped that this list is complete, i.e., that *any* removal function $*$ satisfying these 6 rules is equal to $*_C$ for some semi-modular context C. Indeed we might expect to be able to show $* = *_{C(*)}$, where $C\,(*)$ is the context defined via Definition 3. The following result gives us a good start.

Proposition 3. *Let $*$ be any removal function satisfying ($*$**1**)-($*$**5**) and ($*$**6a**). Then $C\,(*)$ is a context, i.e., satisfies (C**1**)-(C**3**).*

However to get (C**3a**) it seems an extra property is needed:

($*$**C**) If $*(\lambda) \wedge \neg\lambda \vdash *(\chi) \wedge \neg\chi$ then $*(\lambda) \vdash *(\chi)$

We can rephrase this using the *Levi Identity* [14]. Given any removal function $*$ we may define a *revision function* $*^R$ by setting, for each consistent sentence $\lambda \in L$, $*^R(\lambda) = *(\neg\lambda) \wedge \lambda$. Then rule ($*$**C**) may be equivalently written as:

($*$**C'**) If $*^R(\neg\lambda) \vdash *^R(\neg\chi)$ then $*(\lambda) \vdash *(\chi)$

Thus ($*$**C'**) is effectively saying that if revising by $\neg\lambda$ leads to a stronger belief set than revising by $\neg\chi$, then removing λ leads to a stronger belief set than removing χ. The next result confirms that this rule is sound for the removal functions generated by semi-modular contexts, and that this property is enough to show that $C\,(*)$ satisfies (C**3a**).

Proposition 4. *Let C be a semi-modular context. Then $*_C$ satisfies ($*$C). Furthermore if $*$ is any removal function satisfying ($*$C) then the context $C(*)$ satisfies (C3a).*

Rule ($*$C) is actually quite strong. In the presence of ($*$3) it can be shown to imply ($*$4). This means that, in the axiomatisation of $*_C$ we can replace ($*$4) with ($*$C). To show that the list of rules is complete, it remains to prove $* = *_{C(*)}$. It turns out that here we need the following weakening of ($*$6b):

($*$E) $\quad \neg(\lambda \wedge \chi) \wedge *(\lambda) \wedge *(\chi) \vdash *(\lambda \wedge \chi)$

This rule may be reformulated as "$*(\lambda) \wedge *(\chi) \vdash (\lambda \wedge \chi) \vee *(\lambda \wedge \chi)$". In this reformulation, the right hand side of the turnstile may be thought of as standing for all those consequences of the conjunction $\lambda \wedge \chi$ which are *believed* upon its removal. The rule is saying that any such surviving consequence must be derivable from the *combination* of $*(\lambda)$ and $*(\chi)$.

Proposition 5. *Let C be a semi-modular context. Then $*_C$ satisfies ($*$E).*

Theorem 2. *Let $*$ be any removal function satisfying ($*$1),($*$2), ($*$3),($*$C), ($*$5), ($*$6a) and ($*$E). Then $*_{C(*)} = *$.*

Thus, to summarise, the family of removal functions defined from semi-modular contexts is completely characterised by ($*$1)–($*$3), ($*$C), ($*$5), ($*$6a) and ($*$E).

4   Transitivity and Priority

In this section we look at imposing an extra couple of properties on semi-modular contexts $C = (<, \prec)$, both of which were investigated in the case of modular contexts in [5]. There it was shown how the resulting classes of removal functions still remain general enough to include a great many of the classes which have been previously proposed in the context of modular removal.

    The first property is the transitivity of $\preceq$, thus making $\preceq$ a preorder. (Recall $\preceq$ is the converse complement of $\prec$, so this is equivalent to making $\prec$ modular.) According to our above interpretation of $\preceq$ this means *if there is no reason to treat y more favourably than x, and no reason to treat z more favourably than y then there is no reason to treat z more favourably than x.*

Proposition 6. (i). *If $\preceq$ is transitive then $*_C$ satisfies the following strengthening of ($*$C):*

($*$C+) *If $*(\lambda) \wedge \neg\lambda \vdash *(\chi)$ then $*(\lambda) \vdash *(\chi)$*

(ii). *If $*$ satisfies ($*$C+) then the relation $\preceq$ in $C(*)$ is transitive.*

Note this property is a great deal simpler than the one used to characterise transitivity of $\preceq$ in the modular context in [5]. It can be re-written as: If $*^R(\neg\lambda) \vdash *(\chi)$ then $*(\lambda) \vdash *(\chi)$. It says that if the belief set following removal of χ is contained in the belief set following the *revision* by $\neg\lambda$, then it must be contained also in the belief set following the removal of λ. This seems like a reasonable property.

Corollary 1. *For any removal function* $*$, *the following are equivalent:*
(i). $*$ *is generated by a semi-modular context* $\mathcal{C} = (<, \prec)$ *such that* $\preceq$ *is transitive.* (ii). $*$ *satisfies the list of rules given at the end of Sect. 3, with* $(*\mathbf{C})$ *replaced by* $(*\mathbf{C}+)$.

Now consider the following property of a context $\mathcal{C} = (<, \prec)$:

$(\mathcal{C}\mathbf{P})$ If $x \prec y$ and $y \not\prec x$ then $x < y$

This, too, looks reasonable: if $\mathcal{A}$ has an explicit reason to hold x more plausible than y, but not vice versa, then in the final reckoning he should hold x to be strictly more plausible than y. Consider the following property of removal functions:

$(*\mathbf{P})$ If $*(\lambda) \vdash \chi$ and $*(\chi) \not\vdash \lambda$ then $*(\lambda \wedge \chi) \vdash \chi$

This property is briefly mentioned as *Priority* in [3], and is also briefly mentioned right at the end of [7]. It can be read as saying that if λ is excluded following removal of χ, but not vice versa, then χ is strictly more entrenched than λ. **For the case of modular removal,** we can obtain the following exact correspondence between $(\mathcal{C}\mathbf{P})$ and $(*\mathbf{P})$:

Proposition 7. (i). *If* $\mathcal{C}$ *is a modular context satisfying* $(\mathcal{C}\mathbf{P})$ *then* $*_{\mathcal{C}}$ *satisfies* $(*\mathbf{P})$. (ii). *If* $*$ *satisfies* $(*\mathbf{P})$ *then* $\mathcal{C}(*)$ *satisfies* $(\mathcal{C}\mathbf{P})$.

The proof of Proposition 7(i) makes critical use of the modularity of $<$. It turns out that $(*\mathbf{P})$ is *not* sound for general semi-modular contexts, even if we insist on $(\mathcal{C}\mathbf{P})$.

Example 2. Suppose $L = \{p, q\}$ and that $<= \{(01, 11)\}$ while $\preceq = \{(01, 11)\}$ (strictly speaking the reflexive closure of this). One can verify that $\mathcal{C}$ is a semi-modular context and that $(\mathcal{C}\mathbf{P})$ is satisfied. Now let $\lambda = p \vee \neg q$ and $\chi = \neg p$. Then $[*_{\mathcal{C}}(\lambda)] = \{01\}$, $[*_{\mathcal{C}}(\chi)] = \{11, 01, 10\}$ and $[*_{\mathcal{C}}(\lambda \wedge \chi)] = \{01, 10\}$ and we have $*_{\mathcal{C}}(\lambda) \vdash \chi$, $*_{\mathcal{C}}(\chi) \not\vdash \lambda$, and $*_{\mathcal{C}}(\lambda \wedge \chi) \not\vdash \chi$. Hence $(*\mathbf{P})$ is not satisfied.

The question now is, which postulate corresponds to $(\mathcal{C}\mathbf{P})$ for general semi-modular contexts? Here is the answer:

Proposition 8. (i). *If* $\mathcal{C}$ *is a semi-modular context which satisfies* $(\mathcal{C}\mathbf{P})$, *then* $*_{\mathcal{C}}$ *satisfies the following rule:*

$(*\mathbf{P}')$ *If* $*(\lambda) \vdash \chi$ *and* $*(\chi) \vdash *(\lambda \wedge \chi)$ *then* $*(\chi) \vdash \lambda$

(ii). *If* $*$ *satisfies* $(*\mathbf{P}')$, *plus* $(*\mathbf{C})$ *and* $(*\mathbf{1})$, *then* $\mathcal{C}(*)$ *satisfies* $(\mathcal{C}\mathbf{P})$.

It is straightforward to see $(*\mathbf{P}')$ is weaker than $(*\mathbf{P})$ given $(*\mathbf{1})$, while it implies $(*\mathbf{P})$ given $(*\mathbf{6})$.

5   Finite Base-Generated Removal

In this section we mention a concrete and important subfamily of our general family of removal functions, the ideas behind which can be seen already throughout the literature on nonmonotonic reasoning and belief change (see in particular [3] for a general treatment in a belief removal context). Given any, possibly inconsistent, set Σ of sentences, let $cons\,(\Sigma)$ denote the set of all consistent subsets of Σ. We assume agent $\mathcal{A}$ is in possession of a finite set Σ of sentences which are possible *assumptions* or *defaults*, together with a strict preference ordering $\Subset$ on $cons\,(\Sigma)$ (with sets "higher" in the ordering assumed more preferred). We assume the following two properties of $\Subset$:

(**Σ1**) $\Subset$ is a strict partial order
(**Σ2**) If $A \subset B$ then $A \Subset B$

(**Σ2**) is a monotonicity requirement stating a given set of defaults is strictly preferred to all its proper subsets.

Definition 5. *If $\Sigma \subseteq L$ is a finite set of sentences and $\Subset$ is a binary relation over $cons\,(\Sigma)$ satisfying (**Σ1**) and (**Σ2**). Then we call $\mathbb{\Sigma} = \langle \Sigma, \Subset \rangle$ a prioritised default base. If in addition $\Subset$ is modular then we call $\mathbb{\Sigma}$ a modular prioritised default base.*

How does the agent use a prioritised default base $\mathbb{\Sigma} = \langle \Sigma, \Subset \rangle$ to remove beliefs? For $\Sigma \subseteq L$ and $\lambda \in L_*$ let $cons\,(\Sigma, \lambda) \overset{\text{def}}{=} \{S \in cons\,(\Sigma) \mid S \nvdash \lambda\}$. Then from $\mathbb{\Sigma}$ we may define a removal function $*_{\mathbb{\Sigma}}$ by setting, for each $\lambda \in L_*$,

$$*_{\mathbb{\Sigma}}(\lambda) = \bigvee \left\{ \bigwedge S \mid S \in \max_{\Subset} cons\,(\Sigma, \lambda) \right\}.$$

In other words, after removing λ, $\mathcal{A}$ will believe precisely those sentences which are consequences of *all maximally preferred* subsets of Σ which do not imply λ.

We will now show how the family of removal functions generated from prioritised default bases fits into our general family. From a given $\mathbb{\Sigma} = \langle \Sigma, \Subset \rangle$ we may define a context $\mathcal{C}\,(\mathbb{\Sigma}) = (<, \prec)$ as follows. Let $sent_{\Sigma}\,(x) \overset{\text{def}}{=} \{\alpha \in \Sigma \mid x \in [\alpha]\}$. Then

 - $x < y$ iff $sent_{\Sigma}\,(y) \Subset sent_{\Sigma}\,(x)$
 - $x \prec y$ iff $sent_{\Sigma}\,(x) \not\subseteq sent_{\Sigma}\,(y)$

Thus we define x to be more plausible than y iff the set of sentences in Σ satisfied by x is more preferred than the set of sentences in Σ satisfied by y. Meanwhile we have the natural interpretation for $\prec$ that $\mathcal{A}$ has a reason to hold x to be more plausible than y precisely when one of the sentences in Σ is satisfied by x but not y.

Theorem 3. (i). $\mathcal{C}\,(\mathbb{\Sigma})$ *defined above forms a semi-modular context (which is modular if $\Subset$ is modular). (ii). $\preceq$ is transitive and the condition (**CP**) from Sect. 4 holds. (iii). $*_{\mathbb{\Sigma}} = *_{\mathcal{C}(\mathbb{\Sigma})}$.*

Thus we have shown that every removal function generated by a prioritised default base may *always* be generated by a semi-modular context which furthermore satisfies the two conditions on contexts mentioned in the previous section. By the results of the previous sections, this means we automatically obtain a list of sound postulates for the default base-generated removals.

Corollary 2. *Let* Σ *be any prioritised default base. Then* $*_\Sigma$ *satisfies all the rules listed at the end of Sect. 3, as well as* $(*\mathbf{C}+)$ *and* $(*\mathbf{P}')$ *from the last section.*

Note we have shown how every prioritised default base gives rise to a semi-modular context satisfying $\preceq$-transitivity and $(\mathcal{C}\mathbf{P})$. An open question is whether *every* such context arises in this way.

6   AGM Preferential Removal

Recall that three of the basic AGM postulates for contraction do not hold in general for the removal functions generated by semi-modular contexts, namely Inclusion, Recovery and Vacuity. In this section we show how each of these rules can be captured. In [5] it was shown already how they may be captured within the class of modular context-generated removal.

The Inclusion rule is written in our setting as follows:

$(*\mathbf{I})$ $*(\bot) \vdash *(\lambda)$

To capture $(*\mathbf{I})$ for any removal generated from any semi-modular context $\mathcal{C} = (<, \prec)$, we need only to require the following condition on $\mathcal{C}$:

$(\mathcal{C}\mathbf{I})$ $\min_<(W) \subseteq \min_\prec(W)$

According to our interpretation of $\prec$, $(\mathcal{C}\mathbf{I})$ is stating that, for any world x, if $\mathcal{A}$ has some explicit reason favour some world y over x (i.e., $y \prec x$) then in the final reckoning $\mathcal{A}$ must hold *some* world z (not necessarily the same as y) more plausible than x (i.e., $z < x$).

Proposition 9. (i). *If* $\mathcal{C}$ *satisfies* $(\mathcal{C}\mathbf{I})$ *then* $*_\mathcal{C}$ *satisfies* $(*\mathbf{I})$. (ii). *If* $*$ *satisfies* $(*\mathbf{I})$ *then* $\mathcal{C}(*)$ *satisfies* $(\mathcal{C}\mathbf{I})$.

The Recovery rule is written as follows:

$(*\mathbf{R})$ $*(\lambda) \wedge \lambda \vdash *(\bot)$

The corresponding property on contexts $\mathcal{C} = (<, \prec)$ is:

$(\mathcal{C}\mathbf{R})$ If $y \notin \min_<(W)$ and $x \neq y$ then $x \prec y$

Thus the only worlds $\nabla_\preceq(x)$ contains, other than x itself, are worlds in $\min_<(W)$.

Proposition 10. (i). *If* $\mathcal{C}$ *satisfies* $(\mathcal{C}\mathbf{R})$ *then* $*_\mathcal{C}$ *satisfies* $(*\mathbf{R})$. (ii). *If* $*$ *satisfies* $(*\mathbf{R})$ *then* $\mathcal{C}(*)$ *satisfies* $(\mathcal{C}\mathbf{R})$.

Note the combination of $(\mathcal{C}\mathbf{I})$ and $(\mathcal{C}\mathbf{R})$ specifies $\prec$, equivalently $\preceq$, uniquely in terms of $<$, viz. $x \preceq_{agm} y$ iff $x = y$ or $x \in \min_< (W)$, and we obtain the removal recipe of AGM contraction, in which removal of λ boils down to just adding the $<$-minimal $\neg\lambda$-worlds to the $<$-minimal worlds:

$$[\ast_{agm}(\lambda)] = \min_< (W) \cup \min_< ([\neg\lambda]).$$

It is easy to check that the resulting context $\mathcal{C}$ satisfies condition $(\mathcal{C}\mathbf{3a})$ and thus forms a semi-modular context. It is also easy to check $(\mathcal{C}\mathbf{P})$ is satisfied and that the above-defined $\preceq_{agm}$ is transitive. Thus the above $\ast_{agm}$ also satisfies $(\ast\mathbf{C}+)$ and $(\ast\mathbf{P}')$ from Sect. 4. It can also be shown to satisfy $(\ast\mathbf{6b})$.

The Vacuity rule is written as follows:

$(\ast\mathbf{V})$ If $\ast(\bot) \nvdash \lambda$ then $\ast(\lambda) \equiv \ast(\bot)$

Unlike in the modular case, where Vacuity is known to follow from Inclusion for modular removal functions [5], $(\ast\mathbf{V})$ does not even hold in general for the above preferential AGM contraction $\ast_{agm}$. This was essentially noticed, in a revision context, in [2].

Example 3. Let $L = \{p, q\}$ and $<= \{(11, 01)\}$. So $[\ast_{agm}(\bot)] = \{00, 11, 10\}$. Let $\lambda = p$. Then we have $\ast_{agm}(\bot) \nvdash \lambda$ (because $00 \in [\ast_{agm}(\bot)]$), but $\min_< ([\neg\lambda]) = \{00, 01\}$, so $[\ast_{agm}(\lambda)] = \min_< (W) \cup \min_< ([\neg\lambda]) = W \neq [\ast_{agm}(\bot)]$.

In order to ensure $\ast_{agm}$ satisfies $(\ast\mathbf{V})$ it is necessary, as is done in [12], to enforce the following property on $<$.

$(<\mathbf{V})$ $\forall x, y ((x \in \min_< (W) \wedge y \notin \min_< (W)) \to x < y)$.

In other words all $<$-minimal worlds can be compared with, and are below, every world which is not $<$-minimal. For general semi-modular contexts $\mathcal{C} = (<, \prec)$ we also require the following condition, which is weaker than $(\mathcal{C}\mathbf{I})$:

$(\mathcal{C}\mathbf{V})$ If $x, y \in \min_< (W)$ then $x \nprec y$

This property says that for any two of his $<$-minimal worlds, $\mathcal{A}$ will not have explicit reason to hold one to be more plausible than the other.

Proposition 11. (i). *If $\mathcal{C}$ satisfies $(\mathcal{C}\mathbf{V})$ and $(<\mathbf{V})$ then $\ast_{\mathcal{C}}$ satisfies $(\ast\mathbf{V})$.* (ii). *If $\ast$ satisfies $(\ast\mathbf{V})$ then $\mathcal{C}(\ast)$ satisfies $(\mathcal{C}\mathbf{V})$.*

7   Conclusion

In this paper we introduced a family of removal functions, generalising the one given in [5] to allow for incomparabilities in the plausibility relation $<$ between possible worlds. Removal is carried out using the plausibility relation in combination with a second relation $\prec$ which can be thought of as indicating "reasons" for holding one world to be more plausible than another. We axiomatically characterised this general family as well as certain subclasses, and we showed how this family includes some important and natural families of belief removal, specifically those which may be generated from prioritised default bases and the preferential counterpart of AGM contraction. Our results show the central construct used in this paper, i.e., semi-modular contexts, to be a very useful tool in the study of belief removal functions.

For future work we would like to employ semi-modular contexts in the setting of *social belief removal* [6], in which there are several agents, each assumed to have their own removal function, and in which all agents must remove some belief to become consistent with each other. [6] showed that, under the assumption that each agent uses a removal function generated from a *modular* context, certain *equilibrium points* in the social removal process are guaranteed to exist. An interesting question would be whether these results generalise to the *semi-modular* case.

Acknowledgements

Thanks are due to two anonymous reviewers for some helpful comments.

References

1. Alchourrón, C., Gärdenfors, P., Makinson, D.: On the logic of theory change: Partial meet contraction and revision functions. Journal of Symbolic Logic 50(2), 510–530 (1985)
2. Benferhat, S., Lagrue, S., Papini, O.: Revision of partially ordered information: Axiomatization, semantics and iteration. In: Pack Kaelbling, L., Saffiotti, A. (eds.) IJCAI, pp. 376–381. Professional Book Center (2005)
3. Bochman, A.: A Logical Theory of Nonmonotonic Inference and Belief Change. Springer, Heidelberg (2001)
4. Booth, R., Chopra, S., Ghose, A., Meyer, T.: Belief liberation (and retraction). Studia Logica 79(1), 47–72 (2005)
5. Booth, R., Chopra, S., Meyer, T., Ghose, A.: A unifying semantics for belief change. In: Proceedings of ECAI 2004, pp. 793–797 (2004)
6. Booth, R., Meyer, T.: Equilibria in social belief removal. In: KR, pp. 145–155 (2008)
7. Cantwell, J.: Relevant contraction. In: Proceedings of the Dutch-German Workshop on Non-Monotonic Reasoning, DGNMR 1999 (1999)
8. Cantwell, J.: Eligible contraction. Studia Logica 73, 167–182 (2003)
9. Arló Costa, H.: Rationality and value: The epistemological role of indeterminate and agent-dependent values. Philosophical Studies 128(1), 7–48 (2006)
10. Hansson, S.O.: Belief contraction without recovery. Studia Logica 50(2), 251–260 (1991)
11. Hansson, S.O.: Changes on disjunctively closed bases. Journal of Logic, Language and Information 2, 255–284 (1993)
12. Katsuno, H., Mendelzon, A.O.: Propositional knowledge base revision and minimal change. Artif. Intell. 52(3), 263–294 (1992)
13. Kraus, S., Lehmann, D., Magidor, M.: Nonmonotonic reasoning, preferential models and cumulative logics. Artificial Intelligence 44, 167–207 (1991)

14. Levi, I.: The Fixation of Belief and Its Undoing. Cambridge University Press, Cambridge (1991)
15. Maynard-Zhang, P., Lehmann, D.: Representing and aggregating conflicting beliefs. Journal of Artificial Intelligence Research 19, 155–203 (2003)
16. Meyer, T., Heidema, J., Labuschagne, W., Leenen, L.: Systematic withdrawal. Journal of Philosophical Logic 31(5), 415–443 (2002)
17. Rott, H.: Preferential belief change using generalized epistemic entrenchment. Journal of Logic, Language and Information 1, 45–78 (1992)
18. Rott, H.: Change, Choice and Inference: A Study of Belief Revision and Nonmonotonic Reasoning. Oxford University Press, Oxford (2001)
19. Rott, H., Pagnucco, M.: Severe withdrawal (and recovery). Journal of Philosophical Logic 28, 501–547 (1999)
20. Shoham, Y.: A semantic approach to nonmonotic logics. In: LICS, pp. 275–279 (1987)

Computing Compliance

Ivano Ciardelli, Irma Cornelisse, Jeroen Groenendijk, and Floris Roelofsen

Institute for Logic, Language, and Computation, University of Amsterdam

Abstract. Inquisitive semantics (cf. Groenendijk, 2008) provides a formal framework for reasoning about information exchange. The central logical notion that the semantics gives rise to is *compliance*. This paper presents an algorithm that computes the set of compliant responses to a given initiative. The algorithm is sound and complete. The implementation is accessible online via www.illc.uva.nl/inquisitive-semantics.

1 Introduction

Traditionally, logic is concerned with argumentation. As a consequence, the meaning of a sentence is traditionally identified with its *informative* content. In much recent work, this notion is given a dynamic twist, and the meaning of a sentence is taken to be its potential to change the 'common ground' of a conversation. The most basic way to formalize this idea is to think of the common ground as a set of possible worlds, and of a sentence as providing information by eliminating some of these possible worlds.

Of course, this picture is limited in several ways. First, when exchanging information sentences are not only used to provide information, but also—crucially— to *raise issues*, that is, to indicate which kind of information is desired. Second, the given picture does not take into account that updating the common ground is a *cooperative* process. One conversational participant cannot simply change the common ground all by herself. All she can do is *propose* a certain change. Other participants may react to such a proposal in several ways. Changes of the common ground come about by mutual agreement.

In order to overcome these limitations, *inquisitive semantics* (cf. Ciardelli and Roelofsen, 2009; Groenendijk, 2008; Groenendijk and Roelofsen, 2009; Mascarenhas, 2008) starts with an altogether different picture. It views propositions as proposals to enhance the common ground. These proposals do not always specify just one way of enhancing the common ground. They may suggest alternative ways of doing so, among which the responder is then invited to choose. Formally, a proposition consists of one or more *possibilities*. Each possibility is a set of possible worlds and embodies a possible way to enhance the common ground. If a proposition consists of two or more possibilities, it is *inquisitive*: it invites the other participants to respond in a way that will lead to a cooperative choice between the proposed alternatives. Inquisitive propositions raise an issue. They indicate which kind of information is desired. Thus, inquisitive semantics directly reflects the idea that information exchange consists in a cooperative dynamic process of raising and resolving issues.

X. He, J. Horty, and E. Pacuit (Eds.): LORI 2009, LNAI 5834, pp. 55–65, 2009.

Traditional semantics gives rise to the logical notion of entailment, which judges the validity of argumentation. Inquisitive semantics gives rise to the logical notion of *compliance*, which judges whether or not a sentence makes a significant contribution towards resolving a given issue. Extensive motivation for the precise formulation of compliance can be found in (Groenendijk and Roelofsen, 2009). The aim of the present paper is to devise an algorithm that computes the set of compliant responses to a given initiative. Such an algorithm will form the basis for practical applications of inquisitive semantics.

The paper is organized as follows. Section 2 reviews the basic notions of inquisitive semantics and some basic properties of the system, section 3 discusses and illustrates the definition of compliance, and section 4 presents a sound and complete algorithm for computing compliant responses.

2   Inquisitive Semantics

Definition 1 (Language). Let $\mathcal{P}$ be a finite set of proposition letters that we will consider fixed throughout the paper. We denote by $\mathcal{L_P}$ the set of formulas built up from letters in $\mathcal{P}$ and $\bot$ using the binary connectives $\wedge, \vee$ and $\rightarrow$. We will refer to $\mathcal{L_P}$ as the propositional language based on $\mathcal{P}$.

We will also make use of the following abbreviations: $\neg\varphi$ for $\varphi \rightarrow \bot$, $!\varphi$ for $\neg\neg\varphi$, and $?\varphi$ for $\varphi \vee \neg\varphi$.

Definition 2 (Indices). An *index* is a function from $\mathcal{P}$ to $\{0, 1\}$. We denote by ω the set of all indices.

Definition 3 (States). A *state* is a set of indices. We denote by $\mathcal{S}$ the set of all states.

Definition 4 (Support)

1. $s \models p$ iff $\forall w \in s : w(p) = 1$
2. $s \models \bot$ iff $s = \emptyset$
3. $s \models \varphi \wedge \psi$ iff $s \models \varphi$ and $s \models \psi$
4. $s \models \varphi \vee \psi$ iff $s \models \varphi$ or $s \models \psi$
5. $s \models \varphi \rightarrow \psi$ iff $\forall t \subseteq s : \text{if } t \models \varphi \text{ then } t \models \psi$

It follows from the above definition that the empty state supports any formula φ. Thus, we may think of $\emptyset$ as the *inconsistent* state.

Fact 5 (Persistence). If $s \models \varphi$ then for every $t \subseteq s: t \models \varphi$

Fact 6 (Singleton states behave classically). For any index w and formula φ:

$$\{w\} \models \varphi \iff \varphi \text{ is classically true under the valuation } w$$

In particular, $\{w\} \models \varphi$ or $\{w\} \models \neg\varphi$ for any formula φ.

It follows from definition 4 that the support-conditions for $\neg\varphi$ and $!\varphi$ are as follows.

Fact 7 (Support for negation)

1. $s \models \neg\varphi$ iff $\forall w \in s : w \models \neg\varphi$
2. $s \models !\varphi$ iff $\forall w \in s : w \models \varphi$

In terms of support, we define the *possibilities* for a sentence φ and the *proposition* expressed by φ. We also define the *truth-set* of φ, which is the meaning that would be associated with φ in a classical setting.

Definition 8 (Truth sets, possibilities, propositions). Let φ be a formula.

1. A *possibility* for φ is a maximal state supporting φ, that is, a state that supports φ and is not properly included in any other state supporting φ.
2. The *proposition* expressed by φ, denoted by $[\varphi]$, is the set of possibilities for φ.
3. The *truth set* of φ, denoted by $|\varphi|$, is the set of indices where φ is classically true.

Notice that $|\varphi|$ is a state, while $[\varphi]$ is a set of states. The following result guarantees that the proposition expressed by a formula completely determines which states support that formula, and vice versa.

Fact 9 (Support and Possibilities). For any state s and any formula φ:

$$s \models \varphi \iff s \text{ is contained in a possibility for } \varphi$$

Example 10 (Disjunction). Inquisitive semantics crucially differs from classical semantics in its treatment of disjunction. To see this, consider figures 1(a) and 1(b). In these figures, it is assumed that $\mathcal{P} = \{p, q\}$; index 11 makes both p and q true, index 10 makes p true and q false, etcetera. Figure 1(a) depicts the truth set—that is, the classical meaning—of $p \vee q$: the set of all indices that make either p or q, or both, true. Figure 1(b) depicts the proposition associated with $p \vee q$ in inquisitive semantics. It consists of two possibilities. One possibility is made up of all indices that make p true, and the other of all indices that make q true. So, as in the classical setting, $p \vee q$ is *informative*, in that it proposes to eliminate the index where both p and q are false. But it is also *inquisitive*, in that it proposes two alternative ways of enhancing the common ground, and invites a response that is directed at chosing between these two alternatives. This inquisitive aspect of meaning is not captured in a classical setting.

Definition 11 (Inquisitiveness and informativeness)

- φ is *inquisitive* iff $[\varphi]$ contains at least two possibilities;
- φ is *informative* iff φ proposes to eliminate at least one index, that is, iff $\bigcup[\varphi] \neq \omega$

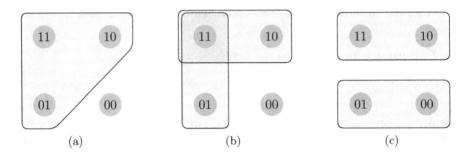

Fig. 1. (a) the classical picture of $p \vee q$, (b) the inquisitive picture of $p \vee q$, and (c) the inquisitive picture of the polar question $?p$

Definition 12 (Questions, assertions, and hybrids)

- φ is a *question* iff it is not informative;
- φ is an *assertion* iff it is not inquisitive;
- φ is a *hybrid* iff it is both informative and inquisitive.

Example 13 (Questions, assertions, and hybrids). We have already seen that $p \vee q$ is both informative and inquisitive, i.e., hybrid. The proposition depicted in figure 1(a) is expressed by $!(p \vee q)$. This proposition consists of exactly one possibility. So $!(p \vee q)$ is an assertion. The proposition depicted in figure 1(c) is expressed by $?p$. This proposition covers the entire logical space, so $?p$ does not propose to eliminate any index. That is, $?p$ is a question.[1]

The following result gives some sufficient syntactic conditions for a formula to be an assertion.

Fact 14. For any proposition letter p and formulas φ, ψ:

1. p is an assertion;
2. $\bot$ is an assertion;
3. if φ, ψ are assertions, then $\varphi \wedge \psi$ is an assertion;
4. if ψ is an assertion, then $\varphi \rightarrow \psi$ is an assertion.

Note that items 2 and 4 imply that any negation is an assertion. In particular, $!\varphi$ is always an assertion. In fact, as a consequence of proposition 7, the possibilities for $\neg\varphi$ and $!\varphi$ can be characterized as follows.

---

[1] Notice that questions do not have to be inquisitive, and assertions do not have to be informative. For instance, the tautology $!(p \vee \neg p)$ is both a question and an assertion, even though (or rather *because*) it is neither inquisitive nor informative. Groenendijk and Roelofsen (2009) give a slightly more involved definition of questions and assertions, which makes sure that the two notions are strictly disjoint. This may be more desirable from a linguistic point of view, but the additional complexity is not quite relevant in the present setting, and is therefore avoided.

Fact 15 (Negation)

1. $[\neg\varphi] = \{|\neg\varphi|\}$
2. $[!\varphi] = \{|\varphi|\}$

Using fact 14 inductively we obtain the following corollary showing that disjunction is the only source of inquisitiveness in our propositional language.

Corollary 16. *Any disjunction-free formula is an assertion.*

In inquisitive semantics, the informative content of a formula φ is captured by the union $\bigcup[\varphi]$ of all the possibilities for φ. For φ proposes to eliminate all indices that are not in $\bigcup[\varphi]$. In a classical setting, the informative content of φ is captured by $|\varphi|$. The following result says that, as far as informative content goes, inquisitive semantics does not diverge from classical semantics. In this sense, inquisitive semantics is a 'conservative extension' of classical semantics.

Fact 17. For any formula φ: $\bigcup[\varphi] = |\varphi|$

We end this section with a definition of inquisitive equivalence.

Definition 18 (Equivalence)
Two formulas φ and ψ are *equivalent*, $\varphi \equiv \psi$, iff $[\varphi] = [\psi]$.

It follows immediately from fact 9 that $\varphi \equiv \psi$ just in case φ and ψ are supported by the same states.

3   Compliance

The notion of compliance judges whether a certain conversational move makes a significant contribution to resolving a given issue. Before stating the formal definition, let us first review some of the basic logico-pragmatical intuitions behind it.

Basic intuitions. Consider a situation where a sentence φ is a *response* to an *initiative* ψ. We are mainly interested in the case where the initiative ψ is inquisitive, and hence proposes several alternatives. In this case, we consider φ to be an optimally compliant response just in case it picks out exactly one of the alternatives proposed by ψ. Such an optimally compliant response is an assertion φ such that the unique possibility α for φ equals one of the possibilities for ψ: $\lfloor\varphi\rfloor = \{\alpha\}$ and $\alpha \in \lfloor\psi\rfloor$. Of course, the responder will not always be able to give such an optimally compliant response. It may still be possible in this case to give a compliant informative response, not by picking out *one* of the alternatives proposed by ψ, but by selecting some of them, and excluding others. The informative content of such a response must correspond with the union of some but not all of the alternatives proposed by ψ. That is, $|\varphi|$ must coincide with the union of a proper non-empty subset of $\lfloor\psi\rfloor$.

If such an informative compliant response cannot be given either, it may still be possible to make a significant compliant move, namely by responding with an inquisitive sentence, replacing the issue raised by ψ with an easier to answer sub-issue. The rationale behind such an inquisitive move is that, if part of the original issue posed by ψ were resolved, it might become possible to subsequently resolve the remaining issue as well.

Summing up, there are basically two ways in which φ may be compliant with ψ:

(a) φ may partially *resolve* the issue raised by ψ;
(b) φ may *replace* the issue raised by ψ by an easier to answer sub-issue.

Combinations are also possible: φ may partially resolve the issue raised by ψ and at the same time replace the remaining issue with an easier to answer sub-issue. What is important is that φ should do nothing *more* than this: it should not provide any information that is not strictly related to the given issue, and it should not raise any issues that are not strictly related to the given issue, or issues that are more difficult to resolve. This means, in particular, that over-informative answers are not compliant. For instance, $p \wedge q$ is not a compliant response to $?p$, because it does not resolve the issue any more than the less informative answer p would do.

These considerations are captured by the following definition:

Definition 19 (Compliance). φ is compliant with ψ, $\varphi \propto \psi$, iff

1. every possibility in $[\varphi]$ is the union of a non-empty set of possibilities in $[\psi]$
2. every possibility in $[\psi]$ restricted to $|\varphi|$ is contained in a possibility in $[\varphi]$

Here, the *restriction* of $\alpha \in [\psi]$ to $|\varphi|$ is defined to be the intersection $\alpha \cap |\varphi|$. To explain the workings of the definition, we will distinguish several cases, depending on whether ψ and φ are assertions, questions or hybrids.

First, consider the case where ψ is an assertion. Then the first clause says that every possibility for φ should coincide with the unique possibility for ψ. This can only be the case if φ is equivalent to ψ. In this case, the second clause is trivially met. Thus, the only way to compliantly respond to an assertion is to confirm it.

Fact 20. If ψ is an assertion, then $\varphi \propto \psi$ iff $[\varphi] = [\psi]$.

If φ is an assertion, then the first clause in the definition of compliance requires that $|\varphi|$ coincides with the union of a set of possibilities for ψ. The second clause is trivially met in this case.

Fact 21. If φ an assertion, then $\varphi \propto \psi$ iff $|\varphi|$ coincides with the union of a non-empty set of possibilities for ψ.

In particular, if φ is an assertion and ψ is *inquisitive*, then fact 21 tells us that φ is compliant with ψ just in case φ partially resolves the issue raised by ψ, without being over-informative. Thus, compliance embodies a strict notion of partial answerhood.[2]

Next, consider the case where φ is a question. Then the first clause in the definition of compliance requires that ψ is a question as well. Moreover, the first clause also requires that every complete answer to φ is at least a partial answer to ψ.

The second clause also plays a role in this case. However, since φ is assumed to be a question, and since questions are not informative, the second clause can be simplified: the restriction of the possibilities for ψ to $|\varphi|$ does not have any effect, because $|\varphi| = \omega$. Hence, the second clause simply requires that every possibility for ψ is contained in a possibility for φ.

Fact 22. If φ is a question, then $\varphi \propto \psi$ iff

1. every possibility in $[\varphi]$ is the union of a non-empty set of possibilities in $[\psi]$
2. every possibility in $[\psi]$ is contained in a possibility in $[\varphi]$

The second constraint prevents φ from being more difficult to answer than ψ. Let us illustrate this with an example. Consider the case where $\psi \equiv ?p \vee ?q$ and $\varphi \equiv ?p$. The propositions expressed by $?p \vee ?q$ and $?p$ are depicted in figure 2.

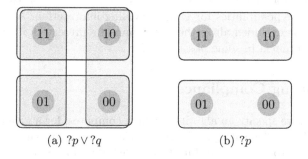

(a) $?p \vee ?q$ (b) $?p$

Fig. 2. Choice question and polar question

Intuitively, $?p \vee ?q$ is a *choice question*. To resolve it, one may either provide an answer to the question $?p$ or to the question $?q$. Thus, there are four possibilities, each corresponding to an optimally compliant response: p, $\neg p$, q and $\neg q$. The question $?p$ is more demanding: there are only two possibilities and thus only two optimally compliant responses, p and $\neg p$. Hence, $?p$ is more difficult to answer than $?p \vee ?q$, and should therefore not count as compliant with it. This is

---

[2] Earlier formal analyses of questions (cf. Groenendijk and Stokhof, 1984) usually characterize partial answerhood in terms of entailment. Such characterizations are satisfactory as long as questions are assumed to *partition* logical space. In inquisitive semantics, questions are no longer associated with partitions: possibilities may overlap. As a consequence, partial answerhood cannot be characterized in terms of entailment anymore (cf. Groenendijk and Roelofsen, 2009).

not taken care of by the first clause in the definition of compliance, since every possibility for $?p$ is also a possibility for $?p \vee ?q$. So the second clause is essential in this case: it says that $?p$ is not compliant with $?p \vee ?q$ because two of the possibilities for $?p \vee ?q$ are not contained in any possibility for $?p$. The fact that these possibilities are, as it were, 'ignored' by $?p$ is the reason that $?p$ is more difficult to answer than $?p \vee ?q$.[3]

Notice that the second clause in the definition of compliance only plays a role in case both φ and ψ are inquisitive. Moreover, the *restriction* of the possibilities for ψ to $|\varphi|$ can only play a role if $|\varphi| \subset |\psi|$, which is possible only if φ is informative. Thus, the second clause can only play a role in its unsimplified form if φ is both inquisitive and informative, i.e., hybrid. If φ is hybrid, just as when φ is a question, the second clause forbids that a possibility for ψ is ignored by φ. But now it also applies to cases where a possibility for ψ is partly excluded by φ. The part that remains should then be fully included in one of the possibilities for φ.

As an example where this condition applies, consider $p \vee q$ as a response to $p \vee q \vee r$. One of the possibilities for $p \vee q \vee r$, namely $|r|$, is ignored by $p \vee q$: the restriction of $|r|$ to $|p \vee q|$ is not contained in any possibility for $p \vee q$. Again, this reflects the fact that the issue raised by $p \vee q$ is more difficult to resolve than the issue raised by $p \vee q \vee r$.

A general characterization of what the second clause says, then, is that φ may only remove possibilities for ψ by providing information. A possibility for ψ must either be excluded altogether, or it must be preserved: its restriction to $|\varphi|$ must be contained in some possibility for φ.

4   Computing Compliance

In this section, we specify an algorithm which computes, for a given sentence ψ, all sentences (up to logical equivalence) that are compliant with ψ. In order to do so, we first introduce a procedure DNF, which determines, for any formula ψ, an equivalent formula $\mathrm{DNF}(\psi)$ which is a disjunction of assertions (a *disjunctive normal form*).

Definition 23. $\mathrm{DNF}(\psi)$ is recursively defined as follows:

1. $\mathrm{DNF}(p) = p$
2. $\mathrm{DNF}(\bot) = \bot$
3. $\mathrm{DNF}(\neg\psi) = \neg\psi$
4. $\mathrm{DNF}(\psi \vee \chi) = \mathrm{DNF}(\psi) \vee \mathrm{DNF}(\chi)$
5. $\mathrm{DNF}(\psi \wedge \chi) = \bigvee_{i,j} (\psi_i \wedge \chi_j)$

   where:
   - $\mathrm{DNF}(\psi) = \psi_1 \vee \ldots \vee \psi_n$
   - $\mathrm{DNF}(\chi) = \chi_1 \vee \ldots \vee \chi_m$

---

[3] Notice that compliance does not hold in the other direction either. That is, $?p \vee ?q$ is not compliant with $?p$ (in this case, the first clause is not satisfied).

  - i ranges over $\{1, \ldots, n\}$
  - j ranges over $\{1, \ldots, m\}$

6. $\mathrm{DNF}(\psi \to \chi) = \bigvee_{k_1, \ldots, k_n} \bigwedge_i (\psi_i \to \chi_{k_i})$

where:

  - $\mathrm{DNF}(\psi) = \psi_1 \vee \ldots \vee \psi_n$
  - $\mathrm{DNF}(\chi) = \chi_1 \vee \ldots \vee \chi_m$
  - i ranges over $\{1, \ldots, n\}$
  - $k_1, \ldots, k_n$ all range over $\{1, \ldots, m\}$

Proposition 24. For all ψ, $\mathrm{DNF}(\psi)$ is a disjunction of assertions.

Proposition 25. For all ψ, $\mathrm{DNF}(\psi) \equiv \psi$

There is a close correspondence between $\mathrm{DNF}(\psi)$ and the possibilities for ψ.

Proposition 26. If π is a possibility for ψ then π is a possibility (the unique possibility) for some disjunct of $\mathrm{DNF}(\psi)$.

The converse, however, is not true. This is because some disjuncts of $\mathrm{DNF}(\psi)$ may be entailed by others. If one disjunct α entails another β, then $|\alpha|$ is *contained* in a possibility for ψ, but it is not identical to any such possibility. To get a full correspondence between the possibilities for ψ and the disjuncts of $\mathrm{DNF}(\psi)$, we must eliminate those disjuncts that entail others. This operation preserves logical equivalence. We call the resulting formula the *clean disjunctive normal form* of ψ, $\mathrm{CDNF}(\psi)$.

Definition 27. $\mathrm{CDNF}(\psi)$ is obtained from $\mathrm{DNF}(\psi)$ by removing any disjunct that classically entails any other disjunct.

Proposition 28. For all ψ, $\mathrm{CDNF}(\psi)$ is a disjunction of assertions.

Proposition 29. For all ψ, $\mathrm{CDNF}(\psi) \equiv \psi$.

Proposition 30. π is a possibility for ψ *if and only if* π is a possibility (the unique possibility) for some disjunct of $\mathrm{CDNF}(\psi)$.

So $\mathrm{CDNF}(\psi)$ gives us, as it were, a syntactic representation of the possibilities for ψ. This is exactly what we need to compute compliant responses. We are now ready to define an algorithm that takes a sentence ψ as its input, and yields a set $\mathrm{COMP}(\psi)$ of sentences that are compliant responses to ψ.

Definition 31 (Algorithm)

1. The algorithm takes as its input a sentence ψ. It first computes $\mathrm{CDNF}(\psi)$. If $\mathrm{CDNF}(\psi)$ consist of a single disjunct, then ψ is an assertion. Then, a sentence is compliant with ψ iff it is equivalent with ψ. So we output $\mathrm{COMP}(\psi) = \{\psi\}$ in this case.

2. If ψ is not an assertion, we first use $\mathrm{CDNF}(\psi) = \psi_1 \vee \ldots \vee \psi_n$ to compute the set $\mathrm{CA}(\psi)$:

$$\mathrm{CA}(\psi) = \{!(\psi_{i_1} \vee \ldots \vee \psi_{i_m}) \mid i_1, \ldots i_m \in \{1, \ldots, n\}, \ m \geq 1\}$$

$\mathrm{CA}(\psi)$ consists of all formulas that are obtained from $\mathrm{CDNF}(\psi)$ by removing some (possibly zero, but not all) disjuncts, and then turning the remaining disjunction into an assertion using the ! operator. The unique possibility for such a formula always coincides with the union of a non-empty set of possibilities for ψ. So all formulas in $\mathrm{CA}(\psi)$ satisfy the first condition in the definition of compliance. Moreover, by fact 21, the second condition does not play a role for assertive responses. So all formulas in $\mathrm{CA}(\psi)$ are compliant with ψ (hence the name CA, short for 'compliant assertions').

3. Oc course, a compliant response to ψ does not have to be an assertion. It may very well be inquisitive. Any inquisitive compliant response, however, must be equivalent with a disjunction of compliant assertions. Thus, we compute the set of potentially compliant responses, PCR, as follows:

$$\mathrm{PCR}(\psi) = \{\chi_1 \vee \ldots \vee \chi_n \mid 1 \leq n \leq |\mathrm{CA}(\psi)| \text{ and } \chi_1 \ldots \chi_n \in \mathrm{CA}(\psi)\}$$

All formulas in PCR satisfy the first condition in the definition of compliance, and vice versa, every formula that satisfies this condition is equivalent with some formula in $\mathrm{PCR}(\psi)$.

4. What remains to be done is to filter out those formulas in $\mathrm{PCR}(\psi)$ that do not satisfy the second condition in the definition of compliance. To do so, we proceed as follows.

   Take a sentence $\chi \in \mathrm{PCR}(\psi)$. We know that $\chi = \chi_1 \vee \ldots \vee \chi_n$, where all χ_i's are assertions. We have to check that every possibility for ψ, when restricted to $|\chi|$, is contained in some possibility for χ.

   To do so, take a disjunct ψ_j of $\mathrm{CDNF}(\psi)$, and check if $\psi_j \wedge !\chi$ classically entails one of the disjuncts of χ. If this works for all ψ_j, then χ is compliant with ψ, otherwise it is not.

   Carrying out this procedure for all $\chi \in \mathrm{PCR}(\psi)$ yields the desired set of sentences $\mathrm{COMP}(\psi)$.

5. Finally, there is some optional 'cleaning up' to do. The formulas in $\mathrm{COMP}(\psi)$ are all disjunctions of assertions, that is, formulas in disjunctive normal form. To allow for a more intelligible output, we would like to bring these formulas into *clean* disjunctive normal form. To do so, we simply apply CDNF to every formula in $\mathrm{COMP}(\psi)$.

We are now ready to state our main result: $\mathrm{COMP}(\psi)$ does not just contain *some* sentences that are compliant with ψ, it actually contains *all* such sentences (up to logical equivalence). The proof of this result is suppressed here for reasons of space. The interested reader is referred to (Cornelisse, 2009).

Theorem 1 (Soundness and Completeness of the Algorithm)
φ is compliant with ψ iff φ is logically equivalent with some sentence in $\mathrm{COMP}(\psi)$.

We end with a remark regarding the implementation of the algorithm. Notice that most of the operations that have to be carried out consist in syntactic manipulation of formulas. The only 'reasoning' steps consist in checking classical entailment. Existing entailment/satisfiability checking algorithms can be used to carry out this task. The algorithm has been implemented, and is accessible through a graphical user interface at www.illc.uva.nl/inquisitive-semantics.

5 Conclusion and Outlook

The established algorithm could serve as the basis for practical applications of inquisitive semantics, and will also aid in further developing and imparting the theoretical framework. Several extensions suggest themselves. In particular, Groenendijk and Roelofsen (2009) discuss some general criteria for *preferring* certain compliant responses over others. Thus, one natural step to take would be to develop an algorithm that, given an initiative ψ and an agent A with information state s_A, determines the *most* compliant response(s) to ψ that A may truthfully utter.

Bibliography

Ciardelli, I., Roelofsen, F.: Generalized Inquisitive Logic: Completeness via Intuitionistic Kripke Models. In: Proceedings of Theoretical Aspects of Rationality and Knowledge (2009)

Cornelisse, I.: A syntactic characterization of compliance in inquisitive semantics. BSc Thesis, Artificial Intelligence, University of Amsterdam (2009)

Groenendijk, J.: Inquisitive semantics: Two possibilities for disjunction. In: Bosch, P., Gabelaia, D., Lang, J. (eds.) Tbilisi 2007. LNCS (LNAI), vol. 5422, pp. 80–94. Springer, Heidelberg (2008)

Groenendijk, J., Roelofsen, F.: Inquisitive semantics and pragmatics. In: Larrazabal, J.M., Zubeldia, L. (eds.) Meaning, Content, and Argument: Proceedings of the ILCLI International Workshop on Semantics, Pragmatics, and Rhetoric (2009), www.illc.uva.nl/inquisitive-semantics

Groenendijk, J., Stokhof, M.: Studies on the Semantics of Questions and the Pragmatics of Answers. Ph.D. thesis, University of Amsterdam (1984)

Mascarenhas, S.: Inquisitive semantics and logic. Manuscript, University of Amsterdam (2008)

Attributing Distributed Responsibility in Stit Logic

Roberto Ciuni and Rosja Mastop

Department of Philosophy
Delft University of Technology
Jaffalaan 5 - 2628 BX Delft
The Netherlands

1 Introduction

One important function of logics of agency (Stit Logic, ATL, CL, Dynamic Logic) is to assess the validity of arguments whereby responsibility is attributed to, or denied by, individuals and groups. Indeed, the vocabulary and computational properties of such logic allow us to express and reason on what agents did, or what empowered them to do (either following a strategy as in ATL or not). It is clear that this kind of reasoning is indispensable when we want to attribute or distribute responsibility to agents, as when we want to ascertain that some agents were not responsible of the events into account: if the group constituted by two persons *did not* or even *could not* for some reasons blow a hospital, then they should not be charged for blowing the hospital. Agency and the formal framework for it then appear to be important for the implementation of a rigourous reasoning on responsibility in (or of) groups.

Various developments have been carried over from a pure logic of agency to frameworks where agency interacts with other notions. Focusing on Stit Logic, the addition of knowledge (Broersen [4] and [5]) and intention (Broersen [6]) operators allows for the expression of statements such as that one did something (un)knowingly or (un)intentionally. Here we make a further step in this direction, by introducing expressions stating that an individual or group did not do more than so-and-so. For instance, if responsibility is attributed to a group for the robbing of a bank, then one member of the group may state that all he did was to drive the others there and back. The setting we propose will enable us to express the conditions for attributing or distributing responsibility.

In Stit Logic any set of individuals counts as a group, and the combination of their individual actions counts as their group action. This means that the group of you and some bank robber can be attributed responsibility for the combination of watering the plants and robbing a bank, although all that you did was water the plants. We shall call "*agents*" the individuals and "*coalitions*" the groups of agents, thus following standard terminology in logics of agency.

If we aim at expressing the fact that an agent did something while another *in fact did not*, it is necessary to expand the basic formal framework of stit logic.

X. He, J. Horty, and E. Pacuit (Eds.): LORI 2009, LNAI 5834, pp. 66–75, 2009.

Indeed, an important feature of stit logic is the so-called coalition monotonicity (discussed below): if coalition A accomplishes φ and coalition B accomplishes ψ, then $A \cup B$ (the two coalitions together) accomplishes $\varphi \wedge \psi$. In other words, coalitions preserve their accomplishments when entering bigger coalitions, and this principle makes sense if we want to formalise cooperative behaviour (as usually logics of agency do). But if $\varphi \wedge \psi$ has been accomplished, this means that the actual course of events has been forced as to include both $\varphi \wedge \psi$, and hence it is not possible to make a difference whether A did only φ or only ψ (or both). The same can be said for B. Yet we want to make this difference very often. This is more evident in a concrete example. Take again the case of the person watering the plants and the robbers. Given the set-theoretical apparatus in the logics for collective actions, the coalition constituted by these persons both water the plants and rob a bank. Obviously we want to dispense with the plant waterer with the charge of robbing a bank, since he *only* watered the plants. However, we cannot do that with the usual apparatus for Stit Logic or other logics for agency. In the present paper, we aim at providing a formal refinement of the Stit apparatus that allows to express the difference we perceive in the above case and similar ones.

2   The Logic of "Seeing to It That"

Stit Logic is named after its primitive operator, the *stit* operator, which in turn is the acronym of "seeing to it that". "$stit_a\varphi$" reads "agent a sees to it that φ". The motivation for the logic has been mainly philosophical: to give a formal semantic for sentences about agents and what they do (these sentences are also called *agentives* in the literature). For instance, it allows a precise formulation of the concept of free action (where one could have done otherwise than he/she actually did). In this fashion a number of systems have been introduced for example by Belnap, Perloff and Xu [2,3] and Horty [10]. However, in recent years *stit logic* has been considered as a useful tool for the logic of agency in the field of computer science, and this as resulted in proposals that are much more sensible to the meta-logical properties we should reasonably want for a stit logic, and more careful to the manageability of the logic itself. The systems proposed by by Broersen [4] and Broersen, Herzig and Troquard [7] go in that direction.

A distinguishing feature of stit logic is that it is able to express what an agent actually *does*, while other prominent logics of agency (Pauly's [13] Coalition Logic and Alur's [1] Alternating-time Temporal Logic) are designed to express just what an agent is *empowered*, or *able*, to do. This superior expressive power has a number of conceptual virtues. For example, it enables us to point out how the agent contributed to the actual state of the world or to what happened. The ability to do this plays an important role in responsibility assignment or distribution. Take the example of a bank that has been robbed. When we want to charge or track down those who are responsible for that, we do not want to single out those agents that were simply *empowered* to do it. We want to charge the agents that *did* it. Coalition Logic and Alternating-time Temporal Logics have the resources to express just the first case (being empowered to do). By contrast, Stit Logic has the expressive resources to discriminate the two cases, since it can express them both. Thus the expressive power of stit logic fits with some important differences we want to make in the *general* context of agency. When it comes to more specific contexts, further expressive power is needed: epistemic operators must be added to express what an agent knowingly does (see Broersen [4]), while deontic ones (or a combined deontic-stit operator) are indispensable if we want to express the notion of what an agent ought to do (see Horty [10]). Similarly, gaining more expressive power is essential if we want to express *collective actions*, i.e. what a *group of agents* does.

The example above makes reference to a set of agents, *i.e.* what we have called a *coalition*. For expressing it, the language of stit logic must be endowed with constants for sets of agents, so that we can form the sentence "$stit_A\varphi$" (reading "coalition A sees to it that φ"). We must extend the language about just one agent with names for set. Such a move allows as to enter the field of multi-agency. As for the reasons for not confining ourselves to the mono-agent case, there is a plenty of them. Most prominently, the most important phenomena concerning agency are *collective*. Suppose we want to single out the responsibility of one agent. In order to do that, we have to contrast *agents* (we have to be able to say that agent a, but not agent b, committed a given crime). Similarly, we must be able to speak of a plurality of agents if we want to individuate those responsibility that are shared by more than one agent (the case where both agent a and agent b committed the given crime). Stit Logic is easily adapted to incorporate such expressions. Horty [10] and Broersen [4] present specific versions of Stit Logic that expresses what *coalitions* do. Notice that cases involving one agent and cases with many agents may be treated uniformly within an apparatus for collective action: $stit_a\varphi$ may be taken as a graphically light version of $stit_{\{a\}}\varphi$, and then actions of a single agent may be taken as actions of the coalition containing just that agent. Other changes are needed beside the linguistic one. For example, Horty [10] uses a function to individuate the alternative options of single agents, and then defines a formal device to single the jointed action of many agents out of the possible actions of each of the agents. However, the systems presented by Broersen [4] avoids this further complication, since the apparatus there introduces applies in a straightforward way both to coalitions and single agents (to be considered singletons). Such a feature is very attractive when it comes to expanding the framework of coalitional Stit. Indeed, it allows to start with a straightforward and manageable apparatus (though a rich one, in any case). Thus, in the present work we shall use a slightly modified version of the system proposed in Broersen [4]. Such a system possess a feature that sets it apart from the classical Stit Logics: it defines its main stit operator (*xstit*) as meaning "seeing to it that in the next state of the world". The reason for such a choice will become clear below.

2.1   The System xstit

The language $\mathcal{L}_{xstit}$ has three modal operators. One of these is $[A\,xstit]\varphi$, which means that the agents in the set A see to it that φ is true directly after their next (collective) action. For instance, {John,Bill} sees to it that the room is vented, when John's next action is to open the window and, simultaneously, Bill's next action is to open the door. A second modal expression in the language is $\Box\varphi$, which says that φ is true (here and now), no matter what any of the agents decide to do. For instance, if the door is locked and no agent has a key, then $\Box\neg[A\,xstit]\,door\,open$ is true: necessarily, after the next actions of the agents the door is (still) closed. The third modality is $X\varphi$, which means that in the state actually resulting from the next action, φ is true.

The language is formally specified as follows:

$$\varphi, \psi, \ldots := p \mid \bot \mid \neg\varphi \mid \varphi \wedge \psi \mid X\varphi \mid \Box\varphi \mid [A\,xstit]\varphi$$

Here $p \in P$ is a propositional variable, and $A \subseteq Ags$ is a set of agents.

The language is interpreted on a Kripke model $\langle S, H, R_X, R_\Box, \{R_A \mid A \subseteq Ags\}, V\rangle$. S is a set of states and H is the set of histories passing through these states. Histories, in turn, are sets of states. A *possibility* is an ordered pair $\langle h, s\rangle$ of a history h and a state s, such that $s \in h$.

The relation $R_\Box$ is an equivalence relation for *historical necessity*, defined by: $\langle h, s\rangle R_\Box\langle h', s'\rangle$ iff $s = s'$. This relation lets you 'see' the alternative possible histories—the decisions that might be taken—to the actual history, relative to the state at which you are in the actual history. The 'next' relation R_X, on the other hand, linearly orders the states in a given history. So $\langle h, s\rangle R_X\langle h', s'\rangle$ only if $h = h'$, and the relation is serial and deterministic.[1]

The relations R_A, for any set A of agents, represent the substance of the (next) actions undertaken by the group A in any given possibility. Various assumptions are made concerning the structure of these relations. They are serial: every coalition acts in any given history-state pair. The results of actions carry over to larger coalitions (*C-mon*). Distinct (non-overlapping) coalitions can always join forces (*Indep*). The empty coalition achieves nothing (*Ineff-0*) and no coalition achieves less than what is necessary to happen next (*X-Eff*), nor more than what next is necessary to happen: no actions constitute a choice between histories that are undivided in next states (*NCUH*). These assumptions are imposed as frame conditions, defining the class of XSTIT frames. Its precise definition can be retrieved straightforwardly from the axioms, because they are standard Sahlqvist formulas. See below for the axioms, and see Broersen [4] for the precise statements of the frame conditions.

A valuation $V : P \mapsto (S \times H)$ assigns truth values to the proposition letters $p \in P$ for every possibility $\langle h, s\rangle$. Given a model $M = \langle S, H, R_X, R_\Box, \{R_A \mid A \subseteq Ags\}, V\rangle$, the semantics is defined as follows:

---

[1] Broersen [4] assumes that the total coalition Ags is deterministic, and so he defines the 'next' relation by equating it with the relation R_{Ags}. Here, we keep the two distinct, in order to make the completeness proof for the logic of "seeing to it that only" to go over smoothly.

$$M, \langle h, s \rangle \models p \qquad \text{iff } \langle h, s \rangle \in V(p)$$
$$M, \langle h, s \rangle \models \neg\varphi \qquad \text{iff not } M, \langle h, s \rangle \models \varphi$$
$$M, \langle h, s \rangle \models \varphi \wedge \psi \qquad \text{iff } M, \langle h, s \rangle \models \varphi \text{ and } M, \langle h, s \rangle \models \psi$$
$$M, \langle h, s \rangle \models X\varphi \qquad \text{iff } \langle h, s \rangle R_X \langle h', s' \rangle \text{ implies that } M, \langle h', s' \rangle \models \varphi$$
$$M, \langle h, s \rangle \models \Box\varphi \qquad \text{iff } \langle h, s \rangle R_\Box \langle h', s' \rangle \text{ implies that } M, \langle h', s' \rangle \models \varphi$$
$$M, \langle h, s \rangle \models [A\,xstit]\varphi \text{ iff } \langle h, s \rangle R_A \langle h', s' \rangle \text{ implies that } M, \langle h', s' \rangle \models \varphi$$

2.2 Axiomatisation

The complete Hilbert calculus Λ_{xstit} is listed below.

PC	propositional calculus
MP	from φ and $\varphi \rightarrow \psi$ infer ψ
Nec	from φ infer $X\varphi, [A\,xstit]\varphi$ and $\Box\varphi$
K$\Box$	$\Box(\varphi \rightarrow \psi) \rightarrow (\Box\varphi \rightarrow \Box\psi)$
T$\Box$	$\Box\varphi \rightarrow \varphi$
5$\Box$	$\neg\Box\varphi \rightarrow \Box\neg\Box\varphi$
(K-X)	$X(\varphi \rightarrow \psi) \rightarrow (X\varphi \rightarrow X\psi)$
(Det-X)	$\neg X\neg\varphi \leftrightarrow X\varphi$
(K-xstit)	$[A\,xstit](\varphi \rightarrow \psi) \rightarrow ([A\,xstit]\varphi \rightarrow [A\,xstit]\psi)$
(Ser)	$[A\,xstit]\varphi \rightarrow \neg[A\,xstit]\neg\varphi$
(Ags-X)	$[Ags\,xstit]\varphi \rightarrow X\varphi$
(C Mon)	$[A\,xstit]\varphi \rightarrow [(A \cup B)\,xstit]\varphi$
(Indep)	$\Diamond[A\,xstit]\varphi \wedge \Diamond[B\,xstit]\psi \rightarrow \Diamond([A\,xstit]\varphi \wedge [B\,xstit]\psi)$, if $(A \cap B) = \emptyset$
(Ineff-0)	$[\emptyset\,xstit]\varphi \rightarrow \Box X\varphi$
(X-Eff)	$\Box X\varphi \rightarrow [A\,xstit]\varphi$
(NCUH)	$[A\,xstit]\varphi \rightarrow X\Box\varphi$

These axiom schemes are all within the Sahlqvist class, so they define a class of frames XSTIT for which the logic is sound and complete (see Broersen [4]).

2.3 Why Do We Use *xstit*?

Working with the operator *xstit* is not the standard option in Stit Logic. So, why are we doing it? The answer is that with the *xstit* operator we have neat technical benefits. Indeed, the usual choice in Stit Logic is defining a *stit* operator by the condition: "$[A\,stit]\varphi$ is true iff for any history h' there is a state s' such that $M, \langle s', h' \rangle \models \varphi$". As a consequence, the states s' in the relevant pairs state-history may be subsequent states whatever of s (w.r.t. h'), and *not* the next states w.r.t. to s. But if we let this go, we lose axiom *K-xstit*. Indeed, we may have the following situation:

1. $[A\,stit](\varphi \rightarrow \psi)$ is true at the pair $\langle s, h \rangle$, and hence for every history h' is in a pair $\langle s', h' \rangle$ that is in R_A with $\langle s, h \rangle$ and forces $\varphi \rightarrow \psi$.
2. All such pairs $\langle s', h' \rangle$ forces $\varphi \rightarrow \psi$ because both φ and ψ fail at such pairs.

3. Every history is also in a pair $\langle s'', h' \rangle$ in R_A with $\langle s, h \rangle$ and such that it forces φ. This makes $[A\,stit]\varphi$ true at the pair $\langle s, h \rangle$. Yet no state that appears in a pair in R_A with $\langle s, h \rangle$ forces ψ. Hence $[A\,stit]\psi$ is false at $\langle s, h \rangle$

As a consequence of the three conditions, $[A\,stit](\varphi \to \psi)$ and $[A\,stit]\varphi$ are true at the pair $\langle s, h \rangle$, but $[A\,stit]\psi$ is not. This makes K-$xstit$ fail.

With the conditions for $xstit$ the above situation is prevented. Indeed, if $[A\,xstit]$ $(\varphi \to \psi)$ $[A\,xstit]\varphi$ are true at $\langle s, h \rangle$, then for every history there is *only one* pair $\langle s', h' \rangle$ (in R_A with $\langle s, h \rangle$) that is relevant for the truth of the two sentences. Such a pair is the one containing the next state of s w.r.t. s'. If $\varphi \to \psi$ and φ are forced at the pair $\langle s', h' \rangle$, then also ψ is, from MP. And if for every history h', $\langle s', h' \rangle$, we have this situation, we have that at $\langle s, h \rangle$ not only $[A\,stit](\varphi \to \psi)$ and $[A\,stit]\varphi$ are true, but also $[A\,stit]\psi$ is (otherwise it should fail in some of the next states s' above, but this contradicts the construction).

3   The Logic of "Seeing to It That Only"

3.1   Language, Models and Semantics

To be able to express that an agent (or coalition) does no more than φ, we extend the language $\mathcal{L}_{xstit}$ with sentences of the form $[A\,istit]\varphi$. The intended meaning of such sentences is that the collective action of group A excludes only possibilities such that φ becomes true. In other words, the group A does not avoid any non-φ possibilities. The addition of the $istit$ operator follows the approach of Humberstone [11] of characterising the 'only' modality as a complex expression:

"Group A sees to it only that φ" $\equiv [A\,xstit]\varphi \wedge [A\,istit]\neg\varphi$

ISTIT frames are tuples $\langle S, H, R_X, R_\square, \{R_A \mid A \subseteq Ags\}, \{R_A^- \mid A \subseteq Ags\} \rangle$, where $\langle S, H, R_X, R_\square, \{R_A \mid A \subseteq Ags\} \rangle$ is an XSTIT frame and such that $R_A^- =_{\mathrm{def}} ((R_\square \circ R_{Ags}) \setminus R_A)$. As is standard, an ISTIT model is obtained by adding a valuation function V to the frame, that assigns truth values for all atomic propositions at each of the history-state pairs of the frame.

The semantics is similarly an extension of the semantics for $\mathcal{L}_{xstit}$. We can therefore suffice with the clause for the additional operator.

$M, \langle h, s \rangle \models [A\,istit]\varphi$ iff $\langle h, s \rangle R_A^- \langle h', s' \rangle$ implies that $M, \langle h', s' \rangle \models \varphi$

3.2   Axiomatisation, Soundness and Completeness

As Humberstone already observed, the property that two relations are disjoint (i.e., $R_A \cap R_A^- = \emptyset$) cannot be expressed in a language with a normal modal semantics. This implies that we cannot identify a complete axiomatisation for the class of ISTIT frames by means of standard correspondence theory. Instead, adapting a proof from Herzig and Gasquet [9], we prove completeness by showing that the frames characterised by the logic are p-morphic images of the ISTIT frames.

The logic Λ_{istit} contains the logic Λ_{xstit} plus the following:

- From φ infer $[A\,istit]\varphi$
- $[A\,istit](\varphi \to \psi) \to ([A\,istit]\varphi \to [A\,istit]\psi)$
- $\Box X\varphi \leftrightarrow ([A\,xstit]\varphi \wedge [A\,istit]\varphi)$
- $\Diamond[A\,istit]\varphi \wedge \Diamond[B\,istit]\psi \to \Diamond([A\,istit]\varphi \wedge [B\,istit]\psi)$ (for $A \cap B = \emptyset$)
- $[A \cup B\,istit]\varphi \to [A\,istit]\varphi$
- $[\emptyset\,istit]\bot$

After necessitation and the scheme K, the third axiom scheme identifies what is necessary to happen with that which is both a necessary consequence of what one does and a necessary consequence of what one (thereby) avoids. This makes the axiom scheme *(X-Eff)* superfluous. The remaining three axiom schemes are straightforward consequences of making R_A^- the complement of R_A: both have independence of agency, *istit* is reverse coalition monotonic, and because the empty coalition is inactive, it avoids no possibility.

Soundness of this logic for the class of ISTIT frames can be proven by showing that the first order conditions corresponding to the axiom schemes are true for all ISTIT models.

- $R_A \cup R_A^- = R_\emptyset$;
- If $\langle h, s\rangle(R_\Box \circ R_A^-)\langle h', s'\rangle$ and $\langle h, s\rangle(R_\Box \circ R_B^-)\langle h'', s''\rangle$ and $A \cap B = \emptyset$, then there is a $\langle h, s\rangle R_\Box \langle h''', s\rangle$ such that both $\langle h''', s\rangle R_A^-\langle h', s'\rangle$ and $\langle h''', s\rangle R_B^-\langle h'', s''\rangle$.
- $R_A^- \subseteq R_B^-$ if $A \subset B$;
- $R_\emptyset^- = \emptyset$.

Checking that these conditions obtain in all ISTIT frames is left to the reader. Let us call the class of frames characterised by this logic $F(\Lambda)$. In this class, the relations R_A and R_A^- are not necessarily disjoint, but they do obey the above four properties.

To prove completeness, we show that any $F(\Lambda)$ model is a p-morphism of some ISTIT model: we can always make the relations R_A and R_A^- disjoint while preserving the truth values of sentences. We do so by constructing, for an arbitrary $F(\Lambda)$ model, the ISTIT model of which it is a p-morphism. The proof is an amendment of the proof given by Herzig and Gasquet for the logic of inaccessible worlds. The construction is by duplication of states and, accordingly, the multiplication of histories.

Let $M = \langle S, H, R_X, R_\Box, \{R_A \mid A \subseteq Ags\}, \{R_A^- \mid A \subseteq Ags\}, V\rangle$ be a point-generated submodel for φ at its possibility $\langle h, s\rangle$. We define the model $M' = \langle S', H', R_X', R_\Box', \{R_A' \mid A \subseteq Ags\}, \{R_A^{-\prime} \mid A \subseteq Ags\}, V'\rangle$ as follows:

- $S' = T_1 \cup T_2$, where S, T_1, and T_2 are mutually disjoint and such that there are two isomorphic mappings $f_1 : S \mapsto T_1$ and $f_2 : S \mapsto T_2$.
- H' is the set of all histories h' such that there is an $h \in H$ and h' contains either $f_1(s)$ or $f_2(s)$ iff $s \in h$. We write $g(h')$ for the unique history $h \in H$ such that all states in h' are mapped onto states in h.
- $\langle h, f_i(s)\rangle R_X'\langle h', f_j(s')\rangle$ iff $\langle g(h), s\rangle R_X\langle g(h'), s'\rangle$ and $h = h'$.
- $\langle h, f_i(s)\rangle R_\Box'\langle h', f_j(s')\rangle$ iff $s = s'$.

- $\langle h, f_i(s) \rangle R'_A \langle h', f_j(s') \rangle$ iff $\langle g(h), s \rangle R_A \langle g(h'), s' \rangle$ and $i = j$, or $\langle g(h), s \rangle (R_A \setminus R_A^-) \langle g(h'), s' \rangle$ and $i \neq j$.
- $\langle h, f_i(s) \rangle R_A^{-'} \langle h', f_j(s') \rangle$ iff $\langle g(h), s \rangle R_A^- \langle g(h'), s' \rangle$ and $i \neq j$, or $\langle g(h), s \rangle (R_A^- \setminus R_A) \langle g(h'), s' \rangle$ and $i = j$.
- $\langle h, f_i(s) \rangle \in V'(p)$ iff $\langle g(h), s \rangle \in V(p)$, for $i = 1, 2$.

The reverse of this construction, the mapping $f_1^{-1} \cup f_2^{-1}$ from M' to M, is a p-morphism.[2] So if φ is satisfiable in M it is also satisfiable in M'. Moreover, it can be observed that M' is an ISTIT model: for all coalitions A, the relations R'_A and $R_A^{-'}$ are disjoint, while their union still completely covers $R_\square \circ R_X$. Also, the determinacy of R_X has been preserved in the construction.

4   Discussion

With this setting at hand, we are able to discriminate between cases where one agent is involved in something and cases where he is not. Coming back to the example of the person watering the plants—call him 'Bob': we have a clear way to express that he did not partake in the bank robbery: in the language of *istit* this is expressed by $[bob\,xstit]water \wedge [bob\,istit]\neg water$. Combined with $[joe\,xstit]rob \wedge [joe\,istit]\neg rob$, this implies that $[\{bob, joe\}\,xstit](water \wedge rob) \wedge [\{bob, joe\}\,istit]\neg(water \wedge rob)$. That is to say, Bob only watered the plants.

In this scenario, Bob is entirely superfluous in the group of him and Joe insofar as the robbery is concerned. So Bob is exculpated by this reasoning. Yet, in other cases, doing *only* a precise thing may be also the *source* of a violation of some sort. For example, suppose that a bank employee convinces me to subscribe an investment with high risk-level. The description of the risk-level is in the second page of the document he shows me. However, he shows me only the first page of such a document. In our apparatus, $[bank\ employee\ xstit]page1 \wedge [bank\ employee\ istit]\neg page1$. He should have shown me the other pages as well, so he misled me. As a consequence, he has to be attributed the responsibility of a mischievous act.

Lastly, as has been noted earlier, in the Stit framework a collective of agents may be said to achieve more than their individual actions combined. A possible example is the killing of a person by a mob, where the 'fatal blow' may not be identifiable, so that legally the killing might be said to be caused by the mob as such, rather than by any of its members. In the Netherlands, some controversy has arisen over this notion of collective responsibility (article 141 of the Dutch Criminal Code). The idea of such irreducibly collective actions is a complex philosophical issue that we shall not address here (see for instance Graham [8]). But it is important that we may express in our language that some harm was caused by a collective of agents without the possibility of attributing it to any individuals.

Using only the *xstit* operator we can express that a group did or did not cause something, but not that their individual actions, as such, combined did or did

---

[2] Or, if preferred, the mapping from each $\langle h, f_i(s) \rangle$ onto $\langle g(h), s \rangle$.

not cause the event. If no individual in a group performs any harmful action, but collectively they do cause harm, then it depends on our concept of collective action whether or not we will want to assign any guilt.

5   Concluding Remarks

An intriguing development in legal research is towards the aim of automated legal reasoning. In that context, a specifically challenging issue is the formalisation of responsibility attribution, which involves more than the attribution of causal relevance (and strict liability), because it includes the attribution of the relevant epistemic attitudes and intentions as well (see for instance Lehmann and Breuker [12]). Our paper is envisaged as part of a larger programme of developing the Stit framework into a formal system suitable for automated attribution of (legal or moral) responsibility. Apart from the reformulation of the Stit framework into a normal (multi-modal and two-dimensional) modal logic by Broersen, the aim is to introduce a variety of modal concepts to the language allowing us to express the relevant conceptual distinctions needed in a fair and correct responsibility attribution (including intention and knowledge). In view of the coalitional approach to the modal operators in the recent Stit-formalisms, the present contribution focusses on the assessment of a *distribution* of responsibility within groups. In simple terms, we need to be able to judge who did what and who refrained from doing what.

The technical work in this paper shows that the extension with the *istit* operator allows us to do this within the scope of a normal modal logic, using the insights of Humberstone and Herzig and Gasquet on the expression of 'only' notions in modal logic and its complete axiomatization.

A further potential application of the present approach, suggested by Broersen, is the analysis of dynamic modal notions (as in dynamic epistemic logic, see [14]) within the Stit framework. An announcement of φ is an action that results in the 'elimination' of *all and only* $\neg\varphi$ worlds in the domain.

References

1. Alur, R., Henzinger, T., Kupferman, O.: Alternating-time temporal logic. In: Proceedings of the 38th IEEE Symposium on Foundations of Computer Science, pp. 100–109 (1997)
2. Belnap, N., Perloff, M.: The way of the agent. Studia Logica 51, 463–484 (1992)
3. Belnap, N., Perloff, M., Xu, M.: Facing the Future: Agents and Choices in our Indeterminist World. Oxford University Press, Oxford (2001)
4. Broersen, J.: A logical analysis of the interaction between 'obligation-to-do' and 'knowingly doing'. In: van der Meyden, R., van der Torre, L. (eds.) DEON 2008. LNCS (LNAI), vol. 5076, pp. 140–154. Springer, Heidelberg (2008)
5. Broersen, J.: A complete stit logic for knowledge and action, and some of its applications. In: Baldoni, M., Son, T.C., van Riemsdijk, M.B., Winikoff, M. (eds.) DALT 2008. LNCS (LNAI), vol. 5397, pp. 47–59. Springer, Heidelberg (2009)

6. Broersen, J.: First steps in the stit-logic analysis of intentional action. In: ESSLLI 2009 workshop on Logical Methods for Social Concepts, LMSC 2009 (2009)
7. Broersen, J., Herzig, A., Troquard, N.: A normal simulation of coalition logic and an epistemic extension. In: Proceedings of TARK XI, pp. 92–101. ACM Digital Library (2007)
8. Graham, K.: Practical Reasoning in a Social World: How We Act Together. Cambridge University Press, Cambridge (2002)
9. Herzig, A., Gasquet, O.: Translating inaccessible worlds logic into bimodal logic. In: Clarke, M., Kruse, R., Moral, S. (eds.) ECSQARU 1993. LNCS, vol. 747, pp. 145–150. Springer, Heidelberg (1993)
10. Horty, J.: Agency and Deontic Logic. Oxford University Press, Oxford (2001)
11. Humberstone, I.L.: Inaccessible worlds. Notre Dame Journal of Formal Logic 24(3), 346–352 (1983)
12. Lehmann, J., Breuker, J.: On automatic causal reasoning. In: Breuker, J., Leenes, R., Winkels, R. (eds.) Legal Knowledge and Information Systems. Jurix 2000, pp. 123–134. IOS Press, Amsterdam (2000)
13. Pauly, M.: Logic for Social Software. PhD thesis, CWI/ILLC, Universiteit van Amsterdam (2001)
14. van Ditmarsch, H., van der Hoek, W., Kooi, B.: Dynamic Epistemic Logic. Synthese Library, vol. 337. Springer, Heidelberg (2007)

Characterizations of Iterated Admissibility
Based on PEGL

Jianying Cui, Meiyun Guo, and Xiaojia Tang

Institute of Logic and Intelligence, Southwest University
Tiansheng road. 2, BeibBei district, 400715 Chongqing, P.R. China

Abstract. Iterated dominance is perhaps the most basic principle in game theory. The epistemic foundation of this principle is based on the assumption that all players are rational. The main contribution of this paper is to characterize the algorithm of iterated admissibility in Probabilistic Epistemic Game Logic (PEGL). Firstly, on the basis of Probabilistic Epistemic Logic we set up a logic PEGL. Secondly, by redefining a concept of rationality, we show that the common knowledge of the rationality characterizes the algorithm of Iterated Admissibility, that is, we provide an epistemic foundation for the solutions or equilibria which are found by the algorithm of Iterated Admissibility(IA). Next, we provide a different characterization of IA using public announcements of the rationality in dynamic logic. The results we obtain can be seen as giving a dynamic epistemic foundation for the algorithm of Iterated Admissibility.

Keywords: strategic-form game, probability logic, iterated admissibility algorithm, rationality, common knowledge.

1 Introduction

The algorithm of iterated elimination and the rational players' decisions in games are research focuses in the fields of game theory and game logics (cf. [1], [2],[3] and [4] etc.) Most of the literatures on the concept of rationality used Bayesian rationality as [5] and [6], that is to say, a strategy of player i is said to be rational if it maximizes player i's expected payoff, given her probabilistic beliefs about the strategies used by her opponents, except for [7], [8] and [9] etc. Although [7] and [8] both give us more complicated descriptions of various iterated deletion algorithms on the basis of different definitions of rationality in epistemic logic respectively, but neither of them give a logical characterization of the algorithm of IA, which is a long-standing and attractive solution concept. Furthermore they both restricted their analyses to pure strategies. In this paper, firstly, building on the contributions of [7], [10] and [8], we put forward a new logic, namely Probability Epistemic Game Logic (PEGL). Secondly, we generalize the result in [11] to cover the mixed strategy, and by redefining rationality we take a syntactic approach to show that common knowledge of this rationality characterizes the algorithm of IA. Therefore, we give an epistemic foundation

X. He, J. Horty, and E. Pacuit (Eds.): LORI 2009, LNAI 5834, pp. 76–89, 2009.

for the algorithm of IA. Subsequently, inspired by [7], we show this rationality can be used as a proper announcement assertion in public announcement logic (PAL). In particular, using our definition of our rationality, the proposition " All players are rational " just fails at the worlds in which the strategies are weakly dominated. And after publicly announcing the proposition for one time, we can remove simultaneously all weakly dominated strategies. This leads to a new sub-game model. In this subgame model, the players may discover that some of their retaining strategies are dominated again owing to the absence of some worlds. By repeating the announcement and removing continuously the irrational worlds, until we get a subgame model in which the proposition holds at every world. For a game with finite strategy spaces, the procedure will stabilize after a finite number of stages. We indicate in this paper that the procedure which is constructed by removing worlds after repeated public announcements of the rationality we defined corresponds to the procedure of iterated admissibility.

The paper is organized as follows. In the next section we review the concept of strategic-form game, admissibility and iterated deletion procedures, and also the Probabilistic Epistemic logic (PEL) introduced by J.Y.Halpern. In Section 3, by combining the PEL with Game Logic, we propose a logic, namely Probabilistic Epistemic Game Logic(PEGL), and show how to provide an epistemic foundation for the algorithm of iterated admissibility based on PEGL. In Section 4, we prove that the announcement of this rationality can also characterize the iterated admissibility. We briefly survey the related works in section 5, and draw the conclusion.

2 Preliminaries

2.1 Game and Dominance

In this paper we restrict our attention to finite two-player strategic games, which are defined as follows.

Definition 1. *A finite strategic-form game is a tuple*
$G = \langle N, \{S_i\}_{i \in N}, \{\Delta(S_i)_{i \in N}, \{u_i\}_{i \in N}, \{U_i\}_{i \in N}, \{\succeq_i\}_{i \in N} \rangle.$[1] *where*

- $N = \{1, 2\}$ *is a set of players;*
- S_i *is a finite set of pure strategies of player* $i \in N$;
- $\Delta(S_i)$ *is a finite set of mixed strategies over* S_i. *If* $S_i = \{s_i^1, ..., s_i^m\}$, *then* $\delta_i = (\delta_i^1, ..., \delta_i^m)(\delta_i \in \Delta(S_i))$ *denotes player* i's *mixed strategy, where* $\delta_i^k (k = 1, ..., m)$ *denotes the probability of player* i *choosing the pure strategy* s_i^k, *and* δ_i^k *satisfies* $0 \leq \delta_i^k \leq 1$, $\sum_{k=1}^m \delta_i^k = 1$;
- u_i *is a payoff function[2]* $: S \to \mathbb{Q}$. *It gives player* i's *utility* $u_i(s)$ *for each pure strategy profile* $s \in S$ (*where* $S = S_1 \times S_2$);
- $U_i : \times_{j \in N} \Delta(S_j) \to \mathbb{Q}$, *which we assume satisfies the expected utility property;*

---

[1] Here we define a game by combing a strategic game $G = \langle N, \{S_i\}_{i \in N}, \{u_i\}_{i \in N},$ $\{\succeq_i\}_{i \in N} \rangle$ with its mixed extension form $\langle N, \{\Delta(S_i)\}_{i \in N}, \{U_i\}_{i \in N}, \{\succeq_i\}_{i \in N} \rangle$.

[2] Here we restrict ourselves to the games where players' utility is rational numbers.

   – $\succ_i$ *is player $i's$ preference on $\times_{j\in N}\Delta(S_j)$;The strict ordering $\delta_i \succ_i \eta_i$ is defined as usual: $\delta_i \succ_i \eta_i$ if and only if $\delta_i \succcurlyeq_i \eta_i$ and not $\eta_i \succcurlyeq_i \delta_i$.*

Where, $\succcurlyeq_i$ is a partial order, satisfied reflexivity, antisymmetry and transitivity. The interpretation of $\delta_i \succcurlyeq_i \eta_i$ is that, according to player i, the expected utility obtained by selecting δ_i is at least as good as the utility of selecting η_i. Usually we specify a player's preference relation by giving a payoff function to represent it. In this paper, we retain the two functions so that we can express games more clearly in our logic.

   Note that, we can take pure strategies as a special case of mixed strategies, for example, pure strategy s_i^1 that the player i chooses in a game is considered as a mixed strategy $\delta_i = \underbrace{(1,0,0...,0)}_{m}$, where $S_i = \{s_i^1,...,s_i^m\}$.

Definition 2. *Given a game G, a strategy s_i^k is weakly dominated (or inadmissible) if there is a $\delta_i \in \Delta(S_i)$, such that $U_i(\delta_i,s_j^l) \geq u_i(s_i^k,s_j^l)$ for all $s_j^l \in S_j$, and $U_i(\delta_i,s_j^h) > u_i(s_i^k,s_j^h)$ for some $s_j^h \in S_j(i \neq j)$.*[3]

In addition, a weakly undominated strategy s_i^k is also called admissible for player i. We can give an elimination procedure called iterated weak dominance or iterated admissibility (cf. [12]). The formal definition is as follows:

Definition 3. *Given a game G, let IAS be the set of iteratively admissible strategies of G defined recursively as follows: $IAS = \times_{i\in N}IAS_i$, where $IAS_i = \bigcap_{m\geq 0}IAS_i^m$, with $IAS_i^0 = S_i$ and for $m \geqslant 1, IAS_i^m = IAS_i^{m-1} \setminus IS_i^{m-1}$, where $IS_i^{m-1} = \{s_i^k \in IAS_i^{m-1} \mid s_i^k$ is inadmissible in the IAS_i^{m-1} with respect to $IAS_{-i}^{m-1}\}$.*

Note that in definition 3, it is assumed that at each stage all dominated strategies are simultaneously deleted. In contrast to most equilibrium concepts, iterative admissibility yields a rectangular set of strategy profiles, i.e. a Cartesian product of sets. Accordingly, whether the choice of a particular player is rational in this sense does not depend on the choices of other players. This IA procedures is illustrated in the following figure 1.where,

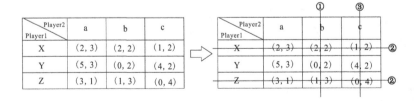

Fig. 1. $IWUS$ procedure

---

[3] Later, we will refer to weakly dominated (or weakly dominate) as dominated(or dominate).

$$IAS_1^0 = \{X, Y, Z\}, \quad IS_1^0 = \varnothing, \quad IAS_2^0 = \{a, b, c\}, \quad IS_2^0 = \{b\};$$
$$IAS_1^1 = \{X, Y, Z\}, \quad IS_1^1 = \{X, Z\}, IAS_2^1 = \{a, c\}, \quad IS_2^1 = \varnothing;$$
$$IAS_1^2 = \{Y\} = IAS_1, \quad IS_1^2 = \varnothing, \quad IAS_2^2 = \{a, c\}, \quad IS_2^2 = \{c\};$$
$$IAS_2^3 = \{a\} = IAS_2. \quad \text{Thus, } IAS = \{(Y, a)\}$$

2.2 Probabilistic Epistemic Logic

The Probabilistic Epistemic logic(PEL)([10]), which allows us to mention the probabilities explicitly in formulas, makes it possible to describe and analyze games with mixed strategies.[4]

Roughly, PEL extends the language of multi-agent $S5$ system by adding a probabilistic operator P_i, and taking a linear inequality $q_1 P(\varphi_1) + ... + q_n P_n(\varphi_n) \geq q$ (i.e. $\sum_{k=1}^m q_k P_i(\varphi_k) \geq q$) as a legal formula in the language of PEL, where $q_1, q_2, ..., q_n, q$ are arbitrary rational numbers, It's an i-probability formula.

$P_i(\varphi) \geq q$ is read as 'the probability agent i assigns to φ is greater than or equal to q', and the probability of a formula φ is the probability of the set of worlds where φ is true.[5]

In addition to axioms and rules for multi-agent $S5$ system, it includes two classes of axioms for reasoning about linear inequalities and probabilities respectively (cf [10]) in PEL. And R.Fagin and J.Y.Halpern showed:

Theorem 1. *PEL is a sound and complete axiomatization for the logic of knowledge and probability.*

3 Probabilistic Epistemic Game Logic

In this section, on the basis of PEL, we construct a new Probabilistic Epistemic Game Logic (PEGL), so that we can study games and discuss the epistemic foundations of the equilibria which are achieved by the algorithm of iterated admissibility.

Definition 4 (Language of PEGL). *Given a game G with $S_i = \{s_i^1, ..., s_i^m\}$, Let $\mathcal{P}$ be a countable set of propositional letters and N a finite set of agents. The language of PEGL denoted by $\mathfrak{L}_{PEGL}$ is given by the following rule in extended Backus-Naur form:*

$$\varphi ::= \bot \mid p \mid \neg \varphi \mid \varphi_1 \vee \varphi_2 \mid K_i \varphi \mid C_N \varphi \mid \sum_{k=1}^m a_k P_i(\varphi_k) \geq a$$

---

[4] Here, we present the PEL which is modified by [13]. It will be more suitable for our purpose.

[5] Note that, here we simplify the probabilistic epistemic models to the models where the σ-algebra of measurable sets is always the powerset of the sample space. Because in this paper we just formalize the concept of the rationality rather than reason about the probabilities, we don't require that the probabilities is assigned on the set of worlds that the agent considers possible, i.e. the property of **CONS**. However, our logical system still satisfies the property **MEAN**, that's to say, all formulas define measurable sets.

where $i \in N, a_1, ..., a_m$, and a are rational numbers, which stand for a utility value of some player i. Intuitively, $K_i\varphi$ says agent i knows that φ, while $E_N\varphi$ says that everyone in group N knows that φ, and $C_N\varphi$ says that φ is common knowledge among the group N. And the set of atomic propositions upon $\mathcal{P}$ contains atomic propositions of the following form:

1. Pure strategy symbols $s_i^1, s_i^2 ..., s_i^m$ etc, S_i is a set of player i's pure strategies; Mixed strategy symbols $\delta_i, \sigma_i, ...$, and $\Delta(S_i)$ is the set of player i's mixed strategies.
2. The symbol Ra_i means player i is rational, Br_i is interpreted as the best response of player, another a symbol NE means it is a Nash Equilibriu.
3. Atomic propositions of the form $\delta_i \succcurlyeq_i s_i^l$ and $\delta_i \succ_i s_i^l$.[6]

We restrict the modal operator P_i only to taking effect on the set of the atomic propositions S_i, and $P_i(s_i^k)$ stands for a probabilistic value that player i assigns to his pure strategy s_i^k, which is somewhat different from PEL.

Definition 5 (The frame of PEGL). *Given a game G, $\mathfrak{F} = \langle W, R_i, P_i, f_i \rangle$ is a frame of PEGL , where*

- $W \neq \emptyset$: consists of all players' pure strategy profiles.
- R_i is an accessibility relation for player i, which is defined as the equivalence relation of agreement of profiles in the i'th coordinate.
- $f_i : W \rightarrow S_i$ is a pure strategic function, which satisfies the following property: $R_i wv$ iff $f_i(w) = f_i(v)$.
- $P : (N \times W) \rightarrow (W \rightharpoonup [0,1])$[7], such that $\forall i \in N, \forall w \in W$, $\sum_{v \in dom(P(i,w))} P(i,w)(v) = 1$. Moreover, We require that $dom(P(i,w)) = \{w' | R_j ww', j \neq i\}$. In addition, we require that if $R_i wv$, then
  $P(i,w)(v) = P(i,v)(w) = P(i,w)(w)$, i.e., for a player i, the same pure strategies should have the same probability.

Thus, a PEGL-frame adds to a two-agent $S5$-Kripke frame a strategic function that associates with every world w a pure strategy profile$(f_i(w), f_j(w)) \in S$ and a function that assigns a probability function to each agent at each world. Given the above the frame for PEGL, we define an initial model based on it.

Definition 6 (The initial model for PEGL). *Given a game G with $S_i = \{s_i^1, ..., s_i^m\}$, an initial probabilistic epistemic game model $M = (W, R_i, V, P_i)$, is defined as follows. Given a world $w \in W$.*

$(M,w) \models p$	iff	$w \in V(p)$
$(M,w) \models s_i^k$	iff	for $\forall v, R_i wv$ implies $f_i(v) = s_i^k$ and $P(i,w)(v) = 1$
$(M,w) \models \delta_i$	iff	for $\forall v, R_i wv$ implies $f_i(v) = s_i^k$ for a $s_i^k \in S_i$ and $0 \leq P(i,w)(v) \leq 1$
$(M,w) \models Br_i$	iff	$\forall \delta_j \in \Delta(S_j), \wedge_{s_i^k \in S_i}(U_i(f_i(w), \delta_j) \geq U_i(s_i^k, \delta_j))$

---

[6] Here we just deal with the relation between mixed strategies and pure strategies, while neglect the preference relations between the mixed strategies.

[7] $\rightharpoonup$ means that it is a partial function.

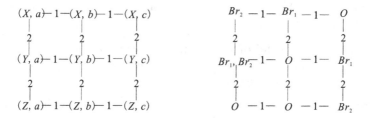

Fig. 2. Epistemic game model(in the lefthand figure) and Distributions of players' best responses (in the righthand figure)

$$(M, w) \models NE \qquad \text{iff} \quad (M, w) \models \wedge_{i \in N} Br_i$$
$$(M, w) \models \neg \varphi \qquad \text{iff} \quad (M, w) \nvDash \varphi$$
$$(M, w) \models \varphi_1 \vee \varphi_2 \qquad \text{iff} \quad (M, w) \models \varphi_1 \text{ or } (M, w) \models \varphi_2$$
$$(M, w) \models K_i \varphi \qquad \text{iff} \quad \text{for } \forall v, R_i wv \text{ implies } (M, v) \models \varphi$$
$$(M, w) \models C_N \varphi \qquad \text{iff} \quad \text{for } \forall v, R_* wv \text{ implies } (M, v) \models \varphi$$
$$(M, w) \models \sum_{k=1}^{m} a_k P_i(\varphi_k) \geq a \quad \text{iff} \quad \sum_{k=1}^{m} a_k P(i, w)(\varphi_k) \geq a^8$$
$$(M, w) \models \delta_i \succsim_i s_i^l \quad \text{iff} \quad (M, w) \models \sum_{k=1}^{m} u_i(s_i^k, f_j(w)) \cdot P_i(s_i^k) \geq u_i(s_i^l, f_j(w))$$
$$(M, w) \models \delta_i \succ_i s_i^l \quad \text{iff} \quad (M, w) \models \sum_{k=1}^{m} u_i(s_i^k, f_j(w)) \cdot P_i(s_i^k) > u_i(s_i^l, f_j(w))$$

In this definition, R_* is the reflexive transitive closure of $\cup_{i \in N} R_i$, and $P(i, w)(\varphi_k) = P(i, w)(\{v \in dom(P(i, w)) | (M, v) \models \varphi_k\})$.[9]

Figure 2(left) represents an initial PEGL model M from the game showed in figure 1. Here player 2's uncertainty relation R_2 runs along columns, because player 2 knows his own action, but not that of player 1. The uncertainty relation of player 1 runs along the rows.

Figure 2(right) describes the distributions of players' best responses in the game of figure 1 limited to pure strategies.

It's easy to prove that $M \models (\neg K_1 Br_1 \wedge \neg K_2 Br_2)$ in figure 2, i.e., neither players knows that he plays the best response. However, rationality means players aim at maximizing their utilities, in other words, a rational player i (Ra_i) always choose her best response (Br_i). Thus, we can deduce : $M \nvDash C_N Ra_i$. But this is in contradiction with the analysis principle of game theory, that is, rationality is common knowledge among players in a game. So it is necessary to modify the definition of rationality, and we need to add epistemic ingredient to the definition of rationality.[10]

Definition 7 (A definition of Rationality). *Player i is rational at a world w if there is not a mixed strategy δ_i, player i knows δ_i to be at least as good as $f_i(w)$, and she considers it is possible that δ_i is better than $f_i(w)$. Formally:*

---

[8] Serving to state, we substitute $\sum_{k=1}^{m} u_i(s_i^k, f_j(w))$ for a_k and $u_i(s_i^l, f_j(w))$ for a in this definition.

[9] We will express that player i assign a probabilistic value to a pure strategy s_i^k by $P(i, w)(s_i^k)$ later.

[10] This part is motivated by [7] and [14].

$M, w \vDash Ra_i \iff \neg \exists \delta_i \in \Delta(S_i), M, w \vDash K_i(\delta_i \succeq_i f_i(w)) \wedge \langle K_i \rangle (\delta_i \succ_i f_i(w))$
$(i \in N, \langle K_i \rangle$ is dual for $K_i).$[11]

Definition 8 (The model of PEGL). *A probabilistic epistemic game model* M_{PEGL} *is a model which is attained by extending* M *with the valuation function satisfies the atomic proposition* Ra_i.

According to this definition, it is easy to verify that Ra_i fails exactly at the rows or the columns with which the weakly dominated strategies correspond for player i in a M_{PEGL}. For instance, in figure 2, Ra_2 fails at the states $(X, b), (Y, b)$ and (Z, b).

Following the research program proposed by [8], we can use an axiom (called RA) to express the notion of rationality: it says that a player is irrational if he chooses a strategy while knowing that a different strategy is at least as good and he considers it is possible that this alternative strategy is actually better than the chosen one:

RA $s_i^k \wedge K_i(\delta_i \succeq_i s_i^k) \wedge \langle K_i \rangle (\delta_i \succ_i s_i^k) \to \neg Ra_i$

Apart from axiom **RA** and all axioms included in PEL, PEGL also has the following additional axioms:[12]

- **G1** $\delta_i \vee \eta_i \vee ... \vee \theta_i$
- **G2** $\neg(\delta_i \wedge \eta_i)$
- **G3** $(\delta_i \succeq_i \eta_i) \vee (\eta_i \succeq_i \delta_i)$
- **G4** $(\delta_i \succ_i \eta_i) \leftrightarrow ((\delta_i \succeq_i \eta_i) \wedge \neg(\eta_i \succeq_i \delta_i))$
- **G5** $\delta_i \to K_i \delta_i$

The intuitive meanings of these axioms is obvious. For example, axioms **G1** and **G2** together imply that each player i chooses exactly one strategy. **G3** and **G4**, on the other hand, means that the ordering of strategies is complete and the corresponding strict ordering is defined as usual respectively. **G5** states player i is aware of her own choice.

We denote by $\mathbf{L}_G$ the PEGL logical system, which satisfies all axioms in PEL, including the axioms **RA** and **G1** to **G5**.

Theorem 2. *Logic* $\mathbf{L}_G$ *is sound with respect to the class of models* M_{PEGL}.

Proof. Owing to the definition of Ra, it's clear that the axiom **RA** is sound in M_{PEGL}. And the verifications about axioms **G1** to **G4** is similar to the proof in [8], so we pass over the details of these verifications. Here, we just prove **G5**.

Let w is an arbitrary world in M_{PEGL}, and $(M_{PEGL}, w) \vDash \delta_i$, this means $\forall v, R_i wv$ implies $f_i(v) = s_i^k$ for a $s_i^k \in S_i$ and for $0 \leq P(i, w)(s_i^k) \leq 1$. Thus, according to $R_i wv$ iff $f_i(w) = f_i(v)$ and the requirement of the Definition 3.2, for $\forall v'$, satisfies $R_i vv'$, we have $f_i(v') = s_i^k$ and for $0 \leq P(i, w)(s_i^k) = P(i, v)(s_i^k) = P(i, v')(s_i^k) \leq 1$. Further, we have $(M_{PEGL}, v) \vDash \delta_i$ for $\forall v'$, satisfies $R_i vv'$, therefore, $(M_{PEGL}, w) \vDash K_i \delta_i$ for $\forall v$, satisfies $R_i wv$.

---

[11] For convenience we replace s_i^k with $f_i(w)$.
[12] This part is motivated by [8].

Next, we will show that the common knowledge of the rationality characterizes the algorithm of Iterated Admissibility, so that we can provide an epistemic foundation for the solutions or equilibria which can be found by the algorithm of Iterated Admissibility. In what follows, we take Ra as $\wedge_{i\in N}Ra_i$,that is, $Ra = \wedge_{i\in N}Ra_i$.

Theorem 3. *Given a probabilistic epistemic game model M_{PEGL} based on a finite strategic-form G and an arbitrary world w, w is in the M_{PEGL}, then, $(M_{PEGL}, w) \models C_N Ra$ iff $f(w) \in IAS$.*

Proof. (a) From left to right: Supposed $(M_{PEGL}, w) \models C_N Ra$. We will prove it by induction.[13]

Firstly, for $\forall v \in W$ satisfied $R_* wv$, we need to show $f(v) \notin IS^0$. If not, then there exists a world w', such that $R_* ww'$. And $f(w') \in IS^0$, this means, there is a player i and one of his pure strategies, let be s_i^l, satisfied $f_i(w') = s_i^l$ and s_i^l is weakly dominated by her some mixed strategy δ_i, as a result, for $\forall s_j^h \in S_j, U_i(\delta_i, s_j^h) \geq u_i(s_i^l, s_j^h)$ and $\exists s_j^g \in S_j, U_i(\delta_i, s_j^g) > u_i(s_i^l, s_j^g)$. Further, for $\forall v \in W, R_i w'v, U_i(\delta_i, f_j(v)) \geq u_i(s_i^l, f_j(v))$, and $\exists v' \in W, R_i w'v', U_i(\delta_i, f_j(v')) > u_i(s_i^l, f_j(v'))$, i.e., $\forall v \in W, R_i w'v, \sum_{k=1}^m u_i(s_i^k, f_j(v)) \cdot P(i,v)(s_i^k) \geq u_i(s_i^l, f_j(v))$, for $\exists v' \in W, R_i w'v', \sum_{k=1}^m u_i(s_i^k, f_j(v')) \cdot P(i,v')(s_i^k) > u_i(s_i^l, f_j(v'))$, where, as R_i is equivalence relation and the requirement of $P(i,w)(v)$ in the definition 3.2., $\delta_i^k = P(i,v)(s_i^k) = P(i,v')(s_i^k)$. Therefore, for $\forall v \in W, R_i w'v, (M_{PEGL}, v) \models \sum_{k=1}^m u_i(s_i^k, f_j(v))P_i(s_i^k) \geq u_i(s_i^l, f_j(v))$, and $\exists v' \in W, R_i w'v', (M_{PEGL}, v') \models \sum_{k=1}^m u_i(s_i^k, f_j(v'))P_i(s_i^k) > u_i(s_i^l, f_j(v'))$. Furthermore, $\forall v \in W, R_i w'v$, $(M_{PEGL}, v) \models \delta_i \succcurlyeq_i s_i^l$, and $\exists v' \in W, R_i w'v', (M_{PEGL}, v') \models \delta_i \succ_i s_i^l$, i.e., $(M_{PEGL}, w') \nvDash Ra_i$, owing to the definition of Ra_i. But this is in contradiction with the hypothesis that $(M_{PEGL}, w) \models C_N Ra$. The reason is we can deduce that $(M_{PEGL}, w') \models Ra_i$ by the hypothesis and $R_* ww'$. So, for $\forall v \in W$ satisfied $R_* wv$, $f(v) \notin IS^0$, furthermore, $f(v) \in IAS^1$.

Next, fix an integer $m \geq 1$, let $f(v) \in IAS^m$, we will prove that $f(v) \notin IS^m$. By contradiction, then, there exists a world w' and a player j and one of her pure strategies s_j^l, satisfied $f_j(w') = s_j^l$, meanwhile, s_j^l is weakly dominated by her some mixed strategy δ_j. By hypothesis, for every player i, $f_i(w') \in IAS_i^m$. It follows that, for $\forall s_i^h \in IAS_i^m, U_j(\delta_j, s_i^h) \geq u_j(s_j^l, s_i^h)$ and $\exists s_i^t \in IAS_i^m, U_j(\delta_j, s_i^t) > u_j(s_j^l, s_i^t)$. Thus, $\forall s_i^h \in IAS_i^m, \sum_{k=1}^m u_j(s_j^k, s_i^h) \cdot P(j,w')(s_j^k) \geq u_j(s_j^l, s_i^h)$, and $\exists s_i^t \in IAS_i^m, \sum_{k=1}^m u_j(s_j^k, s_i^t) \cdot P(j,w')(s_j^k) > u_j(s_j^l, s_i^t)$. Further, according to the above request that if $R_j wv$, then $P(j,w)(v) = P(j,v)(w) = P(j,w)(w)$, and the equivalence relation R_i, we have, $\forall v \in W, R_j w'v$, $\sum_{k=1}^m u_j(s_j^k, f_i(v)) \cdot P(j,v)(s_j^k) \geq u_j(s_j^l, f_i(v))$, and $\exists v' \in W, R_j w'v'$, $\sum_{k=1}^m u_j(s_j^k, f_i(v')) \cdot P(j,v')(s_j^k) > u_i(s_j^l, f_i(v'))$, where, $\delta_j^k = P(j,v)(s_j^k) = P(j,v')(s_j^k)$. So, $(M_{PEGL}, v) \models \delta_j \succcurlyeq_j s_j^l$, and $(M_{PEGL}, v') \models \delta_j \succ_j s_j^l$, accordingly, $(M_{PEGL}, w') \nvDash Ra_j$. But by the hypothesis, we have $(M_{PEGL}, w') \models Ra_j$, contradiction is presented. In additions, because of reflexivity of R_*, $f(w) \in IAS$

---

[13] In this proof, each of the symbols t, h, m, g, l, k refers to a natural number.

is established. Thus,we have shown by induction that if $(M_{PEGL}, w) \models C_N Ra$, then $f(w) \in IAS$.

(b)From right to left: Fix an arbitrary world w and an arbitrary player i, according to the definition of IAS, for $\forall i, s_i^k \in S_i \subseteq IAS_i$, let be $f_i(w) = s_i^k$, then $f_i(w)$ is an admissible strategy for player i in IAS_i. This means, $\neg \exists \delta_i \in \Delta(S_i)$ so that, $U_i(\delta_i, s_j^h) \geq u_i(f_i(w), s_j^h)$ for all $s_j^h \in S_j$, $U_i(\delta_i, s_j^t) > u_i(f_i(w), s_j^t)$ for some $s_j^t \in S_j$. Thus, $(M_{PEGL}, w) \models Ra_i$, further, by the randomicity of w and i, So, we have $(M_{PEGL}, w) \models C_N Ra$.

4   Solving for NE and Public Announce Logic

As a basis for most dynamic epistemic logics, public announcement logic(PAL) can deal with the change of information arising from the action of public announcement by adding a dynamic modality $[\varphi]$ to the standard epistemic logic. $[\varphi]\psi$ means " after a truthful public announcement of φ, formula ψ holds ". And its truth condition is that:

$$M, w \models [\varphi]\psi \text{ iff if } M, w \models \varphi \text{ then } M\mid_\varphi, w \models \psi.^{14}$$

With this language, we can say things like $[\varphi]K_i\psi$: after a truthful public announcement of φ, agent i knows ψ, or $[\varphi]C_N\varphi$: after its announcement, φ has become common knowledge in the group N of agents and so on. As we observed, the issue of " an announcement limit " has close connection with an equilibrium solved by the algorithm of IA. The concept of announcement limit is defined as follows:

Definition 9. *For any model M and formula φ, the announcement limit $\sharp(\varphi, M)$ is the first submodel in the repeated announcement sequence where announcing φ has no further effect.*[15]

Consequently, by repeatedly announcing φ to delete the worlds at which φ is false, and retaining just those worlds where φ holds, it must stop at a finite model, i.e.$\sharp(\varphi, M)$. This yields a sequence of nested decreasing sets. The procedure bears similarities to the above-mentioned procedure of IA. So, it's natural to take a procedure of IA as a procedure of public announcing the rationality by just adding an operator $[\alpha]$ to our language $\mathfrak{L}_{PEGL}$, which provides a dynamic-epistemic analysis for the procedure of game solution. The assertion which players announce publicly must be the statements which they know to be true in PAL. The following theorems guarantee that the rationality which we define can be taken as a suitable assertion for public announcement.

We call the above defined probabilistic epistemic game model M_{PEGL} as a full probabilistic epistemic game model, and take a any submodel of a full probabilistic epistemic game model M_{PEGL} as a general probabilistic epistemic game model M'_{PEGL}.

---

[14] $M\mid_\varphi$ is a submodel of M in which φ is true.
[15] This definition is in [7]

Theorem 4. *Every finite general probabilistic epistemic game model M'_{PEGL} has worlds with Ra true.*[16]

Proof. Note that atomic proposition Ra_i fails exactly at the rows or columns with which weakly dominated strategies correspond for player i in a M'_{PEGL}. Consider any general game model M'_{PEGL}, if there is not a weakly dominated action for all player in the M'_{PEGL}, then Ra is true at every worlds in the model. Thus, iterated announcement of Ra can make no more change on the game model, and get stuck in cycles in this situation. If there is a weakly dominated action for some player in the game, but because of the relativity of the definition of weakly dominated strategy, i.e. if player i has a weakly dominated strategy a, then he must have a strategy which is weakly better than strategy a, let strategy b. Thus, Ra_i holds at all the worlds which belong to the row or the column corresponding to strategy b. On the other hand, for player j, if he has not a weakly dominated action, then Ra_j holds also at all the worlds. Furthermore, Ra_j holds at the worlds which belong to the row or the column corresponding to the strategy b. So, Ra holds in the general game model. But if player j has a weakly dominated action, accordingly she must have a weakly dominant action, suppose that action Y and Ra_j is true at the worlds which belong to the row or the column corresponding to the strategy Y. Therefore, Ra is satisfied at the world (Y, b).

To sum up the above arguments, Every finite general game model has worlds with Ra true.

Theorem 5. *Rationality is epistemically introspective. i.e. The formula $Ra_i \rightarrow K_i Ra_i$ is valid on a general game model M'_{PEGL}.*

Proof. Given a general probabilistic epistemic game model M'_{PEGL}, an arbitrary w in M'_{PEGL}, and $(M'_{PEGL}, w) \vDash Ra_i$, but $(M'_{PEGL}, w) \nvDash K_i Ra_i$. Because $M'_{PEGL}, w \nvDash K_i Ra_i$, which means $\exists v \in W, R_i wv$ so that $(M'_{PEGL}, v) \nvDash Ra_i$, therefore, $\exists \delta_i \in \Delta(S_i)$, satisfied $(M'_{PEGL}, v) \vDash K_i(\delta_i \succcurlyeq_i f_i(v)) \wedge \langle K_i \rangle (\delta_i \succ_i f_i(v))$ i.e., for $\forall v', R_i vv'$ so that $(M'_{PEGL}, v') \vDash (\delta_i \succcurlyeq_i f_i(v))$ and $\exists v'', R_i vv''$ satisfied $(M'_{PEGL}, v'') \vDash (\delta_i \succ_i f_i(v))$. Then, $R_i wv'$ and $R_i wv''$, since R_i is equivalent relation. Thus, $(M'_{PEGL}, w) \vDash K_i(\delta_i \succcurlyeq_i f_i(w)) \wedge \langle K_i \rangle (\delta_i \succ_i f_i(w))$. So, by Definition 3.4., $(M'_{PEGL}, w) \nvDash Ra_i$, this is in contradiction with $(M'_{PEGL}, w) \vDash Ra_i$. The formula $Ra_i \rightarrow K_i Ra_i$ is valid on a general probabilistic game model.

Consequently, these theorems guarantee that we can remove the worlds at which Ra doesn't hold after each player tells others what she know about her behavior at some actual world at the same time.

In figure 3, the left-most model is the model from figure 1. The other models are obtained by public announcements of Ra successively for three times. So, in the last submodel, we have: $(M_{PEGL}, (Y, a)) \vDash [Ra][Ra][Ra]C_N(NE)$. This formula indicates if the players iteratively announce that they are rational, the process of dominated strategy elimination leads them to a solution that is known to be NE.

---

[16] Here, we just prove it on games with pure strategies. It is similarity for proving it on games with mixed strategies.

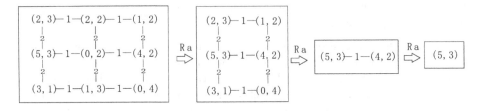

Fig. 3. Announcements of R_a

	Player1	a	b	c			Player1	a	b			Player1	a	b			Player1	a
Player2					Ra	Player2				Ra	Player2				Ra	Player2		
A		(4, 8)	(3, 0)	(1, 3)		A		(4, 8)	(3, 0)		A		(4, 8)	(3, 0)		A		(4, 8)
B		(0, 0)	(2, 6)	(8, 3)		B		(0, 0)	(2, 6)									

Fig. 4. Announcements of R_a

Again, in the following example (as shown in figure 4). Neither of the player 1's strategies is dominated. Both a and b are undominated for player 2, but c is weakly dominated by a $(0.5, 0.5)$ mixture of a and b, i.e., a mixed strategy $(0.5, 0.5, 0)$. Ra_2 is false at the worlds in the column c, thus, after the first public announcement of Ra, we can delete the column c. This leads to the pure strategy C to become dominated, which can also be deleted. Continuing this announcement of rationality and the elimination, we can reach the equilibrium of this game, that is (A, a).

More importantly, based on the Dynamic Epistemic logic we can also characterize the IA.

Theorem 6. *Given a full epistemic game model M_{PEGL} based a finite strategic-form G and an arbitrary world w, w is in a general epistemic game model M'_{PEGL} which is stable by repeated announcement of Ra for all player if and only if $f(w) \in IAS$. That's to say, for $\forall w \in \sharp(Ra, M_{PEGL}) \Leftrightarrow f(w) \in IAS$.*

Proof. By theorem 4, here we only need to prove that if for $\forall w \in \sharp(Ra, M_{PEGL})$, then $(M_{PEGL}, w) \vDash C_N Ra$. And Conversely it is still hold.

Let M'_{PEGL} be a general epistemic game model which is stable by repeated announcement of Ra for all player in a M_{PEGL}, and $\forall w, w \in M'_{PEGL}$ $(i.e., w \in \sharp(Ra, M_{PEGL}))$, by the definition of $\sharp(Ra, M_{PEGL})$, we have:

$(M'_{PEGL}, w) \vDash Ra$. Meanwhile, by the property of a epistemic accessible relation R_i in the M_{PEGL}, it's quite obvious that $(M'_{PEGL}, w) \vDash C_N Ra$.

Conversely, if $\forall w \in W$, $(M'_{PEGL}, w) \vDash C_N Ra$, then, by the theorem in the multi-$S5$ system: $C_N \varphi \rightarrow \varphi$, we have, $(M'_{PEGL}, w) \vDash Ra$. Thus, by the definition of $\sharp(Ra, M_{PEGL})$, $w \in \sharp(Ra, M_{PEGL})$ is true.

5   Related Work and Conclusions

There is a large amount of literatures on the algorithms of iterated elimination in the fields of logic and game theory (cf.[2],[8],[7] [6],[15], [16], [17], [18] etc). In particular, [8]and [7] describe and characterize different algorithms in game theory by redefining rationality based on epistemic logic. Our intellectual debt towards [8] and [7] is clear. Compared to their work, we generalize the pure strategy to the strategic games with mixed strategies, which expands the research field of epistemic analysis of strategic games by adding probabilistic operator into multi-agent epistemic system. What's more important, we give a logic characterization for the algorithm of iterated admissibility, which is known to provide a valuable criterion for selecting among multiple equilibria and to yield sharp predictions in finite games(cf. [2]).

[7] defined two types of rationality, the weak rationality and the strong rationality, which are denoted by WR_i and SR_i, where

$$(M_{PEGL}, w) \models WR_i \Leftrightarrow (M_{PEGL}, w) \models \wedge_{s_i^k \neq f_i(w)} \langle K_i \rangle (f_i(w) \succcurlyeq_i s_i^k),$$

$$(M_{PEGL}, w) \models SR_i \Leftrightarrow (M_{PEGL}, w) \models \langle K_i \rangle \wedge_{s_i^k \neq f_i(w)} (f_i(w) \succcurlyeq_i s_i^k).$$

Repeatedly public announcements of these rationality characterize respectively algorithms of iterated elimination strictly dominated strategy and of rationalizability, which corresponds to Bernheim's version of the rationalizability algorithm in [5]. To compare the definition of our rationality Ra_i to these two rationalities, we can conclude that the worlds removed by announcing WR_i must be deleted by announcing Ra_i, that is,$Ra_i \rightarrow WR_i$ is valid on a general game model M'_{PEGL}. However, there is no relation between Ra_i and SR_i. A NE, which can be solved by iterated announcement of SR, is not necessarily solved by iterated public announcement of Ra, and vice versa. For example, the following games showed in figure 5, G_1 can be solved only by repeated announcements of SR, but for G_2, we can find the NE only by announcement of Ra.

In addition, [7] explored the connection between iterated announcements and epistemic fixed-point logics, proved that the solution zones for repeated announcement are definable in an epistemic fixed-point logic. This links game-theoretic equilibrium theory with current fixed-point logics of computation. From this perspective, when we are considering the function computing the next set for iterated announcement Ra as its update function $F_{M,Ra}(X) = \{w \in X \mid (M|X, w) \models Ra\}$. But because the function $F_{M,Ra}(X)$ is not monotone (cf. [19]), we can't define the announcement limit $\sharp(Ra, M_{PEGL})$ as the greatest

Player1 \ Player2	a	b	c
X	(1, 2)	(1, 0)	(1, 1)
Y	(0, 0)	(0, 2)	(2, 1)

G_1

Player1 \ Player2	a	b	c
X	(1, 2)	(1, 3)	(1, 3)
Y	(0, 2)	(0, 1)	(2, 0)

G_2

Fig. 5. Comparison between SR and Ra

fix-point like SR. Fortunately, in [7], it has been proved that " The iterated announcement limit is an inflationary fixed point ". Therefore, we guess that the iterated announcement Ra limit can also be defined in full generality via a broader sort of procedure in so-called inflationary fixed-point logic [20]. This will be left for our further work.

[8] also put forward two rationalities, namely WR'_i and SR'_i. G.Bonanno examined the implications of common belief and common knowledge of his two rationalities, and proved that the weaker axiom of rationality characterizes the iterated elimination strictly dominated strategy, while the stronger one characterizes the pure-strategy version of the algorithm introduced in [21]. As the same above reason, the rationality we defined, that is, Ra_i is stronger than WR'_i in some sense. Compared to SR'_i, the epistemic game model in this paper is different from the game model defined in [8], which is constructed according to players' belief. As a result, the outcomes of the removing procedure is different too, although they are very close in literal sense.

Meanwhile, what we characterized for the algorithm of iterated admissibility is also based on a dynamic epistemic analysis, while [8] is based on a static epistemic logic. But in [8], frame characterization results are also provided. We left this as our future work.

Acknowledgments. We would like to thank Johan van Benthem and Jonathan A.Zvesper for valuable suggestion and two anonymous referees for their insightful comments. We thank Jeremy Seligman, Fengrong Liu, Yi Wang and Yun Gao for their very helpful remarks and corrections; Those remaining are our responsibilities. This research was supported by National Social Science Foundation of China (No.09CZX033) and funded by the Ministry of Education of China (No. 08JC72040002).

References

1. Aumann, R.J.: Rationality and bounded rationality. Games and Economic Behavior 53, 293–391 (1999)
2. Brandenburger, A., Friedenberg, A., Keisler, H.J.: Admissibility in games. Econometrica 76(2), 307–352 (2008)
3. Gilli, M.: Iterated admissibility as solution concept in game theory. Technical report, University of Milano-Bicocca, Department of Economics (2002)
4. Rubinstein, A.: Modeling Bounded Rationality. MIT Press, Cambridge (1998)
5. Bernheim, D.: Rationalizable strategic behavior. Econometrica 52(4), 1007–1028 (1984)
6. David, P.: Rationalizable strategic behavior and the problem of perfection. Econometrica 52(4), 1029–1050 (1984)
7. van Benthem, J.: Rational dynamics and epistemic logic in games. International Game Theory Review 9(1), 13–45 (2007)
8. Bonanno, G.: A syntactic approach to rationality in games with ordinal payoffs, vol. 3. Amsterdam University Press, Amsterdam (2008)
9. de Bruin, B.: Explaining Games: On the Logic of Game Theoretic Explanations. PhD thesis, Institute for logic, Language and Computation, Universiteit van Amsterdam, Amsterdam, The Netherlands (2004)

10. Fagin, R., Halpern, J.Y.: Reasoning about knowledge and probability. Journal of the ACM 41, 340–367 (1994)
11. Cui, J.Y.: A method for solving nash equilibrium of game based on public announcement logic. Logics for Dynamics of Information and Preferences (2009), http://www.illc.uva.nl/lgc/seminar/?page_id=292
12. Brandenburger, A.: Forward induction. Technical report, Stern School of Business, University of New York (2007)
13. Kooi, B.P.: Knowledge, Chance and Change. PhD thesis, Institute for logic, Language and Computation, Universiteit van Amsterdam, Amsterdam, The Netherlands (2004)
14. Blackburn, P., van Benthem, J., Wolter, F.: Handbook of Modal Logic. Elsevier Science Inc., Amsterdam (2007)
15. Shimoji, M.: On forward induction in money-burning games. Economic Theory 19(3), 637–648 (2002)
16. Apt, K.R., Zvesper, J.A.: Common beliefs and public announcements in strategic games with arbitrary strategy sets. Journal of Logic and Computation (2007), http://www.citebase.org/abstract?id=oai:arXiv.org:0710.3536
17. Christian, E.: Iterated weak dominance in strictly competitive games of perfect information. Games and Economic Behavior (2002)
18. van Benthem, J.: Decisions, actions and game, a logical perspective. Technical report, ILLC, University of Amsterdam and Department of Philosophy, Stanford University (2008), http://staff.science.uva.nl/~johan/Chennai.ICLA3.pdf
19. Apt, K.R.: Relative strength of strategy elimination procedures. Economics Bulletin 3 (2007)
20. Ebbinghaus, H.D., Flum, J.: Finite Model Theory. Springer, Berlin (1995)
21. Stalnaker, R.: Extensive and strategic form: Games and models for games. Research in Economics 21, 2–14 (1997)

Can Doxastic Agents Learn?
On the Temporal Structure of Learning

Cédric Dégremont and Nina Gierasimczuk

Institute for Logic, Language and Computation, Universiteit van Amsterdam
P.O. Box 94242, 1090 GE Amsterdam, The Netherlands
cedric.uva@gmail.com, n.gierasimczuk@uva.nl

Abstract. Formal learning theory formalizes the phenomenon of language acquisition. The theory focuses on various properties of the process of *conjecture-change over time*, and therefore it is also applicable in philosophy of science, where it can be interpreted as a theory of empirical inquiry. Treating "conjectures" as beliefs, we link the process of conjecture-change to doxastic update. Using this approach, we reconstruct and analyze the temporal aspect of learning in the context of temporal and dynamic logics of belief change. We provide a translation of learning scenarios into the domain of dynamic doxastic epistemic logic. Then, we express the problem of finite identifiability as a problem of epistemic temporal logic model checking. Furthermore, we prove a doxastic epistemic temporal logic representation result corresponding to an important theorem from learning theory, that characterizes identifiability in the limit, namely Angluin's theorem. In the end we discuss consequences and possible extensions of our work.

Keywords: Formal learning theory, dynamic epistemic logic, doxastic epistemic logic, temporal logic, epistemic update, belief revision.

1   Introduction

Doxastic epistemic temporal logics and dynamic doxastic logics have been developed and applied both in the context of multi-agent systems and philosophy and can be used to analyze the process of belief change in a temporal perspective. Formal learning theory, on the other hand, is concerned with functions that identify a correct hypothesis from a range of possibilities on the basis of inductively given streams of data. These functions can be viewed as agents that change their beliefs about which hypothesis is correct. In this paper we investigate the connection between formal learning theory and modal logics of belief change and build new bridges between the two frameworks. The motivation for connecting learning theory and modal logics of belief change is two-fold. By analyzing the temporal doxastic structure underlying formal learning theory, we provide additional insight into the semantics of inductive learning. By importing the ideas, problems and methodology from learning theory, logics of epistemic and doxastic change get enriched by new concepts and new problematic perspectives.

X. He, J. Horty, and E. Pacuit (Eds.): LORI 2009, LNAI 5834, pp. 90–104, 2009.
© Springer-Verlag Berlin Heidelberg 2009

Let us outline how the bridging is established (for philosophical discussion see [6]). In what follows we focus on the language learning paradigm, treating languages as sets of positive integers. In formal learning theory (FLT), learning is viewed as a process in which an agent (Learner) considers some range of languages. One of the languages is the actual one, and Learner's aim is to get to know which one it is. Elements of the language are given to Learner one by one. The infinite sequence of data that governs this enumeration includes all and only elements of the language. Several success conditions for Learner can be defined. For instance, we can assume that each time Learner gets a piece of information, she can make a conjecture. We can define the learning process to be successful if Learner's conjectures stabilize on the proper language. This learnability condition is called *identification in the limit* [7]. A more restrictive notion requires that Learner gives an answer only once, at some finite stage of the procedure. This kind of learnability is known as *finite identification* [9]. In Section 2 one can find a formal account of identification in FLT.

Intuitively, our approach to inductive learning in the context of modal logics of belief change (presented in Section 3) is as follows. We take the initial class of languages to be states in an epistemic plausibility model, which mirrors Learner's initial uncertainty and preferences over the range of languages. Each state (language) is assigned a protocol that indicates which sequences of events it allows (which streams of data enumerate that language). The incoming piece of information is taken to be an event that modifies the initial model. The structure resulting from updating the model with a sequence of events generates a doxastic epistemic temporal forest. We formulate the translation in Section 4.1.

We build on this construction in three ways. Firstly, we give a modal characterization of forests generated from a learning situation that satisfies a given learning condition (Section 4.2). Abstracting from this construction, we consider learnability conditions as properties that doxastic epistemic temporal models may or may not satisfy and characterize these classes of frames by a modal formula. (Section 4.3). Finally, we show how FLT characterization theorems have natural counterparts in representation theorems about temporal protocols (Section 4.4). Section 5 concludes and presents directions for further work.

2   Formal Learning Theory

Let $C \subset \mathbb{N}$ be a recursively enumerable set. We will call any $S \subseteq C$ a language. A class of languages $\Omega = \{S_1, S_2, \ldots\}$ is an indexed family of recursive languages if there is a computable function $f : \mathbb{N} \times C \to \{0, 1\}$, such that

$$f(i, w) = \begin{cases} 1 & \text{if } w \in S_i, \\ 0 & \text{if } w \notin S_i. \end{cases}$$

In the rest of this paper we will consider indexed families of recursive languages.

Definition 1. *By a positive presentation (text) of S, ε, we mean an infinite sequence of elements from S such that it enumerates all and only the elements from S allowing repetitions.*

Definition 2 (Notation). *We will use the following notation:*

- $U = \bigcup \Omega$ *is the universal set of Ω;*
- ε_n *is the n-th element of ε; $\varepsilon|n$ is the sequence $(\varepsilon_0, \varepsilon_1, \ldots, \varepsilon_{n-1})$;*
- $set(\varepsilon)$ *is the set of elements that occur in ε;*
- *L is a learning function — a partial map from finite data sequences to indexes of sets, $L : U^* \rightharpoonup \mathbb{N}$.*

Finite identifiability of a class of languages from positive data is defined by the following chain of conditions.

Definition 3 (Finite identification). *A learning function L:*

1. *finitely identifies $S_i \in \Omega$ on ε iff, when inductively given ε, at some point L outputs i, and stops;*
2. *finitely identifies $S_i \in \Omega$ iff it finitely identifies S_i on every ε for S_i;*
3. *finitely identifies Ω iff it finitely identifies every $S_i \in \Omega$.*
4. *Ω is finitely identifiable iff some learning function L finitely identifies Ω.*

Example 1. $\Omega_1 := \{S_i = \{0, i\} | i \in \mathbb{N}\}$. Ω_1 is finitely identifiable by $L : U^* \to \mathbb{N}$:

$$L(\varepsilon|n) = \begin{cases} \text{is undefined if } set(\varepsilon|n) = \{0\}, \\ \max(set(\varepsilon|n)) \text{ otherwise.} \end{cases}$$

In other words, L outputs the correct hypothesis as soon as it receives a number different than 0, and the procedure ends.

Example 2. To see how restrictive the notion of finite identifiability is, take a finite class of finite languages $\Omega_2 = \{S_1, S_2, S_3\}$, where $S_i = \{1, \ldots, i\}$. Ω_2 is not finitely identifiable. To see that, assume that S_2 is the actual language. A learning function can never conclude that S_2 is the actual language. For all it knows, 3 might appear in the future, so it has to leave the S_3-possibility open.

There is a way to deal with this kind of uncertainty. If we allow Learner to answer each time she gets a new piece of data, we can define the success as convergence to the right answer. This leads to the notion of identification in the limit.

Definition 4 (Identification in the limit [7]). *A learning function L:*

1. *identifies S_i in the limit on ε iff for co-finitely many m, $L(\varepsilon|m) = i$;*
2. *identifies S_i in the limit iff it identifies S_i in the limit on every ε for S_i;*
3. *identifies Ω in the limit iff it identifies in the limit every $S_i \in \Omega$.*
4. *Ω is identifiable in the limit iff some learning function identifies Ω in the limit.*

Example 3. First let us consider an example of a finite class of finite sets. Recall the class Ω_2 from Example 2. Ω_2 is identifiable in the limit by the following function $L : U^* \to \mathbb{N}$: $L(\varepsilon|n) = m$, such that $m = max(set(\varepsilon|n))$.

Example 4. The learning function from Example 3 identifies in the limit the following infinite class of finite sets: $\Omega_3 = \{S_i | i \in \mathbb{N} - \{0\}\}$, where $S_n = \{1, \ldots, n\}$.

Example 5. Identifiability in the limit of the class Ω_3 is lost if we enrich it by the set of all natural numbers. Let $\Omega_4 = \{S_i | i \in \mathbb{N}\}$, where $S_0 = \mathbb{N}$ and for $n \geq 1$, $S_n = \{1, \ldots, n\}$. Ω_4 is not identifiable in the limit. To see this, assume that there is a function L that identifies Ω_4. Then, there is a k and n, such that for all $m \geq n$, $L(\varepsilon | m) = k$. Now, if $k \in \{1, 2, 3, \ldots\}$, then L cannot identify the set $\mathbb{N}$. On the other hand, if $k = 0$ then L cannot identify $S_{max(set(\varepsilon | n))}$. So, we get a contradiction, L cannot identify Ω_4.

Another epistemically plausible way to learn is by the elimination of hypotheses that are implausible, e.g. hypotheses that are inconsistent with the incoming data. This paradigm is formalized in the framework of learning by erasing.

Definition 5 (Function stabilization)
In learning by erasing we say that a function stabilizes to number k on environment ε iff for co-finitely many $n \in \mathbb{N}$: $k = min\{\mathbb{N} - \{L(\varepsilon | 1), \ldots, L(\varepsilon | n)\}\}$.

Definition 6 (Learning by erasing [8]). *A learning function L:*

1. *learns $S_i \in \Omega$ by erasing on ε iff L stabilizes to i on ε;*
2. *learns $S_i \in \Omega$ by erasing iff it learns by erasing S_i from every ε for S_i;*
3. *learns Ω by erasing iff it learns by erasing every $S_i \in \Omega$.*
4. *Ω is learnable by erasing iff some learning function learns Ω by erasing.*

It is easy to observe that in this setting learnability heavily depends on the chosen enumeration of languages, since the positive conjecture of the learning function is interpreted as the minimal one that has not been eliminated yet.

3   Modal Logics of Multi-agent Belief Change

We are interested in two logical approaches to multi-agent belief change: the temporal approach [10,5] and the dynamic approach [2]. After introducing models and languages for both approaches we indicate how these two are related.

3.1   The Temporal Approach

Doxastic epistemic temporal logics offer a global view of the evolution of a multi-agent system as events take place, focusing on the information that agents possess and what they believe. We interpret these logics on doxastic epistemic temporal forests [10].

Definition 7. *A* doxastic epistemic temporal model $\mathcal{H}$ *is a tuple:*

$$\langle W, \Sigma, H, (\leq_j)_{j \in A}, (\sim_j)_{j \in A}, V \rangle,$$

where $W \neq \emptyset$ is a countable set of initial states, Σ is a countable set of events, $H \subseteq W(\Sigma^* \cup \Sigma^\omega)$ is a set of histories (sequences of events starting at states from W) closed under non-empty finite prefixes, for each agent $j \in A$, $\leq_j \subseteq H \times H$ is a well-founded pre-order on H, $\sim_j \subseteq H \times H$ is an equivalence relation, and $V : \text{PROP} \to \wp(H)$, $V : \text{NOM} \to W$, i.e. nominals are names for initial states. wh ranges over finite histories starting at w, and $w\varepsilon$ over ω-histories.

ETL models are DETL models without the collection of plausibility pre-orders.

Definition 8. Let $\mathbb{P} : s \mapsto (\{s\}(\Sigma^* \cup \Sigma^\omega)) \cap H$ for $s \in W$. Intuitively, $\mathbb{P}(s)$ is the protocol or bundle of sequences of events associated with s. We refer to the $\langle W, \Sigma, H \rangle$-part of a ETL model as the protocol this model is based on.

We refer to the information of agent i at w by $\mathcal{K}_i[h] = \{h' \in H \mid h \sim_i h'\}$. $\mathcal{B}_i[h] = Min_{\leq_i}\mathcal{K}_i[h]$ are the histories that i considers the most plausible at h.

Let us introduce various *assumptions about doxastic and epistemic agents*.

Definition 9. Let $\mathcal{H} = \langle W, \Sigma, H, (\sim_j)_{j \in \mathcal{A}}, V \rangle$ be an epistemic temporal model.

Perfect Recall. $\mathcal{H}$ satisfies perfect recall iff $\forall he, h'f \in H$ if $K_i[he] = K_j[h'f]$, then $K_i[h] = K_j[h']$. It states that agents do not forget the past information as events take place.

Perfect Observation. $\mathcal{H}$ satisfies perfect observation iff $\forall he, h'f \in H$ if $K_i[he] = K_j[h'f]$, then $e = f$. Perfect observation is satisfied if agents always know exactly what is happening.

Synchronicity. $\mathcal{H}$ satisfies synchronicity iff $\forall h, h' \in H$ if $K_i[h] = K_j[h']$, then $len[h] = len[h']$, where $len(x)$ is the length of sequence x. Synchronicity is satisfied if the agents have access to some external discrete clock and can thus keep track of the time.

Uniform No Miracles. $\mathcal{H}$ satisfies uniform no miracles iff $\forall h, h' \in H$ $\forall e_1$, $e_2 \in E$ with $he_1, h'e_2 \in H$, if there are $h'', h''' \in H$ with $h''e_1, h'''e_2 \in H$ such that $h''e_1 \sim_i h'''e_2$ and $h \sim_i h'$, then $he_1 \sim_i h'e_2$. Uniform no miracles characterizes agents that do not take into account the whole history but that proceed in a step by step way.

Propositional Stability. $\mathcal{H}$ satisfies propositional stability iff for all $h, he \in H$ we have $p \in V(he)$ iff $p \in V(h)$.

Preference Stability. $\mathcal{H}$ satisfies preference stability iff $\forall he, h'f \in H$ we have $he \leq_i h'f$ iff $h \leq_i h'$. It states that agents do not change their mind about the *a priori* plausibility of two histories as events take place. Naturally, it does not mean that the posterior beliefs of the agents might not evolve. Indeed, beliefs are defined as the most plausible states *of an information partition* and the latter might change.

A Hybrid Doxastic Epistemic Temporal Language. The syntax of our hybrid dynamic epistemic temporal language $\mathcal{L}_{\text{DET}}$ is defined inductively as follows:

$$\varphi := p \mid i \mid x \mid \downarrow x.\varphi \mid \neg\varphi \mid \varphi \vee \varphi \mid K_j\varphi \mid B_j\varphi \mid \mathsf{A}\varphi \mid \bigcirc^{-1}\varphi \mid F\varphi \mid P\varphi \mid \forall\varphi$$

p ranges over a countable set of proposition letters PROP, i over a countable set of nominals NOM, x over a countable sets of state variables SVAR, j over $\mathcal{A}$. $K_i\varphi$ ($B_i\varphi$) reads i knows (believes) that φ. F and P stand for future and past. $\forall\varphi$ means: 'in all continuations φ'. Also: $H\varphi := \neg P \neg \varphi$ and $G\varphi := \neg F \neg \varphi$.

$\mathcal{L}_{\text{ETL}}$ is interpreted over a model $\mathcal{H}$, an initial state w, an infinite history $w\epsilon$ and a finite prefix wh of $w\epsilon$ [11,10], with an assignment function $g : \text{SVAR} \rightarrow H$, mapping states variables to nodes. Nominals are satisfied in all histories extending an initial state, while state variables are true in exactly one node.

Definition 10. *We give the semantics of $\mathcal{L}_{\mathrm{DETL}}$. We skip the obvious clauses. We take $e \sqsubseteq e'$ to mean that e is an initial segment of e'.*

$\mathcal{H}, we, wh, g \Vdash p$ iff $wh \in V(p)$

$\mathcal{H}, we, wh, g \Vdash i$ iff $V(i) = w$

$\mathcal{H}, we, wh, g \Vdash x$ iff $g(x) = wh$

$\mathcal{H}, we, wh, g \Vdash \downarrow x.\varphi$ iff $\mathcal{H}, we, wh, g[x := wh] \Vdash \varphi$

$\mathcal{H}, we, wh, g \Vdash K_i\varphi$ iff $\forall vh' \; \forall we$ if $vh' \in \mathcal{K}_i[wh] \& vh' \sqsubseteq ve'$ then $\mathcal{H}, ve', vh' \Vdash \varphi$

$\mathcal{H}, we, wh, g \Vdash B_i\varphi$ iff $\forall vh' \; \forall we$ if $vh' \in \mathcal{B}_i[wh] \& vh' \sqsubseteq ve'$ then $\mathcal{H}, ve', vh' \Vdash \varphi$

$\mathcal{H}, we, wh, g \Vdash \mathsf{A}\varphi$ iff $\forall vh' \; \forall we$ if $vh' \in H \; \& \; vh' \sqsubseteq ve'$ then $\mathcal{H}, ve', vh' \Vdash \varphi$

$\mathcal{H}, we, wh, g \Vdash \bigcirc^{-1}\varphi$ iff $\exists a \in \Sigma \; \exists h' \sqsubseteq \epsilon$ with $h'.a = h$ and $\mathcal{H}, we, wh' \Vdash \varphi$

$\mathcal{H}, we, wh, g \Vdash F\varphi$ iff $\exists e \in \Sigma^* \; \exists h' \sqsubseteq \epsilon$ with $h' = he$ and $\mathcal{H}, we, wh' \Vdash \varphi$

$\mathcal{H}, we, wh, g \Vdash P\varphi$ iff $\exists e \in \Sigma^* \; \exists h' \sqsubseteq \epsilon$ with $h'e = h$ and $\mathcal{H}, we, wh' \Vdash \varphi$

$\mathcal{H}, we, wh, g \Vdash \forall\varphi$ iff $\forall h' \in \mathbb{P}(w)$ s.t. $h \sqsubseteq h$ we have $\mathcal{H}, wh', wh \Vdash \varphi$

3.2 The Dynamic Approach

The dynamic approach of dynamic doxastic and dynamic epistemic logics considers belief change as a step by step operation on models.

Definition 11 ([3]). *An epistemic plausibility model is a tuple: $\mathcal{M} = \langle W, (\sim_i)_{i \in \mathcal{A}}, (\leq_i)_{i \in A}, V\rangle$, where $W \neq \emptyset$, for each agent $i \in \mathcal{A}$, $\sim_i$ is an equivalence uncertainty relation on W, $\leq_i$ is a pre-order on W, and $V : Prop \to \wp(W)$, where PROP is a countable set of propositional letters. We let $\mathcal{K}_i[w] = \{v \in W \mid w \sim_i v\}$ and $\mathcal{B}_i[w] = Min_{\leq_i}\mathcal{K}_i[w]$. We write $v \simeq w$ iff $v \leq w$ and $w \leq v$.*

In DDL epistemic change is viewed as a step by step operation on models, that consists of an update of the epistemic model with an event plausibility model, the latter representing the doxastic and epistemic of what has happened.

Definition 12. *An event model is a triple: $\mathcal{E} = \langle E, (\sim_i^{\mathcal{E}})_{i \in \mathcal{A}}, \mathrm{pre}\rangle$, where $E \neq \emptyset$ is a set of events, for each agent $i \in \mathcal{A}$, $\sim_i^{\mathcal{E}}$ is an equivalence relation on E, and $\mathrm{pre} : E \to \mathcal{L}_{EL}$, is a precondition function and $\mathcal{L}_{EL}$ is an epistemic language. A pointed event model is an event model with one distinguished element from $|\mathcal{E}|$.*

The relation $\sim_i^{\mathcal{E}}$ encodes agent i's epistemic information and uncertainty about the event taking place. The precondition function maps events to epistemic formulas. An event will be executable in some state only if that state satisfies the precondition of this event. We use epistemic event models (without plausibiltity ordering) since they can already capture the setting of finite identifiability.

The effect of updating an epistemic plausibility model $\mathcal{M}$ by an event model $\mathcal{E}$ is computed according to so-called *product update*.

Definition 13. *The product update of epistemic model $\mathcal{M} = \langle W, (\sim_i)_{i \in \mathcal{A}}, V\rangle$ with an event model $\mathcal{E} = \langle E, (\sim_i^{\mathcal{E}})_{i \in \mathcal{A}}, \mathrm{pre}\rangle$ is the model $\mathcal{M} \otimes \mathcal{E}$ whose domain is $\{(w, e) \mid w \in W, e \in E \; \& \; \mathcal{M}, w \Vdash \mathrm{pre}(e)\}$. The epistemic relation in the resulting model is $(w, e) \sim_i' (w', e')$ iff $w \sim_i w'$ and $e \sim_i^{\mathcal{E}} e'$, the plausibility ordering is $(w, e) \leq_i' (w', e')$ iff $w \leq_i w'$, and the valuation is as follows: $(w, e) \in V(p)$ iff $w \in V(p)$.*

An epistemic plausibility model describes what agents currently believe and know, while product update creates the new doxastic epistemic situation after some information event has taken place.

Recently DEL borrowed the crucial idea of *protocol* from the temporal approach [4]. A *protocol* P maps states in an epistemic plausibility model to sets of finite sequences of pointed event models closed under taking prefixes. This defines the admissible runs of some informational process: not every observation may be available, or appropriate. We let $\mathcal{E}$ be the class of all pointed plausibility event models. Let $Prot(\mathcal{E}) = \{P \subseteq (\mathcal{E}^* \cup \mathcal{E}^\omega) \mid P \text{ is closed under finite prefixes}\}$ be the co-domain of protocols, it is the class of all sets of sequences (infinite and finite) of pointed plausibility event models closed under taking finite prefixes.

Definition 14. *Let us take an epistemic plausibility model $\mathcal{M}$, and let $|\mathcal{M}|$ be the domain of $\mathcal{M}$. A local protocol for $\mathcal{M}$ is a function $P : |\mathcal{M}| \to Prot(\mathbb{E})$.*

Definition 15. $P, \epsilon | n$*-generated epistemic model $\mathcal{M}^{P,\epsilon|n}$ is defined inductively in the following way: $\mathcal{M}^{P,\epsilon|0} = \mathcal{M}$; $\mathcal{M}^{P,\epsilon|n+1} = \langle |\mathcal{M}^{P,\epsilon|n+1}|, \sim_{P,\epsilon|n+1}, V_{P,\epsilon|n+1} \rangle$, where:*

1. $|\mathcal{M}^{P,\epsilon|n+1}| := \{s\epsilon|n+1 \mid s\epsilon|n \in \mathcal{M}^{P,\epsilon|n} and \epsilon|n+1 \in P(s)\};$
2. $\sim_{P,\epsilon|n+1} := \sim_{P,\epsilon|n} \cap (|\mathcal{M}^{P,\epsilon|n+1}| \times |\mathcal{M}^{P,\epsilon|n+1}|);$
3. *For every $p \in$ PROP, $V_{P,\epsilon|n+1}(p) := V_{P,\epsilon|n}(p) \cap |\mathcal{M}^{P,\epsilon|n+1}|.$*

3.3 Dynamic Doxastic Language

$\mathcal{L}_{\mathrm{DDE}}$ is a core language that matches dynamic belief update. Its syntax is:

$$\varphi := p \mid \neg\varphi \mid \varphi \vee \varphi \mid \langle \leq_i \rangle\varphi \mid \langle i \rangle\varphi \mid \mathbf{E}\varphi \mid \langle \epsilon, \mathbf{e} \rangle\varphi$$

where i ranges over A, p over a countable set of proposition letters *Prop*, and $(\epsilon, \mathbf{e})$ ranges over a suitable set of symbols for event models.

We use event symbols in the semantic clause and write $\mathbf{pre}(e)$ for $\mathbf{pre}_\epsilon(e)$. We interpret $\mathcal{L}_{\mathrm{DDE}}$ as follows:

Definition 16. *We give the interesting clauses and use usual abbreviations.*

$$\mathcal{M}, w \Vdash \langle \leq_i \rangle\varphi \quad \text{iff } \exists v \text{ such that } w \preceq_i v \text{ and } \mathcal{M}, v \Vdash \varphi$$
$$\mathcal{M}, w \Vdash K_i\varphi \quad \text{iff } \forall v \text{ such that } v \in \mathcal{K}_i[w] \text{ we have } \mathcal{M}, v \Vdash \varphi$$
$$\mathcal{M}, w \Vdash B_i\varphi \quad \text{iff } \forall v \text{ such that } v \in \mathcal{B}_i[w] \text{ we have } \mathcal{M}, v \Vdash \varphi$$
$$\mathcal{M}, w \Vdash \mathbf{E}\varphi \quad \text{iff } \exists v \in W \text{ such that } \mathcal{M}, v \Vdash \varphi$$
$$\mathcal{M}, w \Vdash \langle \epsilon, \mathbf{e} \rangle\varphi \quad \text{iff } \mathcal{M}, w \Vdash \mathbf{pre}(e) \text{ and } \mathcal{M} \times \epsilon, (w, e) \Vdash \varphi$$

3.4 Connection between the Temporal and the Dynamic Approach

There is a connection between the two approaches presented above in Section 3. In fact, the product updaters of the dynamic approach are one interesting type of doxastic (temporal) agents. Indeed iterated product update of an epistemic plausibility model $\mathcal{M}$ according to a uniform line protocol P generates doxastic epistemic temporal forests that validate particular doxastic temporal properties. We will refer to this construction by $For(\mathcal{M}, P)$ and define it as follows.

Definition 17 (DETL forest generated by a DDL protocol). *Each initial epistemic plausibility model* $\mathcal{M} = \langle W, (\sim_i^{\mathcal{M}})_{i \in \mathcal{A}}, (\leq_i^{\mathcal{M}})_{i \in \mathcal{A}}, V^{\mathcal{M}} \rangle$ *and each local protocol* P *yields a generated ETL forest* $For(\mathcal{M}, P)$ *of the form:* $\mathcal{H} = \langle W^{\mathcal{H}}, \Sigma, H, (\sim_i)_{i \in \mathcal{A}}, (\leq_i)_{i \in \mathcal{A}}, V \rangle$*, as follows:*

1. $W^{\mathcal{H}} = |\mathcal{M}|$, $\Sigma = \bigcup_{w \in W} \bigcup_{n \in \mathbb{N}} P(w)(n)$,
2. H *is defined inductively as follows:* $H_1 = W^{\mathcal{H}}$;
   - $H_{n+1} := \{(we_1 \ldots e_{n+1}) \,|\, (we_1 \ldots e_n) \in H_n;$
     $\mathcal{M} \otimes \epsilon_1 \otimes \ldots \otimes \epsilon_n, (w, e_1, \ldots, e_n) \Vdash \mathbf{pre}_n(e_{n+1})$ *and* $e_1 \ldots e_{n+1} \in P(w)\}$;
   - $H = \bigcup_{1 \leq k < \omega} H_k$.
3. *If* $h, h' \in W^{\mathcal{H}}$, *then* $h \sim_i h'$ *iff* $h \sim_i^{\mathcal{M}} h'$;
4. *For* $1 < k \leq m$, $he \sim_i h'e'$ *iff* $he, h'e' \in H_k$, $h \sim_i h'$, e *and* e' *are events from the same event model and* $e \sim_i e'$ *in their event model;*
5. *For* $1 < k \leq m$, $he \leq_i h'e'$ *iff* $he, h'e' \in H_k$ *and* $h \leq_i h'$;
6. *Finally,* $wh \in V(p)$ *iff* $w \in V^{\mathcal{M}}(p)$.

We conclude by mentioning an important representation theorem due to van Benthem et al. [4], that we make use of. It indicates what assumptions we are making about the epistemic agents when working in the dynamic perspective.

Theorem 1 ([4]). *An ETL-model* $\mathcal{H}$ *is isomorphic to the forest generated by the sequential product update of an epistemic model according to some state-dependent DEL-protocol iff it satisfies perfect recall, synchronicity, uniform no miracles and propositional stability.*

4 Analyzing Learnability in a DETL Framework

This section gives first results bridging learning theory and dynamic epistemic temporal logics. We prove that the problem of checking whether a class of sets is finitely identifiable can be reduced to the model-checking problem of $\mathcal{L}_{\mathrm{DET}}$ on doxastic epistemic temporal forests. To start with we show how learning situations can be encoded by an epistemic plausibility model and a local protocol.

4.1 Protocols That Correspond to Set Learning

We now focus on the single agent case, $\mathcal{A} = \{L\}$. We write $\sim$ instead of $\sim_L$.

Definition 18 (Initial epistemic model). *Our initial epistemic model* $\mathcal{M}_\Omega$ *is a triple:* $\langle W_\Omega, \sim_\Omega, V_\Omega \rangle$*, where* $W_\Omega = \Omega$, $\sim_\Omega = W_\Omega \times W_\Omega$*, and for each set* $S_i \in \Omega$*, we take a nominal* i *and we set* $V(i) = \{S_i\}$.

In words, we identify states of the model with sets, we also assume that our agent does not have any particular initial information.

Definition 19 (Single event model). *For each* $e \in U$*, we have a corresponding event model* $\mathcal{E} = \langle \{e\}, \sim^{\mathcal{E}}, \mathbf{pre}_{\mathcal{E}} \rangle$ *where* $\sim^{\mathcal{E}} = \{(e, e)\}$ *and* $\mathbf{pre}_{\mathcal{E}}(e) = \top$.

Given a set S_i, we can transform it into a set of events, we write $\mathbb{E}(S_i) = \{(\mathcal{E}, e) \mid e \in S_i\}$. We trivialize the role of preconditions, the admissible sequences of events are defined by means of protocols.

We now define local protocol. Intuitively, given a state $S_i \in W_\Omega$, our protocol P_Ω should authorize at S_i any ω-sequence that enumerates S_i and nothing more.

Definition 20 (Local protocol). *For every $S_i \in W_\Omega$, $P_\Omega(S_i)$ is the smallest subset of $(\mathbb{E}(U))^* \cup (\mathbb{E}(U))^\omega$ that contains $\{f : \omega \to \mathbb{E}(S_i) \mid f \text{ is surjective}\}$, and that is closed under non-empty finite prefixes.*

4.2   DETL Characterization of Finite Identifiability

Let us first define a DETL version of the notion of belief (resp. knowledge) stabilization to a certain hypothesis.

Definition 21. *j's belief (resp. knowledge) about the initial state stabilizes to w on the history $v\epsilon$ iff there is a finite prefix $e^* \sqsubseteq \epsilon$ such that for any finite sequence e' such that $e^* \sqsubseteq e' \sqsubseteq \epsilon$ we have for all histories sh if $sh \in \mathcal{B}_j[ve']$ then $s = w$ (resp. for $\mathcal{K}_j[ve']$).*

For the case of finite identification we show what follows.

Proposition 1. *The following are equivalent:*

1. *Ω is finitely identifiable.*
2. *In the generated forest $For(\mathcal{M}_\Omega, P_\Omega)$, for all $S_i \in W_\Omega$ and $\epsilon \in P_\Omega(S_i)$ the learner's knowledge about the initial state stabilizes to S_i on $S_i\epsilon$.*
3. *$For(\mathcal{M}_\Omega, P_\Omega) \Vdash \mathtt{A}(\bigcirc^{-1}\bot \to \downarrow x.\forall F\ KH(\bigcirc^{-1}\bot \to x)).$*

Proof. $(1 \Rightarrow 2)$ Assume that Ω is finitely identifiable. We prove the contrapositive. Assume that there is a state $S_i \in W_\Omega$ and ω-sequence $\epsilon \in P_\Omega(S_i)$ such that agent's knowledge does not stabilize to S_i on ϵ. There are two cases.

1. The learner stabilizes to another state, but then by construction of $P_\Omega(S_i)$ and definition of a generated DEL-forest for every finite prefix $h \sqsubseteq \epsilon$, $S_ih \in \mathcal{K}[S_ih]$. Contradiction.
2. After each finite prefix $h \sqsubseteq \epsilon$, there is at least one state different from S_i that remains epistemically possible. Since generated ETL forest satisfies perfect recall (Theorem 1), it follows that there is some state $S_i \neq S_j$ that remains epistemically possible after each finite prefix $h \sqsubseteq \epsilon$. But by construction of $P_\Omega(S_i)$ this is only possible if $S_i \subset S_j$. Every finite subset of S_i is a subset of S_j, and therefore $S_i \in \Omega$ does not have a finite definite tell-tale set. Therefore, from Theorem 7 in [9], Ω is not finitely identifiable.

$(2 \Rightarrow 3)$ We prove the contrapositive. Assume that $For(\mathcal{M}_\Omega, P_\Omega) \nVdash \mathtt{A}(\bigcirc^{-1}\bot \to \downarrow x.\forall FKH(\bigcirc^{-1}\bot \to x)$). This means that some history satisfies $\bigcirc^{-1}\bot$, i.e., there is some initial state in $w \in W_\Omega$, such that for some $\epsilon \in P_\Omega(w)$ and for every finite prefix $h \sqsubseteq \epsilon$ we have $For(\mathcal{M}_\Omega, P_\Omega)w, w\epsilon, wh, g[g(x) := w] \nVdash$

$KH(\bigcirc^{-1}\bot \rightarrow x))$. By truth condition of K and $H(\bigcirc^{-1}\bot \rightarrow x))$, there is some history $vh' \in \mathcal{K}[wh]$ with $v \neq w$. But this means that Learner's knowledge does not stabilize to w on $w\epsilon$ in $For(\mathcal{M}_\Omega, P_\Omega)$. Contradiction.

$(3 \Rightarrow 1)$ By semantics of $A(\bigcirc^{-1}\bot \rightarrow$ we know that in every initial state $S_i \in W_\Omega$: $S_i \Vdash \downarrow x. \forall FKH(\bigcirc^{-1}\bot \rightarrow x)$ (1). Now assume for contradiction that there is some S_i that is not finitely identifiable in Ω. It follows that there is some enumeration ϵ^* of the set such that after any finite prefix of ϵ^*, there is another set S_j that the agent has not excluded (2).

But by (1) we can label S_i by x and for any sequence of events ϵ, there will be a finite prefix $\epsilon|m$ at which $S_i\epsilon|m, \epsilon, g[x := S_i] \Vdash KH(\bigcirc^{-1}\bot \rightarrow x)$ (3). By construction of P_Ω we have a finite prefix $\epsilon^*|n$ such that $S_i\epsilon^*|n, \epsilon^*, g[x := S_i] \Vdash KH(\bigcirc^{-1}\bot \rightarrow x)$ (4). But then the agents knows that the initial state was $g(x) = S_i$ and thus has excluded any other initial state. Contradicting (2). $\quad\square$

The (1-3) equivalence shows that we can characterize finite identifiability by the global satisfaction of a formula from the hybrid doxastic epistemic temporal language. As corollary it shows that the problem of checking whether a class of sets is finitely identifiable can be reduced to the model-checking problem of $\mathcal{L}_{DET}$ on doxastic epistemic temporal forests. (1-2) equivalence indicates that we can abstract away from forests that are actually generated from learning situations and reason directly about DETL models.

4.3   DETL Models for Learnability

Recall the condition from Proposition 1: *In the generated forest $For(\mathcal{M}_\Omega, P_\Omega)$, $S_i \in W_\Omega$ and $\epsilon \in P_\Omega(S_i)$ the learner's knowledge about the initial state stabilizes to S_i on $S_i\epsilon$.* The definitions we use are a natural generalization of that condition.

Definition 22. *A DETL frame $F(\mathcal{H}) = \langle W, \Sigma, H, \leq_L, \sim_L \rangle$ satisfies finite identification (FIN) iff for all $s \in W$ and $s\epsilon \in P(s)$ Learner's knowledge about the initial state stabilizes to s on $s\epsilon$.*

We define what it means for a model to satisfy the learning by erasing property.

Definition 23. *A DETL frame $F(\mathcal{H}) = \langle W, \Sigma, H, \leq_L, \sim_L \rangle$ satisfies learning by erasing (ERASE) iff for all $s \in W$ and $h = s\epsilon \in P(s)$ Learner's belief about the initial state stabilizes to s on $s\epsilon$.*

The preceding definitions indicate that it is possible to abstract away from forests that are actually generated from learning situations and reason directly about DETL frames. We now look into modal characterization of DETL frames satisfying certain learnability conditions. Doing so is a first step to identify the natural modal logics of learning.

The general scheme of such characterizations can be formulated as follows.

A DETL frame $F(\mathcal{H})$ satisfies Learning Condition iff
[Specification of a procedure of choosing the current belief]
$F(\mathcal{H}) \Vdash i \rightarrow$ [Quantifier] F [Epistemic Temporal Condition]i.

The most straightforward is the characterization of finite identifiability.

A DETL frame $F(\mathcal{H})$ satisfies FIN iff $F(\mathcal{H}) \Vdash i \rightarrow \forall F K i$

Learner can finitely identify a class iff for all elements i of the class if i holds, then in the future Learner will know that i.

Further extension of the validity approach demands more expressive power, namely we need to express the existence of an appropriate belief-choosing procedure, which leads to second-order quantification. If we skip the certainty condition, we get the characteristics of limiting identification, similar to learning by erasing.

A DETL frame $F(\mathcal{H})$ satisfies ERASE iff $\exists \leq\ F(\mathcal{H}[\leq]) \Vdash i \rightarrow \forall F G B i$.

The effectiveness of this procedure, in the presence of uncertainty, is guaranteed by the existence of an underlying preference ordering. The temporal condition is weakened, since Learner can not be guaranteed certainty. The success is defined as a stabilization to a correct hypothesis.

In general if we allow some freedom in defining beliefs, we can make an attempt to formalize computable identification in the limit.

A DETL frame $F(\mathcal{H})$ satisfies Comp-LIM iff $\exists \mathfrak{B}\ F(\mathcal{H}[\mathfrak{B}]) \Vdash i \rightarrow \forall F G B i$.

In this expression the $\mathfrak{B}$ is an effective procedure that at each step of the procedure computes the current belief. In general we can make further substitution to our general scheme and see what happens. Let us consider the following example.

Property of $F(\mathcal{H})$ iff $\exists \leq F(\mathcal{H}[\leq])i \rightarrow \exists F G B i$.

Here, we again take a preference ordering to determine the current belief, but we require that the convergence happens only for some environments. We can immediately see that this is an overuse of the scheme. To guarantee an "honest" convergence, we have to insist that it happens for all allowed sequences of events. Otherwise we have to deal with a situation of cheating, when the correct answer is directly "communicated" to the learner by a particular encoding of the answer.

4.4   Characterizing Protocols That Guarantee Learnability

We now prove representation theorems that characterize classes of DETL models in which learnability is guaranteed in terms of properties of the protocol the DETL model is based on. We start by giving two results about finite identification and then we move to a DETL counterpart of Angluin's Theorem.

Proposition 2. *A synchronous, perfect recall, perfect observation DETL model* $\langle W, \Sigma, H, \sim, \leq, V \rangle$ *satisfies* finite identifiability *whenever for all* $w \in W$ *and* *history* $wh \in H \cap \Sigma^\omega$, *there is some natural number* $n \in \omega$ *such that for every* $v \neq w$ *such that* $v \in W$ *and for every* $vh' \in H \cap \Sigma^\omega$ *we have* $(h|n) \neq (h'|n)$.

Proof. Take an arbitrary w. By assumption some $n \in \omega$ such that for every $v \neq w$ such that $v \in W$ and for every $vh' \in H \cap \Sigma^\omega$ we have $(h|n) \neq (h'|n)$. We prove that $w(h|n) \not\sim v(h'|n)$ by induction. Indeed assume that they are in the same information partition. Then by perfect observation the last events were the same. But by perfect recall we also have that the nodes right before were also in the same information partition so we can iterate this argument and apply perfect observation all the way down, proving that $(h|n) \neq (h'|n)$. □

The next result corresponds to the finite identifiability characterization [9].

Proposition 3. *A permutation closed, synchronous, perfect recall, perfect observation DETL model* $\langle W, \Sigma, H, \sim, \leq, V \rangle$ *based on a finite state space satisfies* finite identifiability *whenever for all* $w \in W$ *there is an event* $a \in \mathbb{E}(w)$ *such that for all* $v \in W$ *if* $v \neq w$ $a \in \mathbb{E}(v)$.

Proof. [Sketch] Take an arbitrary $w \in W$. Take $a \in \mathbb{E}$ such that for each $v \neq w$ we have $a \notin \mathbb{E}(v)$. By permutation closure a is included in every $\epsilon \in P(w)$. By the definition of P we know that in every $\epsilon \in P(w)$ event a occurs at some finite stage. Let us then take such $\epsilon \in P(w)$ with $\epsilon_n = a$. Assume for a contradiction that at stage $n + 1$ some state $v \neq w$ is still considered possible. But then it means that $a \in \mathbb{E}(v)$. Contradiction. □

We now turn to a DETL counterpart to a crucial result in learning theory: Angluin's theorem, that characterizes identifiable in the limit classes of sets.

Theorem 2 (Angluin [1]). *A class of sets* Ω *is identifiable in the limit iff for all* $S \in \Omega$ *there is a finite* $D_S \subseteq S$ *such that for all* $S' \in \Omega$, *if* $S \neq S'$ *and* $D_S \in S'$, *then* $S' \not\subseteq S$.

The next result is proved using once more the concept of a *DEL*-generated forest. Before we state the result, let us introduce the following definitions:

Set-driven. A local protocol P for $\mathcal{M}$ is set-driven iff $\forall w \exists S_w \subseteq \mathbb{N}$ such that
   $\forall \varepsilon \in P(w)$ $set(\varepsilon) = S_w$.
A-condition for protocols. A local protocol P satisfies the A-condition iff
   $\forall w \exists e \in P(w) \cap \Sigma^* \forall w \neq v (e \in P(v) \implies P(v) \not\subseteq P(w))$.
Finite identifiability of the incomparable. A local protocol P satisfies the condition of finite identifiability of the incomparable sets iff states whose image under P are $\subseteq$-incomparable constitute finitely identifiable classes.

Let us assume that a local protocol P satisfies finite identifiability of the incomparable. Then we can show the following equivalence.

Theorem 3. *A state space* W *together with a set-driven local protocol* P *satisfies A-condition iff there is a preference ordering* $\leq$ *on* W *and an epistemic plausibility frame* $M = (W, \sim, \leq)$, *where* $\sim = W \times W$ *such that*

(#) *for all $w \in W$ and for all $\varepsilon \in P(w)$ there is some $n \in \omega$ such that for every $m > n$, $w \in |M^{\varepsilon|m}|$ and w is the $\leq$-minimum of $|M^{\varepsilon|m}|$ in the generated doxastic model $M^{\varepsilon|m}$.*

Proof. ($\Rightarrow$) Let us assume that W, P satisfies A-condition, well-foundedness and finite identifiability of the incomparable. Let us define the preference ordering $\leq$ in the following way: $v \leq w$ iff $P(v) \subseteq P(w)$.

Since we deal with an epistemic plausibility model and protocol that corresponds to a set learning situation we have $v \simeq w$ iff $v = w$ (1). We prove that $\leq$ satisfies (#). Take $w \in W$ and choose some environment for w, i.e. some $\varepsilon \in P(w)$. We show that there is some $n \in \mathbb{N}$ such that for every $m > n$, $w \in |M^{\varepsilon|m}|$ and w is the $\leq$-minimum of $|M^{\varepsilon|m}|$ in the generated doxastic model $M^{\varepsilon|m}$. To show that we consider all $v \neq w$ (2) such that $v \leq w$ or such that v is $\leq$-incomparable to w. We show that there is a finite stage of the epistemic update at which v is eliminated, i.e. w is the $\leq$-minimal element of $|M^{\varepsilon|m}|$.

Let us take $v \in W$ such that $v \leq w$. By (1) and (2), if $v \leq w$ then $P(v) \subset P(w)$. Then there is some $e \in \Sigma$ such that $e \in P(w)$ but $e \notin P(W)$. Since protocols allow environments that enumerate all and only elements from the set S_w, e appears at some point at which v is eliminated as inconsistent with e. Since the protocol satisfies the A-condition, i.e. there is no $w \in W$ such that for all $e \in P(w) \cap \Sigma^*$ there is $v \in W$ such that $v \neq w$ and $P(v) \subset P(w)$, then for each $w \in W$ there is only finite number of $v \in W$, such that $v \leq w$.[1] It follows that all $v \leq w$ are going to be eliminated at some finite stage.

If v is $\leq$-incomparable to w, then $P(v) \not\subseteq P(w)$ and $P(w) \not\subseteq P(v)$. Therefore there is an $e \in \Sigma$ with $e \in P(w)$ and $e \notin P(v)$. Since protocols allow environments that enumerate all and only elements from the set S_w, e appears at some point at which v is eliminated as inconsistent with e. Moreover, all $v \in W$ such that v is $\leq$-incomparable to w is eliminated at some finite stage by assumption of finite identifiability of the incomparable.[2] Therefore, at some finite stage m, all $v \in W$ that are either $\leq$-smaller that w or are $\leq$-incomparable to w are eliminated, leaving w the smallest state in $|M^{\varepsilon|m}|$.

($\Leftarrow$) Assume that there is a preference ordering on W that satisfies (#).

To see that the underlying protocols satisfy A-condition for each $w \in W$ we take $\varepsilon_w \in P(w)$ and, from the assumption, for each ε_w there is n such that for all $m \geq n$, $M^{\varepsilon_w|m} = M'$ and in w is minimal wrt $\leq$ in M'. Let us take $\varepsilon_w|m = \sigma_w$. Since for each w, σ_w is finite it is enough to show that for all $v \in W$ such that $v \neq w$ if $\sigma_w \in P(v)$ then $P(v) \not\subseteq P(w)$.

Assume for contradiction that there is $v \in W$ with $\sigma_w \in P(v) \wedge P(v) \subset P(w)$. Let $\tau \in P(v)$ such that $\tau|len(\sigma_w) = \sigma_w$ (there is such because $\tau \in P(v)$). From the assumption, M^τ converges to a model that has w as minimal wrt to $\leq$. But

---

[1] A counterexample is the class of sets $\Omega = \{\{1\}, \{1,2\}, \{1,2,3\}, \ldots, \mathbb{N}\}$. Using the chosen preference relation the set $\mathbb{N}$ cannot be identified.

[2] Otherwise the class of sets $\Omega = \{Even, Even - \{2\} \cup \{3\}, Even - \{4\} \cup \{5\}, \ldots\}$ is allowed, and it is clear we cannot get the "Even" set to become the $\leq$-minimal after any finite number of steps.

$v \neq w$, so for one environment v, namely $\tau \in P(v)$, M^τ converges to a model with w as the minimal and not v. Contradiction.   □

5   Conclusions and Perspectives

We compare the notions of learning theory, doxastic temporal and dynamic doxastic logic. We show that the problem of learnability can be reduced to the model checking (or validity) problem of some doxastic temporal language. By abstracting away from the construction used in this reduction we view learnability from a DETL perspective and provide a representation theorem characterizing identifiability in the limit in terms of properties of temporal protocols. These bridges indicate that the two approaches, learning theory and doxastic temporal logic, can be joined in order to describe the notions of belief and knowledge involved in inductive inference. Also, our representation of initial classes of languages and environments gives an interesting application for the theory of protocols.

Future work includes extending our approach to other types of identification, e.g., identification of functions or learning from positive and negative information; studying the effects of different restrictions on protocols; investigating various constraints one can enforce on learning functions (e.g. consistency, conservatism or set-drivenness) and comparing them to those of epistemic and doxastic agents in the DETL framework. Finally FLT shows that for some classes of problems there are *procedures* of belief change that guarantee success. After reaching the convergence point Learner's beliefs are safe, they will not change under any true information [3]. Belief is fixed and true, but Learner can never be sure about it. Our modal characterizations tend to identify modal logics of learning and the operational concept of 'stable belief' it carries. A next step is to develop complete logics taking these notions of belief as primitives.

References

1. Angluin, D.: Inductive inference of formal languages from positive data. Information and Control 45(2), 117–135 (1980)
2. Baltag, A., Moss, L.S., Solecki, S.: The logic of public announcements, common knowledge, and private suspicions. In: Proc. TARK 1998, pp. 43–56 (1998)
3. Baltag, A., Smets, S.: Dynamic belief revision over multi-agent plausibility models. In: Bonanno, G., et al. (eds.) Proc. LOFT 2006, pp. 11–24 (2006)
4. van Benthem, J., Gerbrandy, J., Hoshi, T., Pacuit, E.: Merging frameworks for interaction. Journal of Philosophical Logic (2009)
5. Fagin, R., Halpern, J.Y., Moses, Y., Vardi, M.Y.: Reasoning About Knowledge. MIT Press, Cambridge (1995)
6. Gierasimczuk, N.: Bridging learning theory and dynamic epistemic logic. Synthese 169(2), 371–384 (2009)
7. Gold, E.: Language identification in the limit. Information and Control 10, 447–474 (1967)
8. Lange, S., Wiehagen, R., Zeugmann, T.: Learning by erasing. In: Arikawa, S., Sharma, A.K. (eds.) ALT 1996. LNCS (LNAI), vol. 1160, pp. 228–241. Springer, Heidelberg (1996)

9. Mukouchi, Y.: Characterization of finite identification. In: Jantke, K.P. (ed.) AII 1992. LNCS, vol. 642, pp. 260–267. Springer, Heidelberg (1992)
10. Parikh, R., Ramanujam, R.: A knowledge based semantics of messages. Journal of Logic, Language and Information 12(4), 453–467 (2003)
11. van der Meyden, R., Wong, K.-s.: Complete axiomatizations for reasoning about knowledge and branching time. Studia Logica 75(1), 93–123 (2003)

Agreement Theorems
in Dynamic-Epistemic Logic
Extended Abstract

Cédric Dégremont[1] and Olivier Roy[2]

[1] Institute for Logic, Language and Computation, Universiteit van Amsterdam,
The Netherlands
cedric.uva@gmail.com
[2] Faculty of Philosophy, University of Groningen, The Netherlands
o.roy@rug.nl

Abstract. In this paper we study Aumann's Agreement Theorem in dynamic-epistemic logic. We show that common *belief* of posteriors is sufficient for agreements in "epistemic-plausibility models", under common and well-founded priors, from which the usual form of agreement results follows, using common knowledge. We do not restrict ourselves to the finite case, and show that in countable structures such results hold if *and only if* the underlying "plausibility ordering" is well-founded. We look at these results from a syntactic point of view, showing that neither well-foundedness nor common priors are expressible in a commonly used language, but that the static agreement result is finitely derivable in an extended modal logic. We finally consider "dynamic" agreement results, show they have a counterpart in epistemic-plausibility models, and provide a new form of agreements via "public announcements". Comparison of the two types of dynamic agreement reveals that they can indeed be different.

1 Introduction

In this paper we study Aumann's Agreement Theorem [1] and some of its subsequent extensions [2] and generalizations [3, 4] in dynamic-epistemic logic [5, 6]. We show that common *belief* of posteriors is sufficient for agreements in "epistemic-plausibility models", under common and well-founded priors, from which the usual form of agreement results follows, using common knowledge. We do not restrict ourselves to the finite case, which thus represents an improvement on known qualitative agreement theorems [4], and show that in countable structures such results hold if *and only if* the underlying "plausibility ordering" is well-founded. We then look at these results from a syntactic point of view, showing that neither well-foundedness nor common priors are expressible in the language proposed in [7], even if it is extended with a common belief operator, but we also show a finitary syntactic derivation of the static agreement result in an extended modal language. We finally consider "dynamic" agreement results. We show that "agreements via dialogues" [3, 4] have a counterpart in

X. He, J. Horty, and E. Pacuit (Eds.): LORI 2009, LNAI 5834, pp. 105–118, 2009.
© Springer-Verlag Berlin Heidelberg 2009

epistemic-plausibility models, and that one also gets agreements via "public announcements", a type of belief update that has so far not been considered in the agreement literature—see [8] and [9]. Comparison of the two types of dynamic agreements reveals that in some situations they are indeed different.

These technical results answer an "internal" question in dynamic-epistemic logic, namely whether agreement results hold in this framework, but they also offer new insights into the contribution of agreement theorems to interactive epistemology. That common belief of posteriors is sufficient for agreements, under common and well-founded priors, strengthens one of the key lessons of agreement theorems, viz. that first-order information is closely dependent on higher-order information in situations of interaction [8]. Our inexpressibility results, on the other hand, support a qualm already voiced in the literature concerning the difficulty for agents to reason about static agreements [10]. The two dynamic results not only make a sharp distinction between two forms of belief changes, they also allow one to capture more adequately the idea that agreements are reached via *public* dialogues. Bringing agreement theorems to dynamic-epistemic logic is thus important both technically and conceptually, and it helps to bridge the existing literature on agreements with the logical approaches to knowledge, beliefs and the dynamics of information.

In this extended abstract all the proofs are omitted, as well as some auxiliary definitions. The reader interested in these details can communicate with the authors.

2   Definitions

In this section we introduce the models in which we study the various agreement results, and the logical language used in [7] to describe them.

2.1   Epistemic Plausibility Models

An epistemic plausibility model [5] is a qualitative representation of the agents' beliefs as well as first- and higher-order information in a given interactive situation.

Definition 1 (Epistemic Plausibility Model). *Given a countable set of atomic propositions* PROP, *an epistemic plausibility model* $\mathcal{M} = \langle W, (\leq_i)_{i \in I},$ $(\sim_i)_{i \in I}, V \rangle$ *has* $W \neq \emptyset$ *and countable,* $I = \{1, 2, \ldots, n\}$ *is a finite set of agents, and for each* $i \in I$, $\leq_i$ *is a total (plausibility) pre-order on* W, $\sim_i$ *is a binary equivalence relation on* W, *and* $V : \text{PROP} \to \wp(W)$. *An epistemic plausibility frame* $\mathcal{F}$ *is an epistemic plausibility model with the valuation* V *omitted.*

The total plausibility pre-order $\leq_i$ induces i's *priors*, and can be viewed as a qualitative counterpart to a prior probability distribution on W. If $w \leq_i w'$ we say that i considers w' at least as plausible as w. Given a set $X \subseteq W$, we say that $w \in X$ is $\leq_i$-minimal in X if $w \leq_i w'$ for all $w' \in X$. The relation $\sim_i$ induces i's *information partition* W. We write $\mathcal{K}_i[w]$ to denote the cell of this partition $\{v \in W \mid w \sim_i v\}$ to which w belongs. $\mathcal{K}_i[w]$ should be regarded as i's (private) information at w. We write $|\mathcal{M}| = W$ for the domain of $\mathcal{M}$.

The next two assumptions are crucial in the following.

Definition 2 ((Local) well-foundedness). *A plausibility pre-order satisfies:*

- **Local well-foundedness.** *If for all $w \in W$ and $i \in I$, for all $X \subseteq \mathcal{K}_i[w]$, X has $\leq_i$-minimal elements.*
- **Well-foundedness.** *If for all $X \subseteq W$ and $i \in I$, X has $\leq_i$-minimal elements.*

$\mathcal{M}$ *satisfies* (Local) Well-foundedness *if every plausibility pre-order has the corresponding property.*

Definition 3 ((A priori/ a posteriori) Most plausible elements)

- *For all $X \subseteq W$, let $\beta_i(X) = min_{\leq_i}(X) = \{w : w$ is $\leq_i$ -minimal in $X\}$.*
- *For all $w \in W$, let $\mathcal{B}_i[w] = \beta_i(\mathcal{K}_i[w])$.*

We write $w \rhd_i^{\mathcal{B}} v$ iff $v \in \mathcal{B}_i[w]$, and $w \to_i^X v$ iff $v \in \beta_i(\mathcal{K}_i[w] \cap X)$.

Intuitively $\beta_i(X)$ are the *a priori* most plausible elements of a set, ignoring the information partitions. $\mathcal{B}_i[w]$ gives the states i considers most plausible, conditional on the information he possesses at w, i.e. conditional on $\mathcal{K}_i[w]$. The relation $w \to_i^X v$ maps w to all states i considers most plausible, conditional on the information he possesses at w *and* on a given subset X. Observe that the set $\{v : w \to_i^X v\}$ might be empty for a given w and a given X, if $X \cap \mathcal{K}_i[w] = \emptyset$ or, in words, if X is already excluded by i's information at w.

Observe that β_i is well-defined if the plausibility pre-order is well-founded, while local well-foundedness is sufficient for $\mathcal{B}_i$ to be well-defined. To draw an analogy with the probabilistic case, this means that local well-foundedness ensures that the conditional beliefs of an agent i are well-defined for all "events" that have a non-empty intersection with the agent's information partition. Well-foundedness, on the other hand, requires i's conditional beliefs to be well-defined for any non-empty subsets of W.

Definition 4 (Common Prior). *There is* common prior beliefs *among group G in an epistemic plausibility model $\mathcal{M}$ when $\leq_i = \leq_j$ for all $i, j \in G$.*

The reflexive-transitive closure of the union of the epistemic accessibility relations $\sim_i$ for all agents i in a group G is the model-theoretic counterpart of the notion of "common knowledge" in G [6, 11]. We define "common belief" analogously.

Definition 5 (Common knowledge). *For each $G \subseteq I$, let $\sim_G^*$ be the reflexive-transitive closure of $\bigcup_{i \in G} \sim_i$. Let $[w]_G^* = \{w' \in W \mid w \sim_G^* w'\}$.*

Definition 6 (Common belief). *For each $G \subseteq I$, let $\rhd_G^*$ be the reflexive-transitive closure of $\bigcup_{i \in G} \rhd_i^{\mathcal{B}}$.*

2.2   Doxastic-Epistemic Logic

The logical language used in [7] to describe epistemic-plausibility models is a propositional modal language with three families of modal operators, which we extend here with "common belief" operators.

Definition 7 (Epistemic Doxastic Language). *The language $\mathcal{L}_{EDL}$ is defined as follows:*

$$\phi := p \mid \neg\phi \mid \phi \wedge \phi \mid K_i\phi \mid B_i^\phi\phi \mid C_G\phi \mid CB_G\phi$$

where i ranges over N, p over a countable set of proposition letters PROP *and $\emptyset \neq G \subseteq I$.*

The propositional fragment of this language is standard, and we write $\perp$ for $p \wedge \neg p$ and $\top$ for $\neg\perp$. A formula $K_i\phi$ should be read as "i knows that ϕ", $C_G\phi$ as "it is common knowledge among group G that ϕ", $CB_G\phi$ as "it is common belief among group G that ϕ." The formula $B_i^\phi\psi$, should be read " conditional on ϕ, i believes that ψ." These formulas are interpreted in epistemic plausibility models as follows:

Definition 8 (Truth definition). *We write $||\phi||^{\mathcal{M}}$ for $\{w \in |\mathcal{M}| : \mathcal{M}, w \Vdash \phi\}$. We omit $\mathcal{M}$ when it is clear from the context.*

$$\begin{aligned}
\mathcal{M}, w \Vdash p \quad &\text{iff} \quad w \in V(p)\\
\mathcal{M}, w \Vdash \neg\phi \quad &\text{iff} \quad \mathcal{M}, w \not\Vdash \phi\\
\mathcal{M}, w \Vdash \phi \wedge \psi \quad &\text{iff} \quad \mathcal{M}, w \Vdash \phi \text{ and } \mathcal{M}, w \Vdash \psi\\
\mathcal{M}, w \Vdash K_i\phi \quad &\text{iff} \quad \forall v \text{ (if } w \sim_i v \text{ then } \mathcal{M}, v \Vdash \phi)\\
\mathcal{M}, w \Vdash B_i^\psi\phi \quad &\text{iff} \quad \forall v \text{ (if } w \rightarrow_i^{||\psi||^{\mathcal{M}}} v\\
&\qquad \text{then } \mathcal{M}, v \Vdash \phi)\\
\mathcal{M}, w \Vdash C_G\phi \quad &\text{iff} \quad \forall v \text{ (if } w \sim_G^* v \text{ then } \mathcal{M}, v \Vdash \phi)\\
\mathcal{M}, w \Vdash CB_G\phi \quad &\text{iff} \quad \forall v \text{ (if } w \rhd_G^* v \text{ then } \mathcal{M}, v \Vdash \phi)
\end{aligned}$$

Simple belief conditional only on i's information at a state w can be defined using the conditional belief operator: $B_i\phi = B_i^\top\phi$, since: $\mathcal{M}, w \Vdash B_i^\top\phi$ iff $\forall v$ (if $w \rhd_i^B v$ then $\mathcal{M}, v \Vdash \phi$).

3   Static Agreements and Well-Foundedness

We first show that well-foundedness is sufficient for agreement on the posteriors under common priors and common *beliefs* of the posteriors. More precisely, we show that if an epistemic plausibility model is well-founded, then common belief that agent i believes that ϕ while j does not believe that ϕ implies that i and j have different priors, which is the contrapositive form of the agreement theorem.

Theorem 1 (Agreement theorem - Common Belief). *If a well-founded epistemic plausibility model $\mathcal{M}$ satisfies $\mathcal{M}, w \Vdash CB_{\{i,j\}}(B_ip \wedge \neg B_jp)$ for some $w \in W$, then i and j have different priors in $\mathcal{M}$.*

This immediately implies the "common knowledge" agreement result below, because $C_G\phi \to CB_G\phi$ is a valid implication in epistemic plausibility models. Note, however, that this result can also have been shown independently, by application of Bacharach's [4] result on qualitative "decision functions", modulo generalization to the countable case.

Corollary 1 (Agreement theorem - Common Knowledge). *If an epistemic plausibility model $\mathcal{M}$ satisfies* well-foundedness *and $\mathcal{M}, w \Vdash C_{\{i,j\}}(B_i p \land \neg B_j p)$ for one $w \in W$, then i and j have different priors in $\mathcal{M}$.*

Well-foundedness is not only sufficient for common priors to exclude the possibility of disagreements when the posterior are common beliefs, it is also necessary, as the Proposition 1 shows. The model behind this result is drawn in figure 1.

W

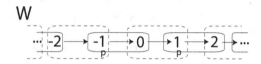

Fig. 1. The epistemic plausibility model constructed in the proof of Proposition 1. The solid and dotted rectangles represent 1's and 2's information partitions on W, respectively. The arrows represent their common plausibility ordering.

Proposition 1. *There exists a pointed epistemic plausibility model $\mathcal{M}, w$ which satisfies* local well-foundedness *and* common prior *such that $\mathcal{M}, w \Vdash C_{\{1,2\}} (B_1 p \land \neg B_2 p)$.*

Well-foundedness is thus necessary for agreement results to hold, and furthermore cannot be weakened to *local well-foundedness*. This condition on the plausibility ordering is thus *the* safeguard against common knowledge of disagreement, once we drop the assumption that the state space is finite.

4   Expressive Power and Syntactic Proofs

$\mathcal{L}_{EDL}$ is a natural choice of language for talking about epistemic-plausibility models, and but we show here that it cannot express Theorem 1 nor Corollary 1, because it cannot express two of their key assumptions, common prior and well-foundedness.

Fact 2. *The class of epistemic plausibility frames that satisfies* common prior *is not definable in $\mathcal{L}_{EDL}$.*

This result, which rests on the two small models drawn in figure 2, confirms the idea that to reason about (common) priors the agents must make "inter-[information]-state comparisons" [10], which they cannot do because their reasonings in $\mathcal{L}_{EDL}$ are local, i.e. they are bounded by the "hard information" [12] they have. This limitation also makes well-foundedness inexpressible, and with it the two static agreement results.

Fig. 2. The two epistemic plausibility model constructed in the proof of Fact 2. 1's and 2's information partitions on W are represented as in figure 1. The arrow in W represents their common plausibility ordering, while in W' the solid arrow and dotted arrows represent 1's and 2's orderings, respectively.

Fact 3. *There is no formula ϕ of $\mathcal{L}_{EDL}$ which is true in a pointed epistemic plausibility model $\mathcal{M}, w$ iff Theorem 1 or Corollary 1 holds in $\mathcal{M}, w$.*

The syntactical counterpart of the model-theoretic agreement results thus resides in more expressive languages. In the full version of the paper we present a finite syntactic derivation of Corollary 1 in $\mathcal{H}(@, \downarrow, C_G, \geq_j, \sim_j)$[1], which extends the hybrid language $\mathcal{H}(@, \downarrow, \geq_j, \sim_j)$ with a common knowledge modality C_G. Formally the language is the following:

$$\phi := p \mid i \mid x \mid \neg\phi \mid \phi \wedge \phi \mid K_j\phi \mid \langle \geq_j \rangle\phi \mid$$
$$@_i\phi \mid @_x\phi \mid \downarrow x.\phi \mid C_G\phi$$

Note that it allows one to scan the plausibility relation directly. The hybrid semantics draws on assignation functions that maps states variables to states and which allows the language to bind a variable to the current state and to refer to it. A detailed presentation of this language and its semantics, together with the syntactic derivation and proof of soundness of the axioms we are using is given in the full version of the paper.

On the positive side this language is able to axiomatize (converse) well-foundedness of the plausibility relation. On the negative side, the satisfiability problem for this language on the class of conversely well-founded frames is Σ_1^1-hard [13], ruling out any finite axiomatization of its validities. The derivation we show, however, is finite and uses only sound axioms. At the time of writing we still do not know whether the agreement results of Section 3 could be derived in a less complex language. The fact that the syntactic derivation reported here pertains to such an expressive language nevertheless shows that reasoning explicitly about agreement results requires onerous expressive resources.

5   Agreements via Dialogues

In this section we turn to "agreements-via-dialogues" [2, 4], which analyze how agents can reach agreement in the process of exchanging information about their beliefs by updating the latter accordingly.

---

[1] In fact we use a language with more primitives, but, as we prove, these are entirely definable in the restricted language.

5.1 Agreements via Conditioning

We first consider agreements by repeated belief conditioning. It is known that if agents repeatedly exchange information about each others' posterior beliefs about a certain event, and update these posteriors accordingly, the posteriors will eventually converge [2, 4]. We show here that this result also holds for the "qualitative" form of beliefs conditionalization in epistemic plausibility models.

We call a *conditioning dialogue about* ϕ [2] at a state w of an epistemic plausibility model $\mathcal{M}$ a sequence of belief conditioning, for each agent, on all other agents' beliefs about ϕ. This sequence can be intuitively described as follows. It starts with the agents' simple belief about ϕ, i.e. for all i: $B_i\phi$ if $\mathcal{M}, w \Vdash B_i\phi$ and $\neg B_i\phi$ otherwise. Agent i's beliefs about ϕ at the next stage is defined by taking his beliefs about ϕ, conditional upon learning the others' belief about ϕ at that stage. Syntactically, this gives, $\mathbb{B}_{1,i} = B_i\phi$ if $\mathcal{M}, w \Vdash B_i\phi$ and $\mathbb{B}_{1,i} = \neg B_i\phi$ otherwise and, for two agents i, j, $\mathbb{B}_{n+1,i} = B_i^{\mathbb{B}_{n,j}\phi}\phi$ if $\mathcal{M}, w \Vdash B_i^{\mathbb{B}_{n,j}\phi}\phi$ and $\neg B_i^{\mathbb{B}_{n,j}\phi}\phi$ otherwise. This syntactic rendering is only intended to fix intuitions, though, since in countable models the limit of this sequence exceeds the finitary character of $\mathcal{L}_{EDL}$. We thus focus on model-theoretic conditioning.

Conditioning on a given event $A \subseteq W$ boils down to refining an agent's information partition by removing "epistemic links" connecting A and non-A states.

Definition 9 (Conditioning by a subset). *Given an epistemic plausibility model $\mathcal{M}$, the collection of epistemic equivalence relation of the agents is an element of $\wp(W \times W)^I$. Given a group $G \subseteq I$, the function $f_G : \wp(W) \to (\wp(W \times W)^I \to \wp(W \times W)^I)$ is a conditioning function for G whenever:*

$$(w, v) \in f_G(A)(i)(\{\sim_i\}_{i \in I}) =$$

$$\begin{cases} (w, v) \in \sim_i \text{ and } (w \in A \text{ iff } v \in A) & \text{if } i \in G \\ (w, v) \in \sim_i & \text{otherwise} \end{cases}$$

Given $\mathcal{M} = \langle W, (\leq_i)_{i \in I}, (\sim_i)_{i \in I}, V \rangle$ we write $f_G(A)(\mathcal{M})$ for the model $\langle W, (\leq_i)_{i \in I}, f_G(A)((\sim_i)_{i \in I}), V \rangle$.

It is easy to see that the relations $\sim_i$ in $f_G(A)(\mathcal{M})$ are equivalence relations. Here we are interested in cases where the agents condition their beliefs upon learning in which *belief state* the others are.

Definition 10 (Belief states). *Let $\mathcal{M}$ an epistemic plausibility model and $A \subseteq W$, we write*

$$B_j^{\mathcal{M}}(A) \text{ for } \{w : \beta_j(\mathcal{K}_j^{\mathcal{M}}[w]) \subseteq A\} \text{ and}$$

$$\neg B_j^{\mathcal{M}}(A) \text{ for } W \setminus B_j^{\mathcal{M}}(A)$$

We define $\mathbb{B}_j^{\mathcal{M},w}(A)$ as follows:

$$\mathbb{B}_j^{\mathcal{M},w}(A) = \begin{cases} B_j^{\mathcal{M}}(A) & \text{if } w \in B_j^{\mathcal{M}}(A) \\ \neg B_j^{\mathcal{M}}(A) & \text{otherwise} \end{cases}$$

Observation 4. *For any plausibility epistemic model $\mathcal{M}$ indexed by a finite set of agents I, $\langle \wp(W \times W)^I, \subseteq \rangle$ is a chain complete poset. Moreover for all $A \subseteq W$, $w \in W$ and $G \subseteq I$, $f_G(A)$ is deflationary.*

Taking $f_I(\bigcap_{j \in I} \mathbb{B}_j^{\mathcal{M},w}(||\phi||^{\mathcal{M}}))$ as a mapping on models, it is easy to see from the preceding observation that conditioning by agents' beliefs about some event is deflationary with respect to the relation of epistemic-submodel. It follows then by the Bourbaki-Witt fixed-point theorem [14] that conditioning by agents' beliefs has a fixed point.

Theorem 5 (Bourbaki-Witt [14]). *Let X be a chain complete poset. If $f : X \to X$ is inflationary (deflationary), then f has a fixed point.*

Given an initial pointed model $\mathcal{M}, w$ and some event $A \subseteq W$, we can construct its fixed point under conditioning by agents' beliefs as the limit of a sequence of models, which are the model-theoretic counterpart of the dialogues described above.

Definition 11. *A conditioning dialogue about ϕ at the pointed plausibility epistemic model $\mathcal{M}, w$, with $\mathcal{M} = \langle W, (\leq_i)_{i \in I}, (\sim_i)_{i \in I}, V \rangle$ is the sequence of pointed epistemic plausibility models $(\mathcal{M}_n, w)$ with*

$$(\mathcal{M}_0, w) = \mathcal{M}, w$$

$$(\mathcal{M}_{\beta+1}, w) = f_I(\bigcap_{j \in I} \mathbb{B}_j^{\mathcal{M}_\beta, w}(||\phi||^{\mathcal{M}}))(\mathcal{M}_\beta), w$$

$$(\mathcal{M}_\lambda, w) = \bigcap_{\beta < \lambda}(\mathcal{M}_\beta, w) \text{ for limit ordinals } \lambda$$

This extends to the countable case the standard representation of a dialogue about ϕ in the literature on dynamic agreements [2, 4]. By observation 4 we know that dialogues cannot last forever, i.e. that each such sequence has a limit.

Corollary 2. *For any pointed epistemic plausibility model $\mathcal{M}, w$ and $\phi \in \mathcal{L}_{EDL}$ there is a α^f such that, for all $i \in I$, $w \in W$ and $\alpha > \alpha^f$, $\mathcal{K}_{\alpha,i}[w] = \mathcal{K}_{\alpha^f,i}[w]$.*

Once the agents have reached this fixed-point α^f—possibly after transfinitely many steps—they have eliminated all higher-order uncertainties concerning the posteriors about ϕ of the others, viz. these posteriors are then common knowledge:

Theorem 6 (Common knowledge of beliefs about ϕ). *At the fixed-point α^f of a conditioning dialogue about ϕ we have that for all $w \in W$ and $i \in I$, if $w \in B_i^{\mathcal{M}_{\alpha^f}, w}(||\phi||^{\mathcal{M}})$ then $w' \in B_i^{\mathcal{M}_{\alpha^f}, w}(||\phi||^{\mathcal{M}})$ for all $w' \in [w]^*_{\alpha^f, I}$, and similarly if $w \notin B_i^{\mathcal{M}_{\alpha^f}, w}(||\phi||^{\mathcal{M}})$.*

With this in hand we can directly apply the static agreement result for common knowledge (Corollary 1, Section 3) to find that the agents do indeed reach agreements at the fixed-point of a dialogue about ϕ.

Corollary 3 (Agreement via conditioning dialogue). *Take any dialogue about ϕ with common and well-founded priors, and α^f as in Corollary 2. Then for all w in W, either $[w]^*_{\alpha^f,I} \subseteq \bigcap_{i \in I} B_i^{\mathcal{M}_{\alpha^f},w}(||\phi||^{\mathcal{M}})$ or $[w]^*_{\alpha^f,I} \subseteq \bigcap_{i \in I} \neg B_i^{\mathcal{M}_{\alpha^f},w}(||\phi||^{\mathcal{M}}).$*

This result brings qualitative dynamic agreement results [3, 4] to epistemic plausibility models, and show that agents can indeed reach agreement via iterated conditioning, even when the finite model assumption is dropped.

5.2   Agreements via Public Announcements

In this section we show that iterated "public announcements" lead to agreements, thus introducing a distinct form of information update to the agreement literature. Public announcements are "epistemic actions" [6] by which truthful, hard information is made public to the members of a group by a trusted source, in such a way that no member is in doubt about whether the others received the same piece of information as he did.

One extends a given logical language with public announcements by operators of the form $[\phi!]\psi$, meaning "after the announcement of ϕ, ψ holds" [15, 16]. A dialogue about ϕ via public announcements among the members of a group G thus starts, as before, with i simple beliefs about ϕ, for all $i \in I$. The agents' beliefs about ϕ at the next stage are then defined as those they would have after a public announcement of all agents' beliefs about ϕ at the first stage. Syntactically, this gives: $\mathbb{B}_{1,i}$ as in Section 5.1, and $\mathbb{B}_{n+1,i}$ as $[\bigcap_{j \in I} \mathbb{B}_{n,j}\phi!]B_i\phi$ if $\mathcal{M}, w \Vdash [\bigcap_{j \in I} \mathbb{B}_{n,j}\phi!]B_i\phi$ and as $[\bigcap_{j \in I} \mathbb{B}_{n,j}\phi!]\neg B_i\phi$ otherwise. For the same reason as in the previous section, we now move our analysis to the level of models.

The A-generated submodel of a given epistemic plausibility model $\mathcal{M}$ is the model that results after the public announcement of A in $\mathcal{M}$. We write $Sub(\mathcal{M}) = \{\mathcal{M}'$ is the A-generated submodel of $\mathcal{M} \mid A \subseteq |\mathcal{M}|\}$ and $\mathcal{M}' \sqsubseteq \mathcal{M}$ whenever $\mathcal{M}' \in Sub(\mathcal{M})$.

Definition 12 (Relativization by agents beliefs). *Let $\mathbb{B}_i(\mathcal{M}, w, \phi)$ be defined as follows:*

$$\mathbb{B}_i(\mathcal{M}, w, \phi) = \begin{cases} ||B_i\phi||^{\mathcal{M}} & \text{if } \mathcal{M}, w \Vdash B_i\phi \\ ||\neg B_i\phi||^{\mathcal{M}} & \text{otherwise} \end{cases}$$

Then given an epistemic-plausibility model $\mathcal{M} = \langle W, (\leq_i)_{i \in I}, (\sim_i)_{i \in I}, V \rangle$, the relativization $!B_w^\phi$ by agents' beliefs about ϕ at w (where $w \in |\mathcal{M}|$), takes $\mathcal{M}$ to $!B_w^\phi(\mathcal{M})$. Here $!B_w^\phi(\mathcal{M})$ is the $\bigcap_{i \in I} \mathbb{B}_i(\mathcal{M}, w, \phi)$-generated submodel $!B_w^\phi(\mathcal{M}) = \langle W^{!B_w^\phi}, \leq_i^{!B_w^\phi}, \sim_i^{!B_w^\phi}, V^{!B_w^\phi} \rangle$ of $\mathcal{M}$ such that:

- $W^{!B_w^\phi} = \bigcap_{i \in I} \mathbb{B}_i(\mathcal{M}, w, \phi)$

  and for each $i \in I$

- $\leq_i^{!B_w^\phi} = \leq_i \cap (W^{!B_w^\phi} \times W^{!B_w^\phi})$

$$- \sim_i^{!B_w^\phi} = \sim_i \cap (W^{!B_w^\phi} \times W^{!B_w^\phi})$$
- *For each* $v \in W^{!B_w^\phi}$, $v \in V^{!B^\phi}(p)$ *iff* $v \in V(p)$

Note that by construction above the actual state w is never eliminated.

Observation 7. *For any plausibility epistemic model $\mathcal{M}$ indexed by a finite set of agents I, $\langle Sub(\mathcal{M}), \sqsubseteq \rangle$ is a chain complete poset. Moreover, for all $\phi \in \mathcal{L}_{EDL}$, $w \in W$, $!B^\phi$ is deflationary.*

It follows then by the Bourbaki-Witt [14] Theorem (see previous subsection) that the process of public announcement of beliefs has a fixed point. Given an initial pointed model $\mathcal{M}, w$ and some formula $\phi \in \mathcal{L}_{EDL}$, we can construct this fixed point by taking the limit of a sequence of models, which we call a public dialogue.

Definition 13. *A public dialogue about ϕ starting in $\mathcal{M}, w$ is a sequence of epistemic-doxastic pointed models $\{(\mathcal{M}_n, w)\}$ such that:*

- $\mathcal{M}_0 = \mathcal{M}$ *is a given epistemic-plausibility model.*
- $\mathcal{M}_{\beta+1} = !B_w^\phi(\mathcal{M}_\beta)$
- $(\mathcal{M}_\lambda)$ *is the submodel of $\mathcal{M}$ generated by $\bigcap_{\beta < \lambda} |\mathcal{M}_\beta|$ for limit ordinals λ*

It is known that such a dialogue need not stop after the first round of announcements, in e.g. the "muddy children" case [17], but by observation 7 we know that it will stop at some point.

Corollary 4 (Fixed-point). *Given an epistemic-plausibility model $\mathcal{M}_0, w$ and a public dialogue about ϕ, there is a α^ϕ such that $(\mathcal{M}_\alpha, w) = (\mathcal{M}_{\alpha^\phi}, w)$ for all $\alpha \geq \alpha^\phi$.*

Moreover at $\mathcal{M}_{\alpha^\phi}, w$, which we call the *fixed point* of the public dialogue about ϕ, the posteriors of the agents about this formula are common knowledge, which means that they will reach an agreement on ϕ if they have common and well-founded priors.

Theorem 8 (Common knowledge at the fixed point). *At the fixed-point of a public dialogue $\mathcal{M}_{\alpha^\phi}, w$ about ϕ, for all $w \in W$ and $i \in I$, if $w \in ||B_i\phi||^{\mathcal{M}_{\alpha^\phi}}$ then $w' \in ||B_i\phi||^{\mathcal{M}_{\alpha^\phi}}$ for all $w' \in [w]_{\alpha^\phi, I}^*$, and similarly if $w \notin ||B_i\phi||^{\mathcal{M}_{\alpha^\phi}}$.*

Corollary 5 (Agreements via Public Announcements). *For any public dialogue about ϕ, if there is common and well-founded priors then at the fixed-point $\mathcal{M}_{\alpha^\phi}, w$ either all agents believe that ϕ or they all do not believe that ϕ.*

This new form of dynamic agreements result is conceptually important because it fits better than iterated conditioning the intuitive idea of a *public* dialogue, or so shall we argue in the next section, by highlighting the differences between the two processes of information exchange.

5.3   Comparing Agreements via Conditioning and Public Announcements

In this section we highlight by way of two examples that public announcements, in comparison with belief conditioning, are indeed *public*. We illustrate this first by comparing how conditioning and public announcements respectively change higher-order information, even in the case of "non-epistemic" facts. We then point out that this difference can indeed lead to different agreements, precisely in cases where the dialogues *are* about epistemic facts.

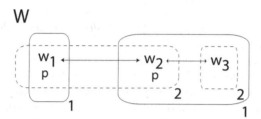

Fig. 3. An epistemic plausibility model where one round of conditioning on p does not remove higher-order uncertainty about p, while a public announcement of p does

Example 9. *Consider the model in Figure 3. The arrows represent 1 and 2's common plausibility ordering, with $w \leq w'$ and $w' \leq w$ for all $w, w' \in W$. The solid and dotted rectangles represent 1 and 2's information partitions, respectively. Take a proposition letter p and assume that $V(p) = \{w_1, w_2\}$. Observe that the agents already agree on p at w_1, but that agent 2 is uncertain about 1's beliefs about p: writing $\Diamond_2 \psi$ for $\neg B_2 \neg \psi$, we have $w_1 \models \Diamond_2 B_1 p \wedge \Diamond_2 \neg B_1 p$. A single public announcement of p at w_1 suffices to remove this higher-order uncertainty: $w_1 \models [p!] C_{1,2} p$. Agent 2's uncertainty about 1's beliefs about p, however, remains after a single conditioning on p. Taking $\Diamond_2^\psi \phi'$ for $\neg B_2^\psi \neg \psi'$, we have $w_1 \models \Diamond_2^p B_1 p \wedge \Diamond_2^p \neg B_1 p$.*

This example illustrates the public character of announcements in comparison with the private character of conditioning. In the first case all agents know that all others have received the same piece of truthful information. This is not necessarily the case for conditioning, even if all agents condition simultaneously on the same piece of information.

Given any pointed epistemic plausibility model $\mathcal{M}, w$ and formula ϕ, the reader can check that both the dialogue about ϕ via public announcements and the dialogue about ϕ via belief conditioning at $\mathcal{M}, w$ lead to the same agreement whenever ϕ is a Boolean combination of propositional letters. This is mainly due to the fact that neither operation changes the "basic facts", i.e. the propositional valuation in a given model. They do, however, treat "informational" facts differently, as the following example shows.

Example 10. *Consider the epistemic plausibility model in Figure 4. The arrows and rectangles are as in example 9. Take a proposition letter p and assume that*

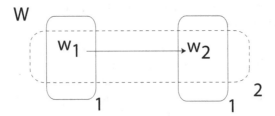

Fig. 4. An epistemic plausibility model where conditioning leads to different agreements than public announcements

$V(p) = \{w_1\}$. Let $\phi := p \land \neg B_2 p$, i.e. *"p but 2 doesn't believes that p"*. *Observe that ϕ holds at w_1, that 1 believes it but that 2 does not. The conditioning dialogue and the dialogue via public announcements, both about ϕ, reach their fixed point n^* after one round in this model, where $[w_1]_{n^*,1} = [w_1]_{n^*,2} = \{w_1\}$. The formula ϕ leads to an "unsuccessful update" by public announcement [6], and at the fixed point of the dialogue neither 1 nor 2 believe that ϕ. In conditioning dialogue, however, both agents do believe that ϕ at the fixed point.*

This example hinges on the fact that public announcement and belief conditioning have a different influence on higher-order information. In conditioning the truth value of the formula under consideration remains fixed. If the formula contains epistemic (K_i or C_G) or doxastic (B_i, CB_G) operators, this means that the conditioning dialogue bears on the knowledge and beliefs of the agents anterior to the information exchange [7]. In dialogues via public announcements the truth value of the formula ϕ is dynamically adapted to the incoming new information, reflecting the fact that knowing that others receive the same piece of information might lead an agent to revise his higher-order information, too.

This highlights the public character of announcements in comparison with belief conditioning, and thus that the former fit well with the intuition of public dialogue that drives the dynamic agreement results.

6   Conclusion

We have studied agreement theorems from the point of view of dynamic-epistemic logic. We have shown that both static and dynamic agreement results hold in epistemic plausibility models, answering an open question in the logic literature. We pointed out the need for rather expressive logical languages to reason explicitly about static agreement results. We have furthermore improved on existing qualitative agreement results by proving that common belief in posteriors is sufficient to ensure agreement, under common and well-founded priors, and so for both finite and countable structures. Finally, we focused on the distinction between conditioning and public announcements to provide two dynamic agreement results, arguing that the later better capture the public character of dialogues. Introducing agreement theorems to dynamic-epistemic logic thus proves to be both technically and conceptually fruitful, and it bridges two important bodies of literature.

Future work should put the full generality dynamic-epistemic logic [6, 18] to use, as well as recent developments in "softer" forms of belief updates [12, 19], to analyze the possibility of agreements in a larger class of situations. It also remains open whether one can finitely axiomatize a logic which can derive the agreement results, in both their static and dynamic forms. Finally, two issues pertaining to the expressibility of the static agreement theorems should be investigated further: first, the definability of the common prior assumption via countable sets of formulas of $\mathcal{L}_{EDL}$, as shown by [20] for the probabilistic case; and second, expressibility of alternative agreement results, as e.g. the one provided in [10].

Acknowledgments

We would like to thank Johan van Benthem, Balder ten Cate, Barteld Kooi, Stig Andur Pedersen and Sonja Smets for helpful comments and discussions. We are also grateful to the referees of TARK, and to the participants at the Research Seminar in Philosophy of Linguistics and the workshop Formal Modeling in Social Epistemology, both at Tilburg University, the Dynamic Logic Working Session at the ILLC in Amsterdam, the LSE-Groningen Exchange Workshop in London and the GroLog seminar in Groningen.

Bibliography

[1] Aumann, R.: Agreeing to disagree. The Annals of Statistics 4(6), 1236–1239 (1976)
[2] Geanakoplos, J., Polemarchakis, H.M.: We can't disagree forever. Journal of Economic Theory 28(1), 192–200 (1982)
[3] Cave, J.A.K.: Learning to agree. Economics Letters 12(2), 147–152 (1983)
[4] Bacharach, M.: Some extensions of a claim of Aumann in an axiomatic model of knowledge. Journal of Economic Theory 37(1), 167–190 (1985)
[5] Baltag, A., Smets, S.: Conditional doxastic models: A qualitative approach to dynamic belief revision. In: Mints, G., de Queiroz, R. (eds.) Proceedings of WOLLIC 2006. Electronic Notes in Theoretical Computer Science, vol. 165 (2006)
[6] van Ditmarsch, H., van de Hoek, W., Kooi, B.: Dynamic Epistemic Logic. Synthese Library Series, vol. 337. Springer, Heidelberg (2007)
[7] Baltag, A., Smets, S.: A qualitative theory of dynamic interactive belief revision. In: Bonanno, G., van der Hoek, W., Wooldridge, M. (eds.) Logic and the Foundation of Game and Decision Theory (LOFT7). Texts in Logic and Games, vol. 3, pp. 13–60. Amsterdam University Press (2008)
[8] Bonanno, G., Nehring, K.: Agreeing to disagree: a survey. Some of the material in this paper was published in [21] (1997)
[9] Menager, L.: Communication, common knowledge, and consensus. PhD thesis, Universite Paris I Pantheon-Sorbonne (2006)
[10] Samet, D.: Agreeing to disagree: The non-probabilistic case. Games and Economic Behavior (in press)
[11] Fagin, R., Halpern, J., Moses, Y., Vardi, M.: Reasoning about Knowledge. MIT Press, Cambridge (1995)
[12] van Benthem, J.: Dynamic logic for belief revision. Journal of Applied Nonclassical Logics 17(2) (2007)

[13] ten Cate, B.: Model theory for extended modal languages. PhD thesis, University of Amsterdam, ILLC Dissertation Series DS-2005-01 (2005)

[14] Bourbaki, N.: Sur le théorème de Zorn. Archiv der Mathematik 2 (1949)

[15] Plaza, J.: Logics of public communications. In: Emrich, M., Pfeifer, M., Hadzikadic, M., Ras, Z. (eds.) Proceedings of the Fourth International Symposium on Methodologies for Intelligent Systems: Poster Session Program, Oak Ridge National Laboratory, pp. 201–216 (1989)

[16] Gerbrandy, J.: Bisimulations on Planet Kripke. PhD thesis, ILLC, Amsterdam (1999)

[17] van Benthem, J.: One is a lonely number. In: Chatzidakis, Z., Koepke, P., Pohlers, W. (eds.) Logic Colloquium 2002. ASL & A.K. Peters, Wellesley (2006)

[18] Baltag, A., Moss, L., Solecki, S.: The logic of public announcements, common knowledge and private suspicions. In: TARK 1998 (1998)

[19] Baltag, A., Smets, S.: Learning by questions and answers: From belief-revision cycles to doxastic fixed points. Under Review (200X)

[20] Heifetz, A.: The positive foundation of the common prior assumption. Games and Economic Behavior 56(1), 105–120 (2006)

[21] Bonanno, G., Nehring, K.: How to make sense of the common prior assumption under incomplete information. International Journal of Game Theory 28(3), 409–434 (1999)

Learning and Teaching as a Game: A Sabotage Approach

Nina Gierasimczuk, Lena Kurzen, and Fernando R. Velázquez-Quesada

Institute for Logic, Language and Computation, Universiteit van Amsterdam.
P.O. Box 94242, 1090GE Amsterdam, The Netherlands
{N.Gierasimczuk, L.M.Kurzen, F.R.VelazquezQuesada}@uva.nl

Abstract. In formal approaches to inductive learning, the ability to learn is understood as the ability to single out a correct hypothesis from a range of possibilities. Although most of the existing research focuses on the characteristics of the learner, in many paradigms the significance of the teacher's abilities and strategies is in fact undeniable. Motivated by this observation, this paper highlights the interactive nature of learning by showing its relation with games. We show how learning can be seen as a *sabotage-type* game between *Teacher* and *Learner*, and we present different variants based on the level of cooperativeness and the actions available to the players, characterizing the existence of winning strategies by formulas of *Sabotage Modal Logic* and analyzing their complexity. We also give a two-way conceptual account of how to further combine games and learning: we propose to use game theory to analyze the grammar inference approach, and moreover, we indicate that existing inductive inference games can be analyzed using learning theory tools. Our work aims at unifying game-theoretical and logical approach to formal learning theory.

Keywords: Formal learning theory, game theory, modal logic, sabotage games, inductive inference games, learning algorithms.

1 Introduction

Formal learning theory (see e.g. [1]) is concerned with the process of inductive inference: it formalizes the process of inferring general conclusions from partial, inductively given information, as in the case of language learning (inferring grammars from sentences) and scientific inquiry (drawing general conclusions from partial experiments).

This general process can be seen as a game between two players: *Learner* and *Teacher*. The game starts with a class of possible worlds from which Teacher chooses the actual one, and Learner has to find out which one it is. Teacher inductively provides information about the world, and whenever Learner receives a piece of information, he picks a conjecture from the initial class, indicating which one he thinks is the case. Different conditions can be defined for the success of the learning process: we can require that after a finite amount of

X. He, J. Horty, and E. Pacuit (Eds.): LORI 2009, LNAI 5834, pp. 119–132, 2009.

data Learner decides on a correct hypothesis (finite identification), or that the sequence of Learner's conjectures converges to a correct hypothesis without him ever being certain about the answer (identification in the limit) [2].

We restrict ourselves to a high-level analysis of the process described above. Our proposal focuses on some important elements of learnability. First of all, we treat learning as a procedure of singling out one correct hypothesis from some range of possibilities. Moreover, we see this procedure not as a one-move choice, but as a sequence of them, and therefore we allow many steps of update before the conclusion is reached. Those two properties make our notion of learning different from the concept of learning formalized in dynamic epistemic logic (see e.g. [3]), where the word "learning" is often used as a synonym of "getting to know" and is usually represented as a one-step epistemic update. Moreover, in our approach we pay attention to the strategies for teaching, highlighting the fact that restricted power and knowledge of Learner can be compensated by additional insights and intentions of Teacher.

Our approach is also motivated by very concrete scenarios of Learner-Teacher interaction as they occur in so called *inductive inference games* such as *Zendo* [4] and *(The New) Eleusis* [5]. We argue that our interactive perspective on Learning and Teaching together with the resulting formal model can also provide tools for game-theoretical and complexity analyses of such concrete games.

The paper is structured as follows. In Section 2 we introduce the sabotage learning framework, showing how sabotage modal logic can express the winning conditions of three versions of Sabotage Learning Games and giving complexity results for them. In Section 3, we analyze Sabotage Learning Games in which players do not need to move in alternation. Section 4 gives additional insights into the relation between learning and games. We first present a refined interactive view on Teaching, based on existing learning algorithms; then, we argue that a mixture of learning theory, game theory and cognitive science tools can be used to analyze existing inductive inference games. Section 5 concludes our considerations.

2   Learning as a Sabotage Game

The main aim of the paper is to highlight the interactive nature of the learning process by showing its relation with games. This is motivated by learning paradigms in which Teacher plays a significant role and, in fact, has a strong influence on whether the process is successful (e.g., *learning from queries and counterexamples* [6]).

Our first step is to show that learning can be seen as a game. We start by considering a simple situation with two players: *Teacher* and *Learner*. From a high-level perspective, we can describe learning as the step-by-step process through which Learner changes his information state. The process is successful if he eventually reaches an information state describing the real state of affairs, a state called the *goal*. The information that Teacher provides can be interpreted as feedback about Learner's current conjecture, allowing him to rule out possible mind changes because they are inconsistent with the received information.

Table 1. Correspondence with Learning Model

Learning Model	Sabotage Games
Hypotheses	States
Correct hypothesis	Goal state
Possibility of a mind change from hypothesis a to hypothesis b	Edge from state a to b
A mind change from hypothesis a to hypothesis b	Transition from state a to b
Giving a counterexample that eliminates the possibility of a mind change from a to b	Removing a transition between a and b

By looking at the Teacher-Learner interaction from this perspective, we can represent the situation as a graph whose vertices represent Learner's possible information states and edges represent transitions between them. Changes in Learner's information state are represented by moves along the edges, and Teacher's feedback is represented as the removal of edges. We say that the learning process has been successful if Learner reaches the goal state. The correspondence between the learning model from formal learning theory and our interactive sabotage approach is described in Table 1.

Note how in our setting, Teacher's information does not rule out states, but *changes* in information states. Then, the removal of one edge does not need to make the target unreachable state since there can be more than one path to it.

2.1 Sabotage Games

The described perspective on learning leads naturally to the framework of Sabotage Games [7,8]. A *Sabotage Game* is played in a directed multi-graph by two players, *Runner* and *Blocker*, which move alternatingly with Runner being the first. Runner moves by making one transition from the current vertex; Blocker moves by deleting an edge from the graph.

In the present paper, we use a variant of the game based on *labelled multi-graphs*. In [9] it is shown to be equivalent to the original game with respect of the existence of a winning strategy.

Definition 1 (Directed Labelled Multi-graph). *Let* $\Sigma = \{a_1, \ldots a_n\}$ *be a finite set of labels. A directed labelled multi-graph is a tuple* $G^\Sigma = (V, \mathcal{E})$ *where* V *is a finite set of vertices and* $\mathcal{E} = (\mathcal{E}_{a_1}, \ldots, \mathcal{E}_{a_n})$ *is a collection of binary relations:* $\mathcal{E}_{a_i} \subseteq V \times V$ *for each* $a_i \in \Sigma$.

In this definition, labels from Σ are used to represent multiple edges between two vertices. The definition of the game is as follows.

Definition 2. *A* Labelled Sabotage Game $SG^\Sigma = \langle V, \mathcal{E}, v, v_g \rangle$ *is given by a directed labelled multi-graph* $(V, \mathcal{E})$ *and two vertices* $v, v_g \in V$. *Vertex* v *represents the position of Runner and* v_g *represents the goal state. Each match is played* *as follows: the initial position $\langle \mathcal{E}^0, v_0 \rangle$ is given by $\langle \mathcal{E}, v \rangle$. Round $k + 1$ from position $\langle \mathcal{E}^k, v_k \rangle$ with $\mathcal{E}^k = (\mathcal{E}_{a_1}^k, \ldots, \mathcal{E}_{a_n}^k)$, consists of Runner moving to some v_{k+1} such that $(v_k, v_{k+1}) \in \mathcal{E}_{a_i}^k$ for some $a_i \in \Sigma$, and then Blocker removing an edge $((v, v'), a_j)$, where $(v, v') \in \mathcal{E}_{a_j}^k$ for some $a_j \in \Sigma$. The new position is $\langle \mathcal{E}^{k+1}, v_{k+1} \rangle$, where $\mathcal{E}_{a_j}^{k+1} = \mathcal{E}_{a_j}^k \setminus \{(v, v')\}$ and $\mathcal{E}_{a_i}^{k+1} = \mathcal{E}_{a_i}^k$ for all $i \neq j$. The match ends if Runner cannot make a move or if he reaches the goal state, with Blocker winning in the first case and Runner winning in the second.*

Note that Blocker cannot get stuck since, whenever it is her turn, there is at least one edge left in the graph (the one that Runner just used).

It is easy to see that the Labelled Sabotage Game has the *history-free determinacy property*: if one of the players has a winning strategy then she also has one that only depends on the current position. Then, we can see each round as a transition from a Sabotage Game $SG^\Sigma = \langle V, \mathcal{E}_k, v_k, v_g \rangle$ to another Sabotage Game $SG^\Sigma = \langle V, \mathcal{E}_{k+1}, v_{k+1}, v_g \rangle$, since previous moves become irrelevant. We will use this perspective throughout the paper. Also, by edges and vertices of the game $SG^\Sigma = \langle V, \mathcal{E}, v, v_g \rangle$, we will mean edges and vertices of $(V, \mathcal{E})$.

Observe that when Blocker removes the edge $(v, v') \in \mathcal{E}_a$, the label a is irrelevant; what matters for who can win the game is how many edges are left from v to v'. Also, vertices and labels are finite, so every match ends after a finite number of rounds.

2.2   Sabotage Learning Games

We now define Sabotage Learning Games with different winning conditions. We will work with Labelled Sabotage Games, using the labelling of the edges to represent different kinds of information changes that take Learner from one state into another.

Definition 3. *A* Sabotage Learning Game (SLG) *is a Labelled Sabotage Game between* Learner *(L, taking the role of Runner) and* Teacher *(T, taking the role of Blocker). We distinguish between three versions, SLGUE, SLGHU and SLGHE. They differ only in their winning conditions, which are provided in Table 2.*

The winning conditions correspond to different levels of Teacher's helpfulness and Learner's willingness to learn. We can have *unhelpful* Teacher and *eager* Learner (*SLGUE*), but there can also be *helpful* Teacher and *unwilling*

Table 2. Sabotage Learning Games

Game	Winning Condition
SLGUE	Learner wins iff he reaches the goal state, Teacher wins otherwise.
SLGHU	Teacher wins iff Learner reaches the goal state, Learner wins otherwise.
SLGHE	Both players win iff Learner reaches the goal state. Both lose otherwise.

Learner (*SLGHU*). The cooperative case corresponds to *helpful* Teacher and *eager* Learner (*SLGHE*). Having provided a formal framework for Teacher-Learner interactions by means of *SLG*s, we now show how we can use Sabotage Modal Logic for reasoning about players' strategic powers in these games.

2.3 Sabotage Modal Logic

Sabotage Modal Logic (SML) has been introduced in [8]. Its language extends the basic modal language by formulas of the form $\diamondsuit\phi$, saying that it is possible to delete a pair from the accessibility relation such that ϕ holds.

Definition 4 (Sabotage Modal Language [8]). *Let* PROP *be a countable set of atomic propositions and* Σ *a finite set of labels. Formulas of the language of SML are given by*

$$\phi ::= p \mid \neg\phi \mid \phi \vee \phi \mid \Diamond_a\phi \mid \diamondsuit_a\phi$$

with $p \in$ PROP *and* $a \in \Sigma$. *We write* $\Diamond\phi$ *for* $\bigvee_{a\in\Sigma} \Diamond_a\phi$ *and* $\diamondsuit\phi$ *for* $\bigvee_{a\in\Sigma} \diamondsuit_a\phi$.

Definition 5 (Sabotage Model [10]). *Given a countable set of atomic propositions* PROP *and a finite set* $\Sigma = \{a_1, \ldots, a_n\}$, *a* Sabotage Model *is a tuple* $M = \langle W, (R_{a_i})_{a_i \in \Sigma}, Val \rangle$ *where* W *is a non-empty set of worlds, each* $R_{a_i} \subseteq W \times W$ *is an accessibility relation and* $Val :$ PROP $\rightarrow \mathcal{P}(W)$ *is a propositional valuation function. The pair* (M, w) *with* $w \in W$ *is called a* Pointed Sabotage Model.

First we give the definition of the model that results from removing an edge.

Definition 6. *Let* $M = \langle W, R_{a_1}, \ldots R_{a_n}, Val \rangle$ *be a Sabotage Model. The model* $M^{a_i}_{(v,v')}$ *that results from removing the edge* $(v, v') \in R_{a_i}$ *is defined as*

$$M^{a_i}_{(v,v')} := \langle W, R_{a_1}, \ldots R_{a_{i-1}}, R_{a_i} \setminus \{(v, v')\}, R_{a_{i+1}}, \ldots R_{a_n}, Val \rangle.$$

Now the semantics of SML is given as follows.

Definition 7. *Given a Sabotage Model* $M = \langle W, (R_a)_{a\in\Sigma}, Val \rangle$ *and a world* $w \in W$, *atomic propositions, negations, disjunctions and standard modal formulas are interpreted as usual. For "transition-deleting" formulas, we have*

$$M, w \models \diamondsuit_a\phi \qquad \text{iff} \qquad \exists\, w, v \in W : (v, v') \in R_a \; \& \; M^a_{(v,v')}, w \models \phi,$$

and $\boxminus_a\phi$ *is defined to be equivalent to* $\neg\diamondsuit_a\neg\phi$.

Theorem 1 ([10]). *Model checking (combined complexity) of Sabotage Modal Logic is PSPACE-complete.*

2.4 Sabotage Learning Games in Sabotage Modal Logic

Sabotage Modal Logic can be used for reasoning about games in which choices of one player restrict the set of possible future moves of herself and the other player. Therefore, this logic is useful for reasoning about players' strategic power in Sabotage Learning Games: for each SLG we can construct a Pointed Sabotage Model in a straightforward way.

Definition 8. *Let* $SG^{\Sigma} = \langle V, \mathcal{E}, v_0, v_g \rangle$ *be a Sabotage Game with* $\mathcal{E} = (\mathcal{E}_a)_{a \in \Sigma}$. *We define the Pointed Sabotage Model* $(M(SG^{\Sigma}), v_0)$ *over* PROP $:= \{goal\}$, *with*

$$M(SG^{\Sigma}) := \langle V, \mathcal{E}, Val \rangle,$$

where $Val(goal) := \{v_g\}$.

For each game in Table 2, we can define a SML formula that is true in the model corresponding to a game if and only if there is a w.s. in the game.

Consider *SLGUE* (the Sabotage Game of [8]). Inductively, we define

$$\gamma_0^{UE} := goal, \qquad\qquad \gamma_{n+1}^{UE} := goal \vee \Diamond \boxminus \gamma_n^{UE}$$

The following result is Theorem 7 of [10] rephrased for Labelled Sabotage Games. The new setting avoids a technical issue present in the original proof.

Theorem 2. *Learner has a winning strategy (w.s.) in the SLGUE* $SG^{\Sigma} = \langle V, \mathcal{E}^0, v_0, v_g \rangle$ *iff* $M(SG^{\Sigma}), v_0 \models \gamma_n^{UE}$, *where* n *is the number of edges of* SG^{Σ}.

Proof. [Sketch] The proof is by induction on n. The base case is immediate.

For the inductive case, Learner has a w.s. in a game with $n + 1$ edges iff he is already at v_g or he can move to a vertex v_1 in which, no matter what edge Teacher removes, he has a w.s. in the resulting SG'^{Σ}. The first case is equivalent to $M(SG^{\Sigma}), v_0 \models goal$. For the second case, note that SG'^{Σ} has n edges, and then by inductive hypothesis we have $M(SG'^{\Sigma}), v_1 \models \gamma_n^{UE}$. Hence, $M(SG^{\Sigma}), v_1 \models \boxminus \gamma_n^{UE}$ and thus $M(SG^{\Sigma}), v_0 \models \Diamond \boxminus \gamma_n^{UE}$. Hence, $M(SG^{\Sigma}), v_0 \models goal \vee \Diamond \boxminus \gamma_n^{UE}$. $\square$

A detailed proof of Theorem 2 can be found in [9].

Next, consider *SLGHU*, the game with helpful Teacher and unwilling Learner. Inductively, we define

$$\gamma_0^{HU} := goal, \qquad\qquad \gamma_{n+1}^{HU} := goal \vee (\Diamond \top \wedge \Box \Diamond \gamma_n^{HU}).$$

We show that this formula corresponds to the existence of a winning strategy for Teacher. Note that Teacher has to make sure that Learner does not get stuck before he has reached the goal state. This is why the conjunct $\Diamond \top$ is needed.

Theorem 3. *Teacher has a w. s. in the SLGHU* $SG^{\Sigma} = \langle V, \mathcal{E}^0, v_0, v_g \rangle$ *iff* $M(SG^{\Sigma}), v_0 \models \gamma_n^{HU}$, *where* n *is the number of edges of* SG^{Σ}.

Proof. Similar to the proof of Theorem 2. $\square$

Finally, for *SLGHE*, the corresponding formula is defined as

$$\gamma_0^{HE} := goal, \qquad\qquad \gamma_{n+1}^{HE} := goal \vee \Diamond \Diamond \gamma_n^{HE}.$$

Theorem 4. *Teacher and Learner have a joint w.s. in the SLGHE* $SG^{\Sigma} = \langle V, \mathcal{E}^0, v_0, v_g \rangle$ *iff* $M(SG^{\Sigma}), v_0 \models \gamma_n^{HE}$, *where* n *is the number of edges of* SG^{Σ}.

Proof. [Sketch] Note that Learner and Teacher have a joint w.s. iff there is a path from v_0 to v_g. Teacher can always remove the edge just used by Learner. $\square$

The above results are summarized in Table 3.

Table 3. Winning Conditions for SLG in SML

Game	Winning Condition in SML	Winner
$SLGUE$	$\gamma_0^{UE} := goal,\quad \gamma_{n+1}^{UE} := goal \vee \Diamond \boxminus \gamma_n^{UE}$	Learner
$SLGHU$	$\gamma_0^{HU} := goal,\quad \gamma_{n+1}^{HU} := goal \vee (\Diamond \top \wedge (\Box \Diamond \gamma_n^{HU}))$	Teacher
$SLGHE$	$\gamma_0^{HE} := goal,\quad \gamma_{n+1}^{HE} := goal \vee \Diamond \Diamond \gamma_n^{HE}$	Both

2.5 Complexity of Sabotage Learning Games

In this section, we investigate the complexity of Sabotage Learning Games. In our framework the complexity of deciding whether a player has a winning strategy in a given SLG corresponds to the complexity of deciding whether learning is possible in a given situation.

Intuitively, learning as in SLG should be easiest with helpful Teacher and eager Learner. This is indeed reflected in the computational complexity of deciding in a given game whether the winning condition is satisfied.

We provided SML formulas expressing the existence of a w.s. for the three versions of SLG. By Theorem 1 (proved in [10]) Model checking (combined complexity) of SML is PSPACE-complete. Thus, deciding whether a player can win a game can be done in PSPACE. For the cases of $SLGUE$ and $SLGHE$, we can also give tight lower bounds.

For $SLGUE$, the standard Sabotage Game, PSPACE-hardness is shown by reduction from QBF [10].

Theorem 5 ([10]). $SLGUE$ is PSPACE-complete.

It remains to show whether $SLGHU$ is also PSPACE-hard. $SLGUE$ seems like the dual of $SLGHU$, but this is not the case due to the different nature of the players' moves. Thus, reducing $SLGUE$ to $SLGHU$ is not straightforward.

Finally, $SLGHE$ is cooperative: both players win or lose together. Here, a w.s. is a *joint strategy*. As mentioned in the proof of Theorem 4, it exists iff the goal is reachable from Learner's position. So, determining if a game can be won

Table 4. Complexity Results for Sabotage Learning Games

Game	Winning Condition	Complexity
$SLGUE$	Learner wins iff he reaches the goal state, Teacher wins otherwise	PSPACE-complete.
$SLGHU$	Teacher wins iff Learner reaches the goal state, Learner wins otherwise.	PSPACE
$SLGHE$	Both players win iff Learner reaches the goal state. Both loose otherwise.	NL-complete

is equivalent to the REACHABILITY (st-CONNECTIVITY) problem which is nondeterministic logarithmic space complete (NL-complete) [11].

Theorem 6. *SLGHE is NL-complete.*

Table 4 summarizes the complexity results. They agree with our intuition that in some sense learning is easiest if Learner and Teacher cooperate.

3   Relaxing Strict Alternation

As mentioned above, the players' moves are asymmetric: Learner moves locally (moving to a vertex accessible *from the current one*) while Teacher moves globally (removing *any* edge from the graph). Thus, Learner's move does not in principle need to be followed by Teacher's move (e.g. Learner can perform several changes of his information state before Teacher can make a restriction).

Definition 9. *A* Sabotage Learning Game without strict alternation *(for Teacher) is a tuple $SLG^* = \langle V, \mathcal{E}, v_0, v_g \rangle$. Moves of Learner are as in SLG and, once he has chosen a vertex v_1, Teacher can choose between removing an edge, in which case the next game is given as in SLG, and doing nothing, in which case the next game is $\langle V, \mathcal{E}, v_1, v_g \rangle$. Again, there are three different versions, now called SLG^*UE, SLG^*HU and SLG^*HE.*

Though in SLG^* Teacher has an additional move, the players' winning abilities do not change. In the rest of this section we show that, for each of the three versions, the games with and without strict alternation are equivalent.

We start with the case of an unhelpful teacher and an eager learner SLG^*UE. Note that even though in this new setting matches can be infinite, in fact if Learner can win the game, he can do so in a finite number of rounds. Consider the case of unhelpful Teacher and eager Learner SLG^*UE. Before we go into the details, note that if Learner can win the game, he can do so in a finite number of rounds.

Theorem 7. *Consider the SLG $\langle V, \mathcal{E}, v_0, v_g \rangle$ with $(V, \mathcal{E})$ a directed labelled multi-graph and v, v_g vertices in it. Learner has a winning strategy in the corresponding SLGUE iff he has a wining strategy in the corresponding SLG^*UE.*

Proof. From left to right, the idea is that in each round L "pretends" that T has removed some edge and then makes the move given by his strategy for $SLGUE$. For the other direction, if L has a w.s. for SLG^*UE, then he can also win the corresponding $SLGUE$ by using the same strategy.   □

Corollary 1. *If Learner has a SLG^*UE-winning strategy in $\langle V, \mathcal{E}, v_0, v_g \rangle$, then he has a SLGUE-winning strategy.*

Proof. If L can respond successfully to all of T's moves in SLG^*, then in particular, he can do so if T removes an edge in every round.   □

The case of helpful Teacher and unwilling Learner is more interesting. One might expect that the additional possibility of skipping a move gives more power to Teacher, since she could avoid removals that would have made the goal unreachable from the current vertex. However, we can show that this is not the case. First, we state the following lemmas.

Lemma 1. *Consider the game $\langle V, \mathcal{E}, v_0, v_g \rangle$ with winning condition SLG^*HU. If there is a path from v_0 to v_g and there is no path from v_0 to a state from where v_g is not reachable, then T has a SLG^*HU-winning strategy.*

Proof. Let us assume that all states reachable from v_0 are on paths to v_g. Then even if T refrains from removing any edge, L will be on a path to the goal. Now, either the path to the goal does not include a loop or it does. If it does not then T can simply wait until L arrives at the goal. If it does, T can remove the edges that lead to the loops in such a way that v_g is still reachable from any vertex. □

Lemma 2. *For all SLG^*HU $\langle V, \mathcal{E}, v_0, v_g \rangle$, if T has a w.s. and there is an edge $(v, v') \in \mathcal{E}_a$ for some $a \in \Sigma$ such that no path from v_0 to v_g uses (v, v'), then T also has a w.s. in $\langle V, \mathcal{E}', v_0, v_g \rangle$, where $\mathcal{E}'$ results from removing (v, v') from $\mathcal{E}_a$.*

Proof. If v is not reachable from v_0, it is easy to see that the claim holds. Assume that v is reachable from v_0. T's w.s. should prevent L from moving from v to v' (otherwise L wins). Hence, T can also win if (v, v') is not there. □

Theorem 8. *If Teacher has a w.s. in the SLG^*HU $\langle V, \mathcal{E}, v_0, v_g \rangle$, then she also has a winning strategy in which she removes an edge in each round.*

Proof. The proof proceeds by induction on the number of edges $n = \sum_{a \in \Sigma} |\mathcal{E}_a|$.

The base case is straightforward. For the inductive case, assume that T has a winning strategy in SLG^*HU $\langle V, \mathcal{E}, v_0, v_g \rangle$ with $\sum_{a \in \Sigma} |\mathcal{E}_a| = n + 1$.

If $v_0 = v_g$, we are done. Otherwise, since T can win, there is some $v_1 \in V$ such that $(v_0, v_1) \in \mathcal{E}_a$ for some $a \in \Sigma$ and for all such v_1 we have:

1. There is a path from v_1 to v_g, and
2. (a) T can win $\langle V, \mathcal{E}, v_1, v_g \rangle$, or
   (b) there is a $((v, v'), a) \in (V \times V) \times \Sigma$ such that $(v, v') \in \mathcal{E}_a$ and T can win $\langle V, \mathcal{E}', v_1, v_g \rangle$ where $\mathcal{E}'$ is the result from removing (v, v') from $\mathcal{E}_a$.

If 2b holds, since $\sum_{a \in \Sigma} |\mathcal{E}'_a| = n$, we are done — we use the inductive hypothesis to conclude that T has a w.s. in which she removes an edge in each round (in particular, her first choice is $((v, v'), a)$). Let us show that 2b holds.

If there is some $(v, v') \in V \times V$ such that $(v, v') \in \mathcal{E}_a$ for some $a \in \Sigma$ and this edge is not part of any path from v_1 to v_g then by Lemma 2, T can remove this edge and 2b holds, so we are done.

If all edges in $(V, \mathcal{E})$ belong to a path from v_1 to v_g, from 1, there are two cases: either there is only one, or there are more than one paths from v_1 to v_g.

In the first case (only one path) (v_0, v_1) can be chosen since it cannot be part of the *unique* path from v_1 to v_g. Assume now that there is more than one path from v_1 to v_g. Let $p = (v_1, v_2, \ldots, v_g)$ be the/a shortest path from v_1 to v_g. This path cannot contain any loops. Then, from this path take v_i such that i is the smallest index for which it holds that from v_i there is a path $(v_i, v'_{i+1}, \ldots v_g)$ to v_g that is at least as long as the path following p from v_i (i.e. $(v_i, v_{i+1}, \ldots, v_g)$). Intuitively, when following path p from v_1 to v_g, v_i is the first point at which one can deviate from p in order to take another path to v_g (recall that we consider the case where every vertex in the graph is part of a path from v_1 to v_g). Now it is possible for T to choose $((v_i, v'_{i+1}), a)$ such that $(v_i, v'_{i+1}) \in \mathcal{E}_a$. Let $\mathcal{E}'$ be the resulting set of edges after removing (v_i, v'_{i+1}) from $\mathcal{E}_a$. Then we are in the game $\langle V, \mathcal{E}', v_1, v_g \rangle$. Note due to the way we chose the edge to be removed, in the new graph it still holds that from v_0 there is no path to a vertex from which v_g is not reachable (this holds because from v_i the goal v_g is still reachable). Then by Lemma 1, T can win $\langle V, \mathcal{E}', v_1, v_g \rangle$, which then implies 2b.

Hence, we conclude that 2b is the case and thus using the inductive hypothesis, T can win $\langle V, \mathcal{E}, v_0, v_g \rangle$ also by removing an edge in every round. □

Corollary 2. *If Teacher has a SLG^*HU-w.s. in $\langle V, \mathcal{E}, v_0, v_g \rangle$, then he has a SLGHU-winning strategy.*

Finally, let us consider the case of helpful Teacher and eager Learner.

Theorem 9. *If Learner and Teacher have a joint SLG^*HE-w.s. in $\langle V, \mathcal{E}, v_0, v_g \rangle$, then they have a joint SLGHE-w.s.*

Proof. If the players have a joint SLG^*HE-w.s., then there is an acyclic path from v_0 to v_g, which L can follow. At each round, T just has to remove the just used edge. □

Let us briefly conclude this section. We have shown that in SLG, allowing Teacher to skip moves does not change the winning abilities of the players. Using these results, both the complexity and definability results from the previous section also apply to the games in which Teacher can skip a move.

4   Interactive Learning and Teaching Scenarios

The view on learning presented above is very general. In this section, we will present two more concrete settings that, we believe, can benefit from an interactive perspective on learning theory as we present it in this paper. First, we present a game theoretic view on the *queries and counterexamples* framework and then we give an outline of how interactive learning models could be used in the analysis of *inductive inference games*.

4.1   Refined View on Teaching: Learning Algorithms

Let us go back to the *queries and counterexamples* paradigm (see [6]). Here, Learner is an algorithm that embodies a winning strategy in the game of learning (the learning procedure succeeds on *all* possible *true* data). Teacher can influence this process by giving counterexamples. Therefore, the game of teaching in such a setting can be formalized in extensive form as in Figure 1.

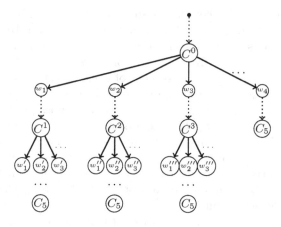

Fig. 1. The tree of the teaching game: dotted lines are Learner's moves, determined by his algorithm; solid lines are Teacher's moves; w_i are counterexamples; C_i are Learner's conjectures; C_5 is the correct hypothesis

A number of game-theoretical issues arise when viewing the run of the learning algorithm as a game. We can e.g. consider the epistemic status of the players, introduce imperfect information and analyze payoff characteristics. Concerning the payoffs, for different classes of teachers such as (un)helpful ones, we can define corresponding preferences or payoffs: the *helpful* teacher might strictly prefer all shortest paths in the game tree, i.e. the paths in which the learner learns the fastest. The *unhelpful* teacher might strictly prefer all the longest paths in the game tree, i.e. the paths in which the learner learns slowly.

We can also provide a choice for Learner in this game. Firstly, we can allow that at each step the learner can choose from different procedures which are part of one algorithm. Secondly, in the beginning Learner can decide which available algorithm to use. Moreover, we can consider another possibility that involves extending the traditional inductive inference paradigm. Usually, learnability of a class is interpreted as the existence of Learner that learns every element from the class independently of the behavior of Teacher — if we introduce the possibility of non-learnability, we can view learning algorithms as winning strategies for eager Learner in the learning game. With the possibility of non-learnability, there are also paths in the game tree in which Learner never makes a correct conjecture. In this framework, a *helpful* teacher would also prefer all (shortest) paths ending in a position in which the learner makes a correct conjecture over all the other paths.

4.2   Inductive Inference Games

Inductive inference games such as *(The New) Eleusis* [12,13,5] and *Zendo* [4] are examples of Learner-Teacher interaction in a very concrete setting. In these games, one player takes the role of Teacher and chooses a secret rule (concerning admissible sequences of cards or configurations of pyramids, in Eleusis and

Zendo, respectively). The other players are Learners whose aim is to inductively discover the rule. They do so by producing instances, and Teacher indicates whether the instances obey the secret rule. These games vary in the additional moves allowed to Learners, from the possibility of judging the correctness of other Learners' instances to the possibility of guessing the actual rule.

The winning conditions for these games allow Learner to win without actually getting to know the rule. Thus, inductive inference games provide a controlled setting in which we can investigate notions of successful learning that seem closer to learning in real life situations than learning as exact identification. In Eleusis, Learner can win by correctly classifying a certain number of instances; in the case of Zendo, Learner can win by conjecturing a rule that cannot be disproved by Teacher (that is, she cannot find an instance that distinguishes between the conjecture and the actual rule).

What makes a game-theoretical analysis of inductive inference games particularly interesting is that it also requires a formal investigation of the notion of *operational knowledge* and the *complexity of hidden rules*. A part of playing the inductive inference games is to be able to converge to an accurate hypothesis about the rule, without ever being sure about the correctness of the own conjectures. The result of this process can be called "operational knowledge", close to true belief based on a reliable strategy of mind changes, but without certainty.

On the other hand, a computational and logical analysis of inductive inference games can give a nice case study of comparing the complexity of logics expressing the winning conditions, the complexity of deciding whether some player has a winning strategy, and the complexity of the rules to be guessed. This can be done by restricting the class of secret rules Teacher can choose from. Then we can use the results from grammar inference to get the complexity of executing the winning strategy (learning algorithm). We can also identify structural properties of counterexamples given by Teacher and conjectures made by the learners that characterize optimal strategies.

Both, epistemic and complexity analyses of inductive inference games can also be investigated empirically. Such results could contribute to the research in epistemic logic by giving an account of real agents dealing with the lack of certainty, and also by introducing the complexity of executing (and not only describing) epistemic actions.

5   Conclusions and Further Work

We have provided a game theoretical approach to learning that takes into account different levels of cooperativeness between Learner and Teacher in a game of perfect information.

Providing a new application of *Sabotage Games*, we have defined *Sabotage Learning Games* with three different winning conditions, each of them representing different levels of cooperativeness between Teacher and Learner. Then, we have shown how *Sabotage Modal Logic* can be used to reason about these games and, in particular, we have identified formulas of the language that characterize each of the three winning conditions, providing also complexity results for each one of them. Our complexity results support the intuitive claim that the cooperation of agents facilitates learning.

From the game-theoretical perspective, Sabotage Learning Games can be extended to more general scenarios by relaxing the strict alternation. As we have mentioned before, there is a difference in the "nature" of moves of the payers in this game. Learner's moves can be seen as internal ones while Teacher's moves can be interpreted externally. Due to this asymmetry, each of Learner's moves does not in principle need to be followed by Teacher's move (e.g. Learner can perform several changes of his information state before Teacher can actually make a restriction). Our results of Section 3 show that if we allow Teacher to skip a move, the winning abilities of the players do not change with respect to the original versions of the games. In the case of helpful Teacher and unwilling Learner, this is quite surprising since it says that if the Teacher can force Learner to learn in the game with non-strict alternation, then even if she is forced to remove edges in each round she can do so without removing edges that are necessary for Learner eventually reaching the goal state. This result crucially depends on the fact that Learner is the first to move; and does not hold in case Teacher starts the game.

From the perspective of Formal Learning Theory, several relevant extensions can be done. We have described the learning process as *changes in information states*, without going further into their epistemic and/or doxastic interpretation. A deeper analysis can give us insights about how the learning process relates belief revision and dynamic epistemic logic.

It can be argued that in some natural learning scenarios, e.g. language learning, the goal of the learning process is concealed from Learner. An extension of the framework of Randomized Sabotage Games [14] could then be used to model the interaction between Learner and Teacher.

In the introduction, we described the concepts of *finite identification* and *identification the limit*. Our work on *SLG*s is closer to the first one, as we understand learning as the ability to reach an appropriate information state, not taking into account what happens afterwards. In particular, we are not concerned with the stability of the resulting belief. *Identification in the limit* extends *finite identification* by looking beyond reachability in order to describe "ongoing behavior". Fixed-point logics, such as the modal μ-calculus [15,16], can provide us with tools to express this notion of learnability. Then epistemic and doxastic interpretations of learning would involve notions of stable belief and a kind of operational, non-introspective knowledge as a result of the process.

References

1. Jain, S., Osherson, D., Royer, J.S., Sharma, A.: Systems that Learn. MIT Press, Chicago (1999)
2. Gold, E.: Language identification in the limit. Information and Control 10, 447–474 (1967)
3. van Ditmarsch, H., van der Hoek, W., Kooi, B.: Dynamic Epistemic Logic. Springer, Netherlands (2007)

4. Looney, A., Cooper, J., Heath, K., Davenport, J., Looney, K.: Playing with Pyramides. Looneylabs (1997)
5. Matuszek, D.: New eleusis (1995), http://matuszek.org/eleusis1.html
6. Angluin, D.: Learning regular sets from queries and counterexamples. Information and Computation 75(2), 87–106 (1987)
7. Löding, C., Rohde, P.: Solving the sabotage game is PSPACE-hard. In: Rovan, B., Vojtáš, P. (eds.) MFCS 2003. LNCS, vol. 2747, pp. 531–540. Springer, Heidelberg (2003)
8. van Benthem, J.: An essay on sabotage and obstruction. In: Mechanizing Mathematical Reasoning, Essays in Honor of Jörg H. Siekmann on the Occasion of His 60th Birthday, pp. 268–276 (2005)
9. Gierasimczuk, N., Kurzen, L., Velázquez-Quesada, F.R.: Games for learning: a sabotage approach (2008),
   http://www.illc.uva.nl/Publications/ResearchReports/
   PP-2009-30.text.pdf
10. Löding, C., Rohde, P.: Solving the sabotage game is PSPACE-hard. Technical report, Aachener Informatik Berichte, Rwth Aachen (2003)
11. Papadimitriou, C.M.: Computational complexity. Addison-Wesley, MA (1994)
12. Dietterich, T.G.: Applying general induction methods to the card game eleusis. In: Proceedings of AAAI 1980, Stanford, CA, pp. 218–220 (1980)
13. Dietterich, T.G., Michalski, R.S.: Learning to predict sequences. In: Michalski, R., Carbonell, J., Mitchell, T. (eds.) Machine Learning: An Artificial Intelligence Approach, vol. II, pp. 63–106. Morgan Kaufmann Publishers, Los Altos (1986)
14. Klein, D., Radmacher, F.G., Thomas, W.: The complexity of reachability in randomized sabotage games. In: Proceedings of FSEN 2009. LNCS. Springer, Heidelberg (2009)
15. Scott, D., de Bakker, W.: A theory of programs. Unpublished manuscript, IBM, Vienna (1969)
16. Kozen, D.: Results on the propositional mu-calculus. Theoretical Computer Science 27, 333–354 (1983)

First-Order Logic Formalisation of Arrow's Theorem

Umberto Grandi and Ulle Endriss

Institute for Logic, Language and Computation
University of Amsterdam

Abstract. Arrow's Theorem is a central result in social choice theory. It states that, under certain natural conditions, it is impossible to aggregate the preferences of a finite set of individuals into a social preference ordering. We formalise this result in the language of first-order logic, thereby reducing Arrow's Theorem to a statement saying that a given set of first-order formulas does not possess a finite model. In the long run, we hope that this formalisation can serve as the basis for a fully automated proof of Arrow's Theorem and similar results in social choice theory. We prove that this is possible in principle, at least for a fixed number of individuals, and we report on initial experiments with automated reasoning tools.

1 Introduction

Social choice theory is a branch of mathematical economics that is concerned with the design and analysis of methods for collective decision making [1]. One of the classical results in the field is Arrow's Theorem [2]; it states that is impossible to aggregate the preferences of a finite set of individuals in a manner that would satisfy a small number of natural properties. In this paper we propose a formalisation of Arrow's Theorem in classical first-order logic (FOL), which may eventually pave the way for an automated proof of this important result.

There have been a number of recent contributions that address the formalisation of theorems in social choice theory (e.g., Pauly [3], Ågotenes et al. [4], Tang and Lin [5], Wiedijk [6], and Nipkow [7]). There are several reasons for this broad interest in applying tools from mathematical logic and automated reasoning to social choice theory. One of them is of course that the full formalisation of a problem domain can help us gain a deeper understanding of that domain. More specifically, in social choice theory, it can clarify the exact nature of the assumptions that are being made to derive, for instance, a characterisation result [3]. Second, a complete formalisation together with an automatically derived (or automatically verifiable) proof can give additional assurances for the correctness of a result. As pointed out by Blau [8], Arrow's original proof contained an error; this has been acknowledged and corrected in the second edition of Arrow's book [2]. While there has been some discussion in the literature whether the standard proofs have been worked out in sufficient detail [7], we certainly do not want to suggest that the major results in social choice theory are not

X. He, J. Horty, and E. Pacuit (Eds.): LORI 2009, LNAI 5834, pp. 133–146, 2009.
© Springer-Verlag Berlin Heidelberg 2009

based on sound foundations. However, for verifying newer and less well studied results, automated reasoning could prove a very useful tool. Finally, the use of automated reasoning in social choice theory has the potential to unveil entirely new results. For instance, we can imagine that it may soon become possible to use automated theorem provers to check whether a known impossibility result persists when we weaken or otherwise alter some of the axioms, or to use model generators to automatically derive counterexamples. To a limited extent, such results have already been achieved in recent work by Tang and Lin [5].

Previous work has discussed formalisations of Arrow's Theorem in modal logic [4], dependence logic,[1] and in the language of set theory [6,7]. Here we explore to what extent it is possible to model the framework underlying Arrow's Theorem in classical FOL. There are two reasons for focusing on FOL: it is a natural language for speaking about linear orders, which are central to the modelling of preferences, and automated theorem proving is more developed for FOL than it is for other systems. We are able to show that it is possible to completely describe the problem within a language of FOL based on the language of linear orders, with one exception: for stating that Arrow's Theorem only applies to the case of a *finite* number of individuals we have to resort to a statement outside the language (we will see that Arrow's Theorem is equivalent to a certain finite theory of FOL axioms not having a finite model). In particular, we will not require any form of second-order quantification, which may seem surprising given that several of the axioms used in Arrow's Theorem certainly have a "second-order flavour". Our axiomatisation draws on several ideas from an important recent paper by Tang and Lin [5], but goes beyond that work in providing a complete axiomatisation of the Arrovian framework of social welfare functions in classical FOL.

The remainder of the paper is organised as follows. In Section 2 we recall Arrow's Theorem and prove a useful lemma. Section 3 presents our axioms and ends with the restatement of Arrow's Theorem in our framework. The models of our axiomatisation are studied in detail in Section 4, with particular attention being paid to the issue of an infinite number of individuals. Related work is discussed in Section 5; and in Section 6 we discuss our preliminary results with an automated theorem prover and conclude. For the rest of the paper we shall assume familiarity with the basic concepts of first-order logic (see e.g. [10]).

2   Social Welfare Functions and Arrow's Theorem

In this section we review Arrow's Theorem and the framework of social welfare functions in which it is stated. We also discuss a recent contribution by Tang and Lin [5], who give a new proof of Arrow's Theorem based on an inductive argument, in which the base case can be checked automatically using automated reasoning tools, and we show how to generalise a lemma proved by these authors so as to also cover the case where there are an infinite number of alternatives that need to be ranked.

---

[1] J. Väänänen (personal communication, 2009); see also [9].

Let I be a set of individuals expressing preferences over a set A of alternatives. We assume that these preferences are represented by linear orders[2] P_i, so that aP_ib holds if individual i strictly prefers a to b. We denote with $\mathcal{L}(A)$ the set of all linear orders on A, and call a *social welfare function* (SWF) for A and I a function $w : \mathcal{L}(A)^I \to \mathcal{L}(A)$. A SWF associates with every *preference profile* $\underline{P} = (P_1, \ldots, P_n) \in \mathcal{L}(A)^I$ a linear order $w(\underline{P})$, that in most interpretations is taken to represent the aggregation of the preferences of the individuals into a "social preference order" over A.

There are several properties that such an aggregation mechanism may satisfy, and some of them have been argued to be natural requirements for a SWF. The fact that in our definition w is defined on *all* preference profiles in $\mathcal{L}(A)^I$ represents what is often stated as a first such property, the *universal domain* condition. The three additional properties that lead to the statement of Arrow's Theorem are the following:

- **UN:** A SWF w satisfies *unanimity* if, whenever every individual prefers alternative a to alternative b, so does society. Formally, if aP_ib for every individual $i \in I$, then $a\,w(\underline{P})\,b$.
- **IIA:** w satisfies *independence of irrelevant alternatives* if the social ranking of two alternatives a and b depends only on their relative ranking by every individual. The formal condition is that, given two preference profiles $\underline{P}$ and $\underline{P}'$, if for every individual $i \in I$ we have that aP_ib if and only if $aP_i'b$, then $a\,w(\underline{P})\,b$ if and only if $a\,w(\underline{P}')\,b$.
- **ND:** w is *non-dictatorial* if there is no individual $i \in I$ such that for every profile $\underline{P}$ the social preference order $w(\underline{P})$ is equal to P_i.

Arrow's Theorem [2] states that:

Theorem 1. *If A and I are finite and non-empty, and if $|A| \geq 3$, then there exists no SWF for A and I that satisfies* **UN**, **IIA** *and* **ND**.

Several proofs of the theorem are known (see e.g. [11]), and most of them give a general argument that works for any number of individuals and any number of alternatives. A new inductive proof has recently been given by Tang and Lin [5]: the authors prove two lemmas to reduce the general statement to a base case with 3 alternatives and 2 individuals, and verify this last step with a computer, using either constraint programming or a satisfiability solver. The first lemma is the inductive step on the number of alternatives: "if there exists a SWF for $m+1$ alternatives and n individuals that satisfies Arrow's conditions, then there exists a SWF for m alternatives and the same number of individuals that still satisfies Arrow's conditions." The contrapositive of this lemma spreads the impossibility from the base case to every finite set of alternatives: "if Arrow's Theorem holds for the case of 3 alternatives and n individuals, then it holds for every finite set of m alternatives and n individuals." We now prove a generalisation of this lemma that also covers the case of an *infinite* number of alternatives:

---

[2] The original statement of Arrow's Theorem assumes weak orders, although many proofs in the literature are restricted to this simpler case. We will discuss how our framework can be extended to the more general case in Section 6.

Lemma 1. *If there exists a SWF w for A and I, with $|A| \geq 3$, that satisfies* **UN**, **IIA** *and* **ND**, *then there exists a set A' with $|A'| = 3$ and a SWF for A' and I that satisfies the same properties.*

Note that the contrapositive of Lemma 1 reads: "if Arrow's Theorem holds for the case of 3 alternatives and n individuals, then it also holds for any larger set A (including the infinite case) and n individuals".

Proof. Let $A' = \{a_1, a_2, a_3\}$ be any set containing three different alternatives in A; every linear order P over A' can be extended to a linear order P^e over the whole set A (though not in a unique way). Define a SWF w' for A' and I in the following way:

$$x \, w'(\underline{P}) \, y \; :\Leftrightarrow \; x \, w(\underline{P}^e) \, y$$

where $\underline{P}$ is a preference profile over A' and $\underline{P}^e$ any extension to a preference profile over A. By **IIA** this definition does not depend on the extension chosen; w' remains unanimous and independent of irrelevant alternatives by definition.

It remains to show that w' is non-dictatorial. Suppose the contrary: we prove that w would then be dictatorial too, in contradiction with the assumptions. Let i be the dictator for w', and x and y two different alternatives in A, and suppose that $x P_i y$ in a certain profile $\underline{P}$. We now show that also $x \, w(\underline{P}) \, y$ must hold, thus i is a dictator on every pair of alternatives in A. The case where both x and y are in A' is trivial. We can therefore restrict ourselves to the case where there are at least two distinct elements in A' different from x and y, a_1 and a_2. Let individual i change her preference relation such that $a_1 P_i a_2$, obtaining profile $\underline{P}'$. Let now every individual (including i) rearrange her preference such that $x P_j a_1$ and $a_2 P_j y$, and call this profile $\underline{P}''$. Both steps can be done without affecting the initial ranking of x and y, thus by **IIA** $x \, w(\underline{P}) \, y$ if and only if $x \, w(\underline{P}'') \, y$. By unanimity of w we have $x \, w(\underline{P}'') \, a_1$ and $a_2 \, w(\underline{P}'') \, y$. Since i is a dictator relative to A', it must be the case that $a_1 \, w(\underline{P}'') \, a_2$ holds, and thus by transitivity also $x \, w(\underline{P}'') \, y$, which as previously observed implies $x \, w(\underline{P}) \, y$. $\square$

3   Axiomatisation

In this section we present a formal system that can model the social choice framework of Arrow's Theorem. Our approach borrows several ideas from Tang and Lin [5], whose main concern, however, is a different one and who do not provide a complete axiomatisation. Arrow's conditions suggest a formalisation in second-order logic, due to the quantification over preference profiles. Following Tang and Lin [5], we instead introduce a set of "situations" and consider them as names for different preference profiles. In our case the set of situations will be (a subset of the domain) marked by a unary predicate, thus allowing us to quantify over this set, which in turn enables us to give a first-order axiomatisation. We will indicate with $\underline{P}^u$ the preference profile associated to situation u. We first define the following first-order signature $\mathcal{L} = \{a_1, a_2, a_3, i_1, s_1, A^{(1)}, I^{(1)}, S^{(1)}, p^{(4)}, w^{(3)}\}$:

- a_1, a_2, a_3 are constants indicating three alternatives, i_1 indicates an individual, and s_1 indicates a situation;

- the three unary predicates mark alternatives (A), individuals (I), and situations (S);
- the predicate p represents, given an individual z and a situation u, the linear order P_z^u associated with situation u; and
- w stands for the social welfare function, representing with a ternary predicate the social preference relation $w(\underline{P}^u)$ for every situation u.

Using this language, we start by introducing the axioms of linear order for p:

LINp:
- $I(z) \wedge S(u) \wedge A(x) \wedge A(y) \rightarrow (p(z,x,y,u) \vee p(z,y,x,u) \vee x = y)$
- $I(z) \wedge S(u) \wedge A(x) \rightarrow \neg p(z,x,x,u)$
- $I(z) \wedge S(u) \wedge A(x_1) \wedge A(x_2) \wedge A(x_2) \wedge$
  $\quad p(z,x_1,x_2,u) \wedge p(z,x_2,x_3,u) \rightarrow p(z,x_1,x_3,u)$

All axioms presented in this paper are to be considered universally closed; therefore the first axiom should be read as: "for all z, u, x and y, if z is an individual, if u is a situation and if x and y are alternatives, then either individual z in situation u prefers x to y, or she prefers y to x, or x is equal to y." This is the completeness (or connectedness) axiom, and the second and the third are the irreflexivity and transitivity axioms.

The analogous axioms for $w(\cdot, \cdot, u)$ follow:

LINw:
- $S(u) \wedge A(x) \wedge A(y) \rightarrow (w(x,y,u) \vee w(y,x,u) \vee x = y)$
- $S(u) \wedge A(x) \rightarrow \neg w(x,x,u)$
- $S(u) \wedge A(x) \wedge A(y) \wedge A(t) \wedge w(x,y,u) \wedge w(y,t,u) \rightarrow w(x,t,u)$

The next two sets of axioms guarantee that there are at least 3 different alternatives, that i_1 is an individual, s_1 is a situation, and that A, I and S form a partition of the universe of a model:

MIN:
- $A(a_1) \wedge A(a_2) \wedge A(a_3) \wedge I(i_1) \wedge S(s_1)$
- $\neg(a_1 = a_2) \wedge \neg(a_1 = a_3) \wedge \neg(a_2 = a_3)$

PART:
- $A(x) \rightarrow (\neg I(x) \wedge \neg S(x))$
- $I(x) \rightarrow (\neg A(x) \wedge \neg S(x))$
- $S(x) \rightarrow (\neg I(x) \wedge \neg A(x))$
- $A(x) \vee I(x) \vee S(x)$

The next two axioms restrict the arguments of p and w to be of the correct type:

DEF:
- $p(z,x,y,u) \rightarrow (I(z) \wedge A(x) \wedge A(y) \wedge S(u))$
- $w(x,y,u) \rightarrow (A(x) \wedge A(y) \wedge S(u))$

The next axiom guarantees that two distinct situations cannot encode the same preference profile, thus the encoding of situations into preference profiles must be injective:

INJ:
- $S(u) \wedge S(v) \wedge u \neq v \rightarrow$
  $\quad \exists z.\exists x.\exists y.[I(z) \wedge A(x) \wedge A(y) \wedge p(z,x,y,u) \wedge p(z,y,x,v)]$

To express the condition of universal domain in our language, and to be able to quantify over the entire set of situations, we use another idea from the same paper by Tang and Lin [5]: identify the set $\mathcal{L}(A)$ with the symmetric group $S(A)$ of all permutations over A and generate it via transpositions. This is the job of the next axiom:[3]

PERM: • $p(z,x,y,u) \to \exists v. \{S(v) \wedge p(z,y,x,v) \wedge$
$\forall x_1.[p(z,x,x_1,u) \wedge p(z,x_1,y,u) \to p(z,x_1,x,v) \wedge p(z,y,x_1,v)] \wedge$
$\forall x_1.[(p(z,x_1,x,u) \to p(z,x_1,y,v)) \wedge (p(z,y,x_1,u) \to p(z,x,x_1,v))] \wedge$
$\forall x_1.\forall y_1.[x_1 \neq x \wedge x_1 \neq y \wedge y_1 \neq y \wedge y_1 \neq x \to (p(z,x_1,y_1,u) \leftrightarrow p(z,x_1,y_1,v))] \wedge$
$\forall z_1.\forall x_1.\forall y_1. [z_1 \neq z \to (p(z_1,x_1,y_1,u) \leftrightarrow p(z_1,x_1,y_1,v))]\}$

The complexity of this axiom is largely due to the fact that linear orders are being represented as binary relations. Given our representation of P_i not as a complete sequence of elements in A but as a subset of A^2, we have to require that, given a situation u, an individual z, and two alternatives x and y, there exists another situation v such that (the following five items correspond to the five lines of the axiom):

 - the relative positions of x and y have been switched in P_z^v;
 - if an alternative x_1 was between x and y in P_z^u, then its relation with respect to x and y is switched in P_z^v;
 - if x_1 was more preferred than x in P_z^u, then in v it is more preferred than y (and thereby also x); if it was less preferred than y in P_z^u, then in v it is less preferred than x (and thereby also y).
 - for every pair of alternatives different from x and y the relative ranking is copied;
 - $P_{z'}^v = P_{z'}^u$ for every individual $z' \neq z$.

Call T_{SWF} the theory composed of all the axioms above, as it summarises the properties of social welfare functions. Adding the next three axioms we obtain a theory that we shall call T_{ARROW}:

UN: • $S(u) \wedge A(x) \wedge A(y) \to [(\forall z.(I(z) \to p(z,x,y,u))) \to w(x,y,u)]$
IIA: • $S(u_1) \wedge S(u_2) \wedge A(x) \wedge A(y) \to$
$[\forall z.(I(z) \to (p(z,x,y,u_1) \leftrightarrow p(z,x,y,u_2))) \to (w(x,y,u_1) \leftrightarrow w(x,y,u_2))]$
ND: • $I(z) \to \exists x.\exists y.\exists u.[S(u) \wedge A(x) \wedge A(y) \wedge p(z,x,y,u) \wedge w(y,x,u)]$

Arrow's Theorem can now be restated as:[4]

Theorem 2. T_{ARROW} *has no finite models.*

It is worth noting that some of our axioms, such as **PART** or **INJ**, are not strictly required. Including these axioms permits to have more "control" in the resulting models and improves the readability of the axiomatisation.

---

[3] Observe that in this axiom the variables x_1, y_1, and z_1 must be explicitly quantified, because they are within the scope of an existential quantifier; the other variables x, y, z, and u are instead implicitly bound by the universal closure of the axiom.

[4] This equivalence is a straightforward consequence of Proposition 1 that will be stated in the following section. Once we have proved that every model of T_{SWF} is associated with a SWF, it will be sufficient to check that our last three axioms correspond to Arrow's conditions to prove that Theorem 2 is equivalent to Arrow's Theorem.

4 Dealing with the Infinite

In Section 3 we have referred to T_{SWF} as the theory of social welfare functions, and in this section we justify this choice of words by proving that T_{SWF} axiomatises this class.[5] We will do so by associating with every SWF w a model $\mathcal{M}_w$ of T_{SWF}, and proving a completeness result. This enables us to determine precisely to what extent Arrow's Theorem can be proved automatically. Special attention will be devoted to the issue of an infinite domain, where Arrow's Theorem does not hold. We will present two different approaches to overcome this difficulty, first by fixing the number of individuals directly in the language, and then a second one based on results by Kirman and Sondermann [12]. From now on we shall assume that the set of alternatives is non-empty and contains at least 3 elements, and that the set of individuals is non-empty.

A model of T_{SWF} is a structure $\mathcal{M} = (M, a_1, a_2, a_3, i_1, s_1, A, I, S, p, w)$, specifying the interpretation of every symbol in the language presented in Section 3.

Definition 1. *If w is a SWF for A and I, then $\mathcal{M}_w$ is the following $\mathcal{L}$-model:*
  (i) *the universe $M = A \sqcup I \sqcup \mathcal{L}(A)^I$; the disjoint union of the sets corresponds to the three unary predicates A, I and S (in particular the set S is equal to the set of all preference profiles $\mathcal{L}(A)^I$);*
  (ii) *a_1, a_2, a_3 are three different alternatives, i_1 is an individual, and s_1 is a preference profile;*
  (iii) *$(z, x, y, u) \in p \Leftrightarrow x\, P_z^u\, y$, where P_z^u is the preference relation of z in profile u; and*
  (iv) *$(x, y, u) \in w \Leftrightarrow x\, w(\underline{P}^u)\, y$.*

If A is finite, then the resulting model $\mathcal{M}_w$ is in some sense unique, depending only on the choice of the constants. In the case where A is infinite, on the other hand, this is not the only model that can be built from w. To obtain a full characterisation we need the following definition:

Definition 2. *Given a set A, let $S(A)$ denote the set of all permutations over A. A transposition is a permutation that switches just two elements of the set. $G \subseteq S(A)$ is closed under transpositions if whenever $g \in G$, $g \circ \tau \in G$ for every transposition τ.*

Observe that if A is finite, then the only subset of $S(A)$ closed under transpositions is $S(A)$ itself.

Let now w be a SWF on an infinite set of alternatives A. We have already remarked that we can identify the set $\mathcal{L}(A)$ with the set $S(A)$ of all permutations over A. With every choice of $G_i \subset S(A)$ closed under transpositions for every individual $i \in I$ we can associate a model of T_{SWF}, using the same construction as in Definition 1, except that the set of situations is now the Cartesian product

---

[5] Using the terminology introduced by Pauly [3], we will prove that T_{SWF} *absolutely axiomatises* the set of partial SWFs satisfying a condition of closure on the domain. This translates in the finite case into an absolute axiomatisation of all SWFs.

$S = \prod_{i \in I} G_i$. In the finite case this definition boils down to Definition 1, because $\mathcal{L}(A)$ is the only possible choice for every individual. The following completeness result shows that these are all possible models of T_{SWF}:

Proposition 1. $\mathcal{M} \models T_{\text{SWF}}$ *if and only if there exist two non-empty sets A and I, with $|A| \geq 3$, and a SWF w for A and I such that $\mathcal{M} = \mathcal{M}_w$.*

Proof. It is easy to prove that $\mathcal{M}_w$ is a model of T_{SWF}. By definition, for every z and u the relations $p(z, \cdot, \cdot, u)$ and $w(\cdot, \cdot, u)$ are linear orders over A, so the **LINp** axioms are satisfied as well as **LINw**. The axioms **MIN**, **PART** and **INJ** are valid thanks to (i) and (ii) in Definition 1. The set of situations S is either the set of all preference profiles or a Cartesian product $\prod_{i \in I} G_i$ of subsets of $\mathcal{L}(A)$ closed under transpositions. This is sufficient to validate axiom **PERM**: given a situation u in S, for every individual and for every pair of alternatives the linear order obtained by switching these two alternatives is the composition of an element in G_i with a transposition. Therefore the new profile is still an element of S, i.e., there exists a situation v that represents this profile.

Suppose now that $\mathcal{M} \models T_{\text{SWF}}$. We can define the two sets I and A as the subsets of the universe indicated by the unary predicates. To every element in S we can associate a preference profile, the one encoded in the relation $p^{\mathcal{M}}$. From the relation $w^{\mathcal{M}}$ we can define a *partial* SWF, whose domain is the set of all preference profiles encoded in S, a subset $G \subseteq \mathcal{L}(A)^I$. By **PERM**, if we take the projection of G on every component i, denoted with G_i, we obtain a set of linear orders that is closed under transpositions: for every individual i, if $g \in G_i$ then g composed with every transposition (a swap of a pair of alternatives) is still in G_i. Thus G is of the form $\prod_{i \in I} G_i$, and $\mathcal{M} = \mathcal{M}_w$ as defined in Definition 1. $\square$

In view of our ultimate goal of using automated reasoning in social choice theory, a result like Theorem 2 is of little practical use, despite its theoretical interest. What should be sought is a formalisation of Arrow's theorem in a sentence that can be derived formally from our theory. The first attempt of proving the inconsistency of T_{ARROW} fails, because Arrow's Theorem does *not* hold in the case of an infinite number of individuals, as has first been pointed out by Fishburn [13]. (The issue of an infinite number of *alternatives*, on the contrary, is fully resolved by Lemma 1.) Fishburn's result translates in our framework into the existence of an infinite model $\mathcal{M}$ of T_{SWF} such that $\mathcal{M} \models (\textbf{UN} \wedge \textbf{IIA} \wedge \textbf{ND})$. Since there is no first-order formula that characterises finite models (see e.g. [10]), we have to somehow circumvent this problem.

One possibility is to give up some generality and to fix the number of individuals in the language. Let therefore the new language $\mathcal{L}_n$ be $\mathcal{L} \cup \{i_2, \ldots, i_n\}$ with $n-1$ new constants, and call T_{SWF}^n the theory composed of all axioms of T_{SWF} plus the following axioms:

- $i_k \neq i_j$ for every $k \neq j$
- $I(i_2) \wedge \cdots \wedge I(i_n)$
- $I(z) \rightarrow (z = i_1) \vee \cdots \vee (z = i_n)$

With a proof analogous to that of Proposition 1 we obtain a completeness result for T_{SWF}^n with respect to SWFs defined for a set I of n individuals. Now the following automated-reasoning friendly proposition holds:

Proposition 2. *If w is a SWF for A and I with $|A| \geq 3$ and $|I| = n$, and if $\mathcal{M}_w$ is the corresponding model, then $\mathcal{M}_w \models \neg(\boldsymbol{UN} \wedge \boldsymbol{IIA} \wedge \boldsymbol{ND})$. Therefore, for every n there exists a proof of $\neg(\boldsymbol{UN} \wedge \boldsymbol{IIA} \wedge \boldsymbol{ND})$ in T_{SWF}^n.*

Proof. The proof follows closely that of Lemma 1. In that proof, we never used the condition of universal domain in its full generality: every time we defined a new profile, it was always constructible with a finite sequence of switches between pairs of alternatives. The condition of closure under transpositions therefore guarantees that the result extends to every $\mathcal{M}_w$ defined on a finite set I.   □

The second approach we present is an indirect one: derive a consequence of T_{ARROW} that forces the resulting models to be infinite. Following the presentation of Arrow's Theorem in the case of an infinite number of individuals given by Kirman and Sondermann [12], this statement is the following: if a SWF satisfies **UN** and **IIA**, then the collection of "winning coalitions", those subsets $J \subseteq I$ such that if $x P_j y$ for every $j \in J$ then $x\, w(\underline{P})\, y$, is an ultrafilter over I. A full axiomatisation of this statement can be given in the same language of T_{SWF} and is sketched in Appendix A. The condition of non-dictatorship corresponds to requiring the ultrafilter to be free: an unsatisfiable requirement if the set of individuals is finite. This finally formalises the argument of Fishburn [13] we presented in this section: if a SWF satisfies **UN**, **IIA** and **ND**, then the number of individuals must be infinite.

In conclusion, we have proved that an automated proof of Arrow's Theorem is possible, despite not in its most general form: for every finite number of individuals there is a (possibly different) first-order proof of the theorem.[6] The general case can be proved indirectly by deducing a set of statements about the sets of winning coalitions that force the set of individuals to be infinite. We report on our preliminary results with automated theorem prover in the last section.

5   Related Work

While we are not aware of any other work exploring the limits of classical first-order logic in expressing the Arrovian framework of social welfare functions, there have been several contributions to the literature making proposals for a full formalisation of Arrow's Theorem, using a variety of logical frameworks. In this section, we briefly review some of them.

As mentioned before, Tang and Lin [5] have shown that Arrow's Theorem in its general form (for finite A and I) follows from Arrow's Theorem for 3 alternatives and 2 individuals. For this base case, these authors give a formalisation in *propositional logic*. This is possible, because the number of possible situations

---

[6] And since the set of theorems of a first-order theory is recursively enumerable it will eventually be found by an automated theorem prover.

(preference profiles) is finite (namely $3! \times 3! = 36$) for this scenario. While the number of SWFs is already prohibitively large in this case (namely $6^{36} \approx 10^{28}$), a complete instantiation of Arrow's conditions for 36 situations is still feasible, and Tang and Lin [5] report that unsatisfiability can be verified using a state-of-the-art SAT-solver in less than 1 second. While our implementation of the same base case in FOL cannot compete with this performance, it arguably has the advantage of being more easily extended. The propositional language presented in [5] has the advantage of being rapidly solved, but can only be used to verify a base case. Building on this language, we aim instead at providing a fully automated proof of Arrow's Theorem without relying on any inductive lemma. (Note that the role of Lemma 1 is that of a theoretical guarantee for the existence of such a proof, at least for a fixed number of individuals, and it would not be part of any eventual automated derivation.) Also, our axiomatisation in Prover9 syntax is human-readable and easily fits on a single page (see Section 6 and Appendix B), while Tang and Lin's input to the SAT-solver is very large and has to be computer-generated (it consists of 106354 clauses).

Kaneko and Suzuki [14] discuss bounds on the size of a potential proof of Arrow's Theorem in a Gentzen-style *sequent calculus*, for the special case of 2 individuals and 3 alternatives.

Ågotnes et al [4] develop a *modal logic* for expressing concepts from social choice theory, including Arrow's Theorem. This logic is specifically designed for this purpose, and to date no automated procedure has been developed. The potential of the approach is limited by the fact that the number of individuals as well as the number of alternatives is fixed in the language.

Yet another approach is the one adopted by Nipkow [7] and Wiedijk [6]. These authors verify formally two proofs of Arrow's Theorem given by Geanakoplos [11] using *proof checkers* (Isabelle and Mizar, respectively). Their language is the language of set theory and their objects are sets; the condition of finiteness of the set of individuals is expressible in this language and this makes it possible to formalise and check the full statement of Arrow's Theorem. However, this approach requires a substantial amount of work in the process of rewriting an existing proof and then allows us to check every single simple step automatically.

The FOL framework developed by Rubinstein [15], while working with FOL, is different from ours. It aims at proving the existence of *single-profile analogues* of various results in social choice theory using social welfare functions defined on models of a suitable first-order theory. The single-profile approach avoids quantification over preference profiles from the outset. The exact relationship between these two frameworks certainly deserves future investigation.

6   Conclusions and Future Work

In this work we have given a first-order axiomatisation of social welfare functions, formalising the framework in which Arrow's Theorem is stated. We have been able to reduce non-trivial conditions to first-order statements, such as the universal domain condition and **IIA**. The issue of an infinite number of alternatives has been solved by proving a lemma that reduces the impossibility to the case of 3 alternatives. We have proved that, if the number of individuals is fixed in our language, then there is a formal derivation of Arrow's Theorem from our axioms, and we have suggested an indirect approach to formalise the general case with a possibly infinite number of individuals.

All these results support the belief that automated reasoning can play a role in proving theorems of social choice theory, and we carried out some preliminary experiments using an automated theorem prover. The system we used is Prover9, the successor of the well-known and widely used Otter theorem prover [16]. The task of writing an input file containing our axiomatisation does not pose any challenge, thanks to the simplicity of the syntax and the high readability of our axioms (see Appendix B). However, to date we have not been able to generate an automated proof of Arrow's Theorem. We designed a step-by-step proof of the simplest case of Arrow's Theorem for 2 individuals and 3 alternatives, following the formalisation of a simple proof of Arrow's Theorem by Nipkow [7]. At each step we received a negative response, with the prover exceeding the search space limits or not providing an answer in a reasonable amount of time.

A critical point, that may go some way towards explaining the difficulty of automatically deriving a proof, is that all of the intermediate lemmas we formalised rely on some steps where the existence of a particular preference profile has to be shown, using the condition of universal domain. This seems to require a clever use of the axiom of permutation, guessing the correct sequence of swaps to get from a profile to another, and it is likely to be the cause of the failure of Prover9 on these tasks. It is very likely that a suitable reformulation of the axioms, in a way that can help and guide the work of the theorem prover, would prove successful in increasing its speed and efficiency.

Despite these difficulties we were able to obtain some simple results, mainly by restricting the domain to the case of 2 individuals and 3 alternatives. For instance, we were able to generate an automated proof for the fact that the unanimity condition entails a weaker condition known as the *non-imposition* property [1]. A SWF satisfies non-imposition if for every pair of distinct alternatives x and y there is a profile $\underline{P}$ such that $x\,w(\underline{P})\,y$. In the syntax of Prover9 this condition can be written as follows:

```
A(x) & A(y) & x!=y -> (exists u (S(u) & w(x,y,u))).
```

We added two axioms to those in Appendix B in order to fix the number of alternatives and individuals, and we instantiated the axiom of permutation to this restricted domain. This still produces a readable axiomatisation, and Prover9 succeeds in providing a proof after about 3 hours on a standard desktop machine (with a memory limit of 500 Mb). The proof consists of 193 steps, and the number of clauses generated is 20623974, of which 257685 have been kept to arrive at the proof. We have also run the same problem using another automated theorem prover, the E prover [17], and we obtained a positive response in a few seconds (on the other hand, for more difficult problems, E tends to run out of memory faster than Prover9).

This work can be extended in a number of ways. First, it is likely that a reformulation of the axioms and a guided use of the theorem prover will significantly improve performance and lead to the creation of a usable tool for social choice theorem proving. Second, it would be interesting to extend the axiomatisation to allow for preferences that are weak orders, allowing both the individual and the social order to express ties between alternatives. This can be achieved by replacing the irreflexivity axiom for both p and w with a reflexivity axiom, and adjusting the axiom of permutation to entail the condition of universal domain: starting from a linear order over alternatives, added "by default" in a model, it is possible to generate all weak orders requiring that in every situation every two alternatives can not only be swapped, but also ranked the same in the preference relation of every individual. Third, a large number of other results in social choice theory are likely to also be expressible in first order-logic. Examples include Sen's theorem on the impossibility of a Paretian Liberal and the Gibbard-Satterthwaite Theorem on the impossibility of strategy-proof voting rules that are non-dictatorial. In this direction, as already remarked, lies the main potential of this method: the use of automated reasoning as a tool for an easier exploration of new results in social choice theory.

Acknowledgements. We would like to thank Daniele Porello, Joel Uckelman, Stéphane Airiau, and two anonymous reviewers for their useful comments. A particular thanks to Joel for helping us getting started with Prover9.

References

1. Arrow, K.J., Sen, A.K., Suzumura, K. (eds.): Handbook of Social Choice and Welfare. North-Holland, Amsterdam (2002)
2. Arrow, K.J.: Social Choice and Individual Values, 2nd edn. John Wiley & Sons, Chichester (1963)
3. Pauly, M.: On the role of language in social choice theory. Synthese 163(2), 227–243 (2008)
4. Ågotnes, T., van der Hoek, W., Wooldridge, M.: Reasoning about judgment and preference aggregation. In: Proc. 6th International Joint Conference on Autonomous Agents and Multiagent Systems (AAMAS 2007), IFAAMAS (2007)
5. Tang, P., Lin, F.: Computer-aided proofs of Arrow's and other impossibility theorems. Artificial Intelligence 173(11), 1041–1053 (2009)
6. Wiedijk, F.: Arrow's impossibility theorem. Formalized Mathematics 15(4), 171–174 (2007)
7. Nipkow, T.: Social choice theory in HOL: Arrow and Gibbard-Satterthwaite. Technical report, Technische Universität München (2008)
8. Blau, J.H.: The existence of social welfare functions. Econometrica 25(2), 302–313 (1957)
9. Väänänen, J.: Dependence Logic. Cambridge University Press, Cambridge (2007)
10. Shoenfield, J.R.: Mathematical Logic. Addison-Wesley, Reading (1967)
11. Geanakoplos, J.: Three brief proofs of Arrow's impossibility theorem. Economic Theory 26(1), 211–215 (2005)
12. Kirman, A., Sondermann, D.: Arrow's theorem, many agents, and invisible dictators. Journal of Economic Theory 5(2), 267–277 (1972)
13. Fishburn, P.C.: Arrow's impossibility theorem: Concise proof and infinite voters. Journal of Economic Theory 2(1), 103–106 (1970)

14. Kaneko, M., Suzuki, N.Y.: A proof-theoretic evaluation of Arrow's theorem. Working Paper presented at the 9th International Meeting of the Society for Social Choice and Welfare, Montréal (2008)
15. Rubinstein, A.: The single profile analogues to multi profile theorems: Mathematical logic's approach. International Economic Review 25(3), 719–730 (1984)
16. McCune, W.: OTTER 3.3 Reference Manual. Technical Memo ANL/MCS-TM-263, Argonne National Laboratory, Argonne, IL (2003)
17. Schulz, S.: System Description: E 0.81. In: Basin, D., Rusinowitch, M. (eds.) IJCAR 2004. LNCS (LNAI), vol. 3097, pp. 223–228. Springer, Heidelberg (2004)

Appendix A: Axioms for Kirman-Sondermann Theorem

The set $\mathcal{J}$ of "winning coalitions" is an ultrafilter:[7]

- $I \in \mathcal{J}$ (**UN**):
  $\exists u.\exists x.\exists y.(\forall z.(I(z) \to p(z,x,y,u)) \to w(x,y,u))$
- $J \in \mathcal{J}$ and $J \subseteq K$ then $K \in \mathcal{J}$:
  $w(x,y,u) \to [\forall z.((I(z) \wedge p(z,x,y,u)) \to p(z,x,y,v)) \to w(x,y,v)]$
- $J_1, J_2 \in \mathcal{J}$ then $J_1 \cup J_2 \in \mathcal{J}$:
  $w(x,y,u_1) \wedge w(x,y,u_2) \to$
  $[\forall z.(I(z) \wedge p(z,x,y,u_1) \wedge p(z,x,y,u_2) \leftrightarrow p(z,x,y,v)) \to w(x,y,v)]$
- $J \subset I$ then $J \in \mathcal{J}$ or $J^c \in \mathcal{J}$:
  $\forall z.(I(z) \to (p(z,x,y,u) \leftrightarrow \neg p(z,x,y,v))) \to (w(x,y,u) \vee w(x,y,v))$
- Free ultrafilter (**ND**):
  $\neg\exists z.(I(z) \wedge \forall x.\forall y.\forall u.(w(x,y,u) \leftrightarrow p(z,x,y,u)))$

Call **FUF** the conjunction of these axioms. With an analogous proof to that of Proposition 2 we obtain that $T_{\text{ARROW}} \vdash$ **FUF**. This gives a formal proof that the set of winning coalitions under Arrow's conditions must be a free ultrafilter (i.e., the Kirman-Sonderman Theorem). Since it is not possible to build a free ultrafilter over a finite set, this formal proof is an indirect formalisation of Fishburn's generalisation of Arrow's Theorem.

Appendix B: T_{ARROW} in Prover9 Syntax

```
% LINp
(I(z) & S(u) & A(x) & A(y)) -> (p(z,x,y,u) | p(z,y,x,u) | x=y).
(I(z) & S(u) & A(x)) -> -p(z,x,x,u).
(I(z) & S(u) & A(x) & A(y) & A(v) & p(z,x,y,u) & p(z,y,v,u)) -> p(z,x,v,u).
```

---

[7] The axioms that follow formalise the notion of ultrafilter in this particular case only. Their formulation use a definition of "winning coalitions" that strongly relies on Lemma A by Kirman and Sondermann [12].

```
% LINw
(S(u) & A(x) & A(y)) -> (w(x,y,u) | w(y,x,u) | x=y).
(S(u) & A(x) & A(y)) -> -w(x,x,u).
(S(u) & A(x) & A(y) & A(v) & w(x,y,u) & w(y,v,u)) -> w(x,v,u).

% MIN
A(a1) & A(a2) & A(a3) & I(b1) & S(c1) & a1!=a2 & a2!=a3 & a1!=a3.

% PART
A(x) -> (-I(x) & -S(x)).
I(x) -> (-A(x) & -S(x)).
S(x) -> (-I(x) & -A(x)).
A(x) | I(x) | S(x).

% DEF
p(z,x,y,u) -> (I(z) & A(x) & A(y) & S(u)).
w(x,y,u) -> (A(x) & A(y) & S(u)).

% INJ
(S(u) & S(v) & u!=v) ->
exists z exists x exists y (I(z) & A(x) & A(y) & p(z,x,y,u) & p(z,y,x,v)).

% PERM
p(z,x,y,u) -> exists v (S(v) & p(z,y,x,v) &
(all x1 (p(z,x,x1,u) & p(z,x1,y,u) -> p(z,x1,x,v) & p(z,y,x1,v))) &
(all x2 (p(z,x2,x,u) -> p(z,x2,y,v))) &
(all x3 (p(z,y,x3,u) -> p(z,x,x3,v))) &
(all x4 all y1 (x4!=x & x4!=y & y1!=y & y1!=x ->
   (p(z,x4,y1,u) <-> p(z,x4,y1,v)))) &
(all z1 all x5 all y2 (z1!=z ->
   (p(z1,x5,y2,u) <-> p(z1,x5,y2,v))))).

% UN
(S(u) & A(x) & A(y)) -> ((all z (I(z) -> p(z,x,y,u))) -> w(x,y,u)).

% IIA
(S(u1) & S(u2) & A(x) & A(y)) ->
((all z (I(z) -> (p(z,x,y,u1)<->p(z,x,y,u2)))) -> (w(x,y,u1)<->w(x,y,u2))).

% ND
I(z) ->
exists x exists y exists u (A(x) & A(y) & S(u) & p(z,x,y,u) & w(y,x,u)).
```

Twelve Angry Men:
A Study on the Fine-Grain of Announcements

Davide Grossi and Fernando R. Velázquez-Quesada

Institute for Logic, Language and Computation, Universiteit van Amsterdam.
P.O. Box 94242, 1090 GE Amsterdam, The Netherlands
{D.Grossi, F.R.VelazquezQuesada}@uva.nl

Abstract. By moving from a suggestive example, the paper analyzes how information flows among agents involved in a deliberation. By deliberating, agents become aware of details, draw the attention of the group to some issues, perform inferences and announce what they know. The proposed framework—which builds on the paradigm of dynamic logic—captures how, during a deliberation, information results from step-wise multi-agent interaction.

Keywords: epistemic logic, dynamic epistemic logic, awareness, inference, interaction, deliberation.

1 Introduction

A jury faces the following task:

> "You've listened to the testimony [. . .] It's now your duty to sit down and try and separate the facts from the fancy. One man is dead. Another man's life is at stake. If there's a reasonable doubt [. . .] as to the guilt of the accused [. . .], then you must bring me a verdict of not guilty. If there's no reasonable doubt, then you must [. . .] find the accused guilty. However you decide, your verdict must be *unanimous*."

This is the setting of Sydney Lumet's 1957 movie "12 Angry Men", and represents a paradigmatic example of a collective decision-making scenario. These kind of scenarios, where a group of agents have to establish whether a given state-of-affairs (e.g., guiltiness beyond any reasonable doubt) holds or not [1], have been object of extensive study in the literature on judgment aggregation [2]. However, while judgment aggregation focuses on the social-theoretic aspects of such decision-making processes (viz. properties of voting rules, possibility of reaching collective judgments with 'desirable' properties, etc.), this paper looks at the deliberation phase that typically precedes the very act of voting, and in particular at its knowledge-related aspects.[1]

To best illustrate the problem, we consider the following example from the mentioned movie. It will be our running example throughout the whole paper.

---

[1] Recent literature on judgment aggregation has recognized the formal analysis of the deliberation that precedes voting as a key open research question in the field [3].

X. He, J. Horty, and E. Pacuit (Eds.): LORI 2009, LNAI 5834, pp. 147–160, 2009.
© Springer-Verlag Berlin Heidelberg 2009

Example 1 (12 Angry Men). The jury members are engaged in the deliberation that will lead to their unanimous verdict. An excerpt of the discussion follows.

A: *Now, why were you rubbing your nose like that?*

H: *If it's any of your business, I was rubbing it because it bothers me a little.*

A: *Your eyeglasses made those two deep impressions on the sides of your nose.*

A: *I hadn't noticed that before.*

A: *The woman who testified that she saw the killing had those same marks on the sides of her nose.*

. . .

G: *Hey, listen. Listen, he's right. I saw them too. I was the closest one to her. She had these things on the side of her nose.*

. . .

D: *What point are you makin'?*

D: *She had dyed hair, marks on her nose. What does that mean?*

A: *Could those marks be made by anything other than eyeglasses?*

. . .

D: *How do you know what she saw? How does he know all that? How do you know what kind of glasses she wore? Maybe they were sunglasses! Maybe she was far-sighted! What do you know about it?*

C: *I only know the woman's eyesight is in question now.*

. . .

C: *Don't you think the woman may have made a mistake?*

B: *Not guilty.*

Agent *A*, convinced that the defendant cannot be proven guilty beyond any doubt, tries to substantiate its claim in the face of *B*'s opposing claim. When *H* rubs her nose, he becomes aware of an issue which has not been considered: marks on the nose. Moreover, when considering the issue, *A* remembers that the woman actually had such marks, and he truthfully announces it. Now he can draw the conclusion that she should wear glasses, announcing it to everyone. After the announcement, it is *D* who infers that the woman's eyesight is in question now. Finally, *B* draws the necessary conclusion and announces it: the defendant is not guilty beyond any reasonable doubt.

In Dynamic Epistemic Logic (DEL) [4], the example could be distilled into a public announcement to the effect that the witness is known to be unreliable. Such representation, however, abstracts from a number of interesting subtleties involved in the example, where the final knowledge —the non-guiltiness of the defendant—is reached through a complex multi-agent procedure. The present paper assumes this more fine-grained perspective on how knowledge flows and is shared in decision-making scenarios such as the one sketched, proposing a dynamic logic framework to cope with such situations.

2   The Static Framework

We start with an informal introduction to the epistemic notions the paper will use to analyze our running example. We talk about "information" in order not to commit, at this stage, to loaded notions such as "knowledge" or "belief".

We use the term *explicit information* for the agent's information that is directly available to her without performing an inference or any other process. *Implicit information*, on the other hand, contains *every piece of information* the agent can obtain after performing the adequate *inferences*, provided that she has enough resources to do it. Finally, *awareness of* refers to the topics the agent can talk about. In fact, it defines the agent's particular language.

Intuitively, *implicit knowledge* extends *explicit knowledge* and, as a superset of them, we have *awareness of*, since the agent cannot have implicit knowledge about something without being able to talk about it.

2.1   The Formal Definitions

Definition 1 (Language $\mathcal{L}$). *Let* P *be a set of atomic propositions and let* A *be a set of agents. Formulas* φ *and rules* ρ *of the language* $\mathcal{L}$ *are given by*

$$\varphi ::= p \mid {}^{[i]}p \mid \mathrm{Ac}_i\,\varphi \mid \mathrm{R}_i\,\rho \mid \neg\varphi \mid \varphi \vee \psi \mid \Box_i\varphi$$
$$\rho ::= (\{\varphi_1, \ldots, \varphi_{n_\rho}\}, \psi)$$

with $p \in \mathsf{P}$ *and* $i \in \mathsf{A}$. *We denote by* $\mathcal{L}_f$ *the formulas of* $\mathcal{L}$ *and by* $\mathcal{L}_r$ *its rules.*

Formulas of the form $\mathrm{Ac}_i\,\varphi$ *(access formulas) indicate that* agent i can access formula φ, *and formulas of the form* $\mathrm{R}_i\,\rho$ *(rule-access formulas) indicate that* agent i can access rule ρ. *Formulas of the form* ${}^{[i]}p$ *indicate that* agent i has proposition p at her/his disposal *for expressing her information. Other boolean connectives* $(\wedge, \rightarrow, \leftrightarrow)$ *as well as the diamond modalities* $\Diamond_i$ *are defined as usual* $(\Diamond_i\varphi := \neg\Box_i\neg\varphi$, *for the last case).*

The following definitions will be useful.

Definition 2 (Premises, conclusion and translation). *Let* ρ *be a rule in* $\mathcal{L}_r$ *of the form* $(\{\varphi_1, \ldots, \varphi_{n_\rho}\}, \psi)$. *We define its* premises *and its* conclusion *as* $\mathrm{prem}(\rho) := \{\varphi_1, \ldots, \varphi_{n_\rho}\}$ *and* $\mathrm{conc}(\rho) := \psi$, *respectively. Moreover, we define its* translation $\mathrm{TR}(\rho) \in \mathcal{L}_f$ *as an implication whose antecedent is the (finite) conjunction of the rule's premises and whose consequent is the rule's conclusion, that is,* $\mathrm{TR}(\rho) := \bigwedge \mathrm{prem}(\rho) \rightarrow \mathrm{conc}(\rho)$.

Formulas of the form ${}^{[i]}p$ allow us to express the availability of atomic propositions. The extension to express availability of formulas of the whole language is defined as follows.

Definition 3. *Let* i, j *be agents in* A.

$$
\begin{aligned}
{}^{[j]}(\mathrm{Ac}_i\,\varphi) &:= {}^{[j]}\varphi & {}^{[j]}(\varphi \vee \psi) &:= {}^{[j]}\varphi \wedge {}^{[j]}\psi \\
{}^{[j]}(\mathrm{R}_i\,\rho) &:= {}^{[j]}\mathrm{TR}(\rho) & {}^{[j]}(\Box_i\varphi) &:= {}^{[j]}\varphi \\
{}^{[j]}(\neg\varphi) &:= {}^{[j]}\varphi & {}^{[j]}({}^{[i]}\varphi) &:= {}^{[j]}\varphi
\end{aligned}
$$

In words, φ *is available to agent* j *iff all the atoms in* φ *are available to her.*[2]

Here is the semantic model in which formulas of $\mathcal{L}$ will be interpreted.

---

[2] A related notion, considering also availability of agents, is studied in [5].

Definition 4 (Semantic model). *With the sets* P *and* A *as before, a* semantic model *is a tuple* $M = \langle W, R_i, V, \mathsf{PA}_i, \mathsf{AC}_i, \mathsf{R}_i \rangle$ *where:*

- $\langle W, R_i, V \rangle$ *is a standard Kripke model with* W *the non-empty set of worlds,* $R_i \subseteq W \times W$ *an* accessibility relation *for each agent* $i \in$ A *and* $V : W \to \wp(\mathsf{P})$ *the* atomic valuation.
- $\mathsf{PA}_i : W \to \wp(\mathsf{P})$ *is the* propositional availability function, *returning those atomic propositions agent* $i \in$ A *has at her/his disposal at each possible world.*
- $\mathsf{AC}_i : W \to \wp(\mathcal{L}_f)$ *is the* access set function, *returning those formulas agent* $i \in$ A *can access at each possible world.*
- $\mathsf{R}_i : W \to \wp(\mathcal{L}_r)$ *is the* rule set function, *returning those rules agent* $i \in$ A *can access at each possible world.*

The pair (M, w) *with* M *a semantic model and* w *a world in it is called a* pointed semantic model. *We denote by* ***M*** *the class of all semantic models.*

Formulas of $\mathcal{L}$ are interpreted in pointed semantic models as follows.

Definition 5 (Semantic interpretation). *Let* $M = \langle W, R_i, V, \mathsf{PA}_i, \mathsf{AC}_i, \mathsf{R}_i \rangle$ *be a semantic model, and take a world* $w \in W$. *The* satisfaction relation $\models$ *between formulas of* $\mathcal{L}$ *and the pointed semantic model* (M, w) *is given by.*

$$
\begin{array}{llll}
(M, w) \models p & \text{iff} & p \in V(w) & \\
(M, w) \models \mathsf{Ac}_i \varphi & \text{iff} & \varphi \in \mathsf{AC}_i(w) & \\
(M, w) \models \neg\varphi & \text{iff} & \text{it is not the case that } (M, w) \models \varphi \\
(M, w) \models \varphi \vee \psi & \text{iff} & (M, w) \models \varphi \text{ or } (M, w) \models \psi \\
(M, w) \models \Box_i \varphi & \text{iff} & \text{for all } u \in W, R_i w u \text{ implies } (M, u) \models \varphi
\end{array}
$$

$$
\begin{array}{llll}
(M, w) \models {}^{[i]}p & \text{iff} & p \in \mathsf{PA}_i(w) \\
(M, w) \models \mathsf{R}_i \rho & \text{iff} & \rho \in \mathsf{R}_i(w)
\end{array}
$$

As it becomes evident from Definitions 4 and 5, our logic is based on a sorted language where special atoms are introduced to represent signatures in the object language (i.e., ${}^{[i]}p$) and direct access to information (i.e., $\mathsf{Ac}_i \varphi$ and $\mathsf{R}_i \rho$) which are then interpreted by the dedicated valuation functions PA_i and, respectively, AC_i and R_i. These sorts are used to capture the "fine-grain" we allude to in the title. We will discuss this choice in more detail in Section 2.2, but first we provide a sound and complete axiom system.

Theorem 1 (Sound and complete axiom system for $\mathcal{L}$ w.r.t. M). *The axiom system of Table 1 is sound and strongly complete for formulas of $\mathcal{L}$ with respect to models in* ***M***.

Table 1. Axiom system for $\mathcal{L}$ w.r.t. ***M***

$\vdash \varphi$	for φ a propositional tautology
$\vdash \Box_i(\varphi \to \psi) \to (\Box_i\varphi \to \Box_i\psi)$	for every agent i
$\vdash \Diamond_i\varphi \leftrightarrow \neg\Box_i\neg\varphi$	for every agent i
If $\vdash \varphi \to \psi$ and $\vdash \varphi$, then $\vdash \psi$	
If $\vdash \varphi$, then $\vdash \Box_i\varphi$	for every agent i

Proof (Sketch). *Soundness is proved by showing that axioms are valid and rules preserve validity. For completeness, construct the standard modal canonical model M and, for each maximal consistent set of formulas w, define the propositional availability, access set and rule set function in the following way:*

$$\mathsf{PA}_i(w) := \{p \in \mathsf{P} \mid {}^{[i]}p \in w\}, \qquad \mathsf{AC}_i(w) := \{\varphi \in \mathcal{L}_f \mid \mathrm{Ac}_i\,\varphi \in w\},$$
$$\mathsf{R}_i(w) := \{\rho \in \mathcal{L}_r \mid \mathrm{R}_i\,\rho \in w\}$$

With these definitions, we get completeness from the fact that the novel formulas, ${}^{[i]}p$, $\mathrm{Ac}_i\,\varphi$ *and* $\mathrm{R}_i\,\rho$, *satisfy the* Truth Lemma:

$$(M, w) \models {}^{[i]}p \quad \text{iff} \quad {}^{[i]}p \in w, \qquad (M, w) \models \mathrm{Ac}_i\,\varphi \quad \text{iff} \quad \mathrm{Ac}_i\,\varphi \in w,$$
$$(M, w) \models \mathrm{R}_i\,\rho \quad \text{iff} \quad \mathrm{R}_i\,\rho \in w$$

$\square$

Note how there are no axioms for formulas of the form ${}^{[i]}p$, $\mathrm{Ac}_i\,\varphi$ and $\mathrm{R}_i\,\rho$. As mentioned, such formulas are simply special *atoms* for the dedicated *valuation functions* PA_i, AC_i and R_i. Moreover, these functions do not have any special property[3] and there is no restriction in the way they interact with each other[4]. Just like axiom systems for Epistemic Logic (*EL*) do not require axioms for atomic propositions, our system does not require axioms for these special atoms.

2.2   Defining Awareness, Implicit and Explicit Information

We now formalize the notions introduced at the beginning of Section 2.

Definition 6. *The notions of* awareness, implicit information *and* explicit information *for an agent i are defined as in Table 2.*

Table 2. Definitions of awareness, implicit and explicit information for agent i

i is *aware of* formula φ	$\mathrm{Aw}_i(\varphi) := \Box_i\,{}^{[i]}\varphi$
i is *aware of* rule ρ	$\mathrm{Aw}_i(\rho) := \Box_i\,{}^{[i]}\,\mathrm{TR}(\rho)$
i is *implicitly informed* about formula φ	$\mathrm{Im}_i(\varphi) := \Box_i({}^{[i]}\varphi \wedge \varphi)$
i is *implicitly informed* about rule ρ	$\mathrm{Im}_i(\rho) := \Box_i({}^{[i]}\,\mathrm{TR}(\rho) \wedge \mathrm{TR}(\rho))$
i is *explicitly informed* about formula φ	$\mathrm{Ex}_i(\varphi) := \Box_i({}^{[i]}\varphi \wedge \varphi \wedge \mathrm{Ac}_i\,\varphi)$
i is *explicitly informed* about rule ρ	$\mathrm{Ex}_i(\rho) := \Box_i({}^{[i]}\,\mathrm{TR}(\rho) \wedge \mathrm{TR}(\rho) \wedge \mathrm{R}_i\,\rho)$

What about the reading of other combinations of access to worlds with access to formulas and propositional availability? They are better read in terms of what they miss in order to become explicit information. For example the *EL* definition of information, $\Box_i\varphi$, expresses now information that will become explicit as soon as the agent considers the atoms occurring in φ and have access to φ itself. In the same way, $\Box_i(\varphi \wedge \mathrm{Ac}_i\,\varphi)$ describes information that will become explicit when the agent considers the atoms in φ.

By unfolding the definitions, we can show that the notions behave as desired.

---

[3] For example, we do not restrict AC_i to formulas *true* in the corresponding world, a restriction that would need $\mathrm{Ac}_i\,\varphi \to \varphi$ as an axiom of the system.

[4] See the discussion on strong and weak unawareness at the end of Section 2.2.

Proposition 1. *Let ξ be either a formula or else a rule of $\mathcal{L}$. The formulas $\mathrm{Ex}_i(\xi) \to \mathrm{Im}_i(\xi)$ and $\mathrm{Im}_i(\xi) \to \mathrm{Aw}_i(\xi)$ are valid in $\boldsymbol{M}$-models.*

In the remaining of the section our definitions are put in perspective with other proposals available in the literature. We argue for information about formulas, but similar cases can be done for information about rules.

Other Possibilities for Defining Explicit Information. The formal definition of explicit information/knowledge has several variants in the literature, in particular, the $\square_i\varphi \wedge \mathrm{Ac}_i\,\varphi$ of [6], and the $\mathrm{Ac}_i\,\varphi$ of [7,8,9,10].

A simple inspection of Table 2 shows that we opted here for yet another definition, along the lines presented in [11]. According to this definition, all the ingredients of explicit information fall under the scope of the modal box. This choice guarantees, once information is interpreted as knowledge, a weak form of both positive and negative introspection which otherwise fails for the definitions sketched above:

$$\mathrm{Ex}_i(\varphi) \to \mathrm{Im}_i(\mathrm{Ex}_i(\varphi)) \qquad \text{and} \qquad \neg\mathrm{Ex}_i(\varphi) \to \mathrm{Im}_i(\neg\mathrm{Ex}_i(\varphi))$$

The proof can be obtained by simple modal principles. Intuitively, if i has explicit knowledge that φ then she implicitly knows—she should be in principle able to infer—that she has explicit knowledge that φ. Conversely, if she does not have explicit knowledge that φ then she implicitly knows that she does not have explicit knowledge.

Syntactic Awareness vs. Semantic Awareness. The proposed formalization of awareness builds on the intuition that, at each state, each agent has only a particular subset of the language at disposal for phrasing her knowledge, so to say. This intuition is modeled via dedicated atoms $^{[i]}p$ and via the inductive extension of the $[i]$ superscript to any formula by Definition 3. This is an eminently syntactic way to look at the availability of bits of language to agents and, thus, to look at awareness.

An alternative model-theoretic approach can be obtained via a relation holding between states equivalent up to a signature $\mathrm{P}_i \subseteq \mathrm{P}$ [12].

Definition 7 (Equivalence up to a signature). *Let P be a countable set of atoms, W a set of states and V a valuation function. The states $w, w' \in W$ are equivalent up to a signature $\mathrm{P}_i \in \wp(\mathrm{P})$ (P_i-equivalent) iff for any $p \in \mathrm{P}_i$, we have $p \in V(w)$ iff $p \in V(w')$. If $w, w' \in W$ are P_i-equivalent, we write $\sim_i$.*

Intuitively, the relation $\sim_i$ links states that are indistinguishable if the atoms in $\mathrm{P} - \mathrm{P}_i$ are neglected. Such relation can then be used as accessibility relation, thereby yielding an extension of S5 in which the availability of certain formulas to agents gets a semantics in terms of a notion of indistinguishability yielded by the set of atoms considered. So, the fact that agent i can make use of φ in eliciting her information means that she can distinguish, thanks to the atoms she has at disposal, between φ-states and $\neg\varphi$-states which, in turn, boils down to the truth of $[\sim_i]\varphi \vee [\sim_i]\neg\varphi$, where $[\sim_i]$ gets the obvious interpretation.[5]

We have the following result generally relating $^{[i]}\varphi$ and $[\sim_i]\varphi$ formulas.

---

[5] We refer the interested reader to [12] for more details.

Proposition 2 (Syntactic vs. semantic propositional availability). *Let* P
*be a countable set of atoms, W a set of states and V a valuation function. For
each propositional availability function PA_i there exists a propositional equivalence relation $\sim_i$ modulo signature $\mathrm{P}_i \subseteq \mathrm{P}$ such that, for any state w and boolean
formula φ:*

$$(W, \mathsf{PA}_i, V), w \models {}^{[i]}\varphi \implies (W, \sim_i, V), w \models [\sim_i]\varphi \vee [\sim_i]\neg\varphi.$$

The implication is strict.

*Proof. Our proof is by construction. Each formula ${}^{[i]}\varphi$ is either true or false at
a pointed model $((W, \mathsf{PA}_i, V), w)$ and, by Definition 3, that depends only on the
truth values of its atoms in that model. Now, put $\mathrm{P}_i := \{p \mid (W, \mathsf{PA}_i, V), w \models
{}^{[i]}p\}$. By Definition 7 we obtain the relation $\sim_i$. The desired implication is proved
via a simple induction on the complexity of φ. That the implication is strict,
follows directly from the fact that the truth of $[\sim_i]\varphi$-formulas is preserved under
substitution of equivalents, while this does not hold for ${}^{[i]}\varphi$-formulas.* $\quad\square$

In other words, the syntactic representation of the availability of atoms to
agents is stronger than the model-theoretic one. In particular, the failure of
the inverse implication in Proposition 2 constitutes the main reason for the
assumption of a syntactic view in the present paper.

Strong and Weak Unawareness. Our semantic models do not impose any
restriction for formulas in access sets. In particular, they can contain formulas
involving atomic propositions that are not in the corresponding propositional
availability set, that is, $\mathsf{Ac}_i\,\varphi \wedge \neg\,{}^{[i]}\varphi$ is satisfiable. If we ask for formulas in
access sets to be built only from available atoms (if we ask for $\mathsf{Ac}_i\,\varphi \to {}^{[i]}\varphi$ to
be valid), we can represent only *strong unawareness*: if the agent is unaware of
φ, then becoming aware of it does not give her any explicit information about
φ, simply because φ (or any formula involving it) cannot be in her access set.

On the other hand, our unrestricted setting allow us to represent also *weak
unawareness*: becoming aware of atoms in φ can give the agent explicit information about φ because φ can be already in her access set. This allow us to model
situations where the agent *remembers*: I am looking for the keys in the bedroom,
and then when someone introduces the possibility for them to be in the kitchen,
I remember that actually I left them next to the oven. A similar notion can be
represented with the model-theoretical approach to awareness described above.

2.3 Working with *Knowledge*

Our current definitions do not guarantee us that the agent's information is *true*.
This is simply because the real world does not need to be among those that the
agents consider possible. We now define the class $\mathbf{M}_K$, containing those models
describing *true* information, that is, *knowledge*.

Definition 8 (Class $\mathbf{M}_K$). *A semantic model $M = \langle W, R_i, V, \mathsf{PA}_i, \mathsf{AC}_i, R_i \rangle$ is
in the class $\mathbf{M}_K$ if and only if R_i is an* equivalence *relation for all agents i.*

Given the reflexivity of every R_i, it is easy to show that in $\mathbf{M}_K$-models, implicit and explicit information are *true* at the evaluation point.

Proposition 3. *In $\mathbf{M}_K$-models, implicit and explicit information are true information, that is, $\mathrm{Im}_i(\varphi) \rightarrow \varphi$ and $\mathrm{Ex}_i(\varphi) \rightarrow \varphi$ are valid.*

When working with models in $\mathbf{M}_K$, we will use the term *knowledge* instead of the term *information*. Instead of implicit and explicit information, we will talk about *implicit* and *explicit* knowledge. A sound and complete axiom system for validities of $\mathcal{L}$ in $\mathbf{M}_K$-models is given by the standard S5 system.

Theorem 2 (Sound and complete axiom system for $\mathcal{L}$ w.r.t. $\mathbf{M}_K$). *The axiom system of Table 1 plus the axioms of Table 3 (for every agent i) is sound and strongly complete for formulas of $\mathcal{L}$ with respect to models in $\mathbf{M}_K$.*

Table 3. Extra axioms for $\mathcal{L}$ w.r.t. $\mathbf{M}_K$

$\vdash \Box_i\varphi \rightarrow \varphi$	$\vdash \Box_i\varphi \rightarrow \Box_i\Box_i\varphi$	$\vdash \neg\Box_i\varphi \rightarrow \Box_i\neg\Box_i\varphi$

2.4 The Example

We are now in the position to start a formal analysis of Example 1 by representing the information state of the relevant members of the jury at the beginning of the conversation, which is done in Table 4. The relevant atomic propositions are gls (*the woman wears glasses*), mkns (*she has marks in the nose*), esq (*her eyesight is in question*) and glt (*the accused is guilty beyond any reasonable doubt*). The relevant rules, abbreviated as $\varphi \Rightarrow \psi$ with φ the (conjunction of the) premise(s) and ψ the conclusion, are $\sigma_1 : \text{mkns} \Rightarrow \text{gls}$, $\sigma_2 : \text{gls} \Rightarrow \text{esq}$ and $\sigma_3 : \text{esq} \Rightarrow \text{glt}$.

Table 4. Information state of the agents in Example 1

A	$\Box_A(\mathrm{TR}(\sigma_1) \wedge \mathrm{R}_A\,\sigma_1)$ $\Box_A(\mathrm{TR}(\sigma_2) \wedge \mathrm{R}_A\,\sigma_2)$ $\Box_A(\mathrm{TR}(\sigma_3) \wedge \mathrm{R}_A\,\sigma_3)$	$\Box_A(\text{mkns} \wedge \mathrm{Ac}_A\,\text{mkns})$	$\mathrm{Aw}_A(\text{glt})$ $\mathrm{Aw}_A(\text{esq})$
B	$\Box_B(\mathrm{TR}(\sigma_1) \wedge \mathrm{R}_B\,\sigma_1)$ $\Box_B(\mathrm{TR}(\sigma_2) \wedge \mathrm{R}_B\,\sigma_2)$ $\Box_B(\mathrm{TR}(\sigma_3) \wedge \mathrm{R}_B\,\sigma_3)$		$\mathrm{Aw}_B(\text{glt})$
C	$\Box_C(\mathrm{TR}(\sigma_1) \wedge \mathrm{R}_C\,\sigma_1)$ $\Box_C(\mathrm{TR}(\sigma_2) \wedge \mathrm{R}_C\,\sigma_2)$ $\Box_C(\mathrm{TR}(\sigma_3) \wedge \mathrm{R}_C\,\sigma_3)$	$\Box_C(\text{mkns} \wedge \mathrm{Ac}_C\,\text{mkns})$	$\mathrm{Aw}_C(\text{glt})$
G	$\Box_G(\mathrm{TR}(\sigma_1) \wedge \mathrm{R}_G\,\sigma_1)$ $\Box_G(\mathrm{TR}(\sigma_2) \wedge \mathrm{R}_G\,\sigma_2)$ $\Box_G(\mathrm{TR}(\sigma_3) \wedge \mathrm{R}_G\,\sigma_3)$		$\mathrm{Aw}_G(\text{glt})$

In words, all the agents know—in the standard epistemic sense—that if somebody has some signs on her nose that means she wears glasses, that if she wears glasses then we can question her eyesight, and that someone with questioned eyesight cannot be a credible eye-witness. Also, all the agents can in principle follow this line of reasoning because each one of them has access to these rules in all the worlds each one considers possible. However, only A and C have access to the bit of information which is needed to trigger the inference, namely, that the witness had those peculiar signs on her nose. This is, nonetheless, not enough since no agent is considering the atoms mkns and gls in their "working languages". The only bit of language they are considering concerns the defendant being guilty or not and, in A's case, the concern about the witness eyesight.

All in all, the key aspect here is that the bits of information that can possibly generate explicit knowledge are spread across the group. The effect of the deliberation is to share this bits through dedicated announcements, which is the topic of the next section.

3   Dynamics of Information

Our framework allow us to describe the information of agents at some given point in time. It is time to provide the tools that allow us to describe how this information changes.

Three are the fundamental informational operations considered in our paper. The first one, the *awareness* operation, makes the agent aware of a given atomic proposition q; it is the processes through which the agent extends her awareness and it can represent the introduction of completely unknown information or a "remembering" action (see the discussion about *weak* and *strong unawareness* above). The second one, *inference*, allows the agent to extend the information she can access by the application of a rule; it is the process through which the agent extends her explicit information. The third one, *announcement*, represents agents' communication. These operations are defined as follows.

Definition 9. *Let* $M = \langle W, R_i, V, \mathsf{PA}_i, \mathsf{AC}_i, \mathsf{R}_i \rangle$ *be a semantic model.*

- *Take* $q \in \mathsf{P}$ *and* $j \in \mathsf{A}$. *The* awareness *operation yields the model* $M_{\ulcorner q \urcorner;j} = \langle W, R_i, V, \mathsf{PA}'_i, \mathsf{AC}_i, \mathsf{R}_i \rangle$, *differing from* M *just in the propositional availability function of agent* j, *which is given by*

$$\mathsf{PA}'_j(w) := \mathsf{PA}_j(w) \cup \{q\} \qquad \text{for every } w \in W$$

  In words, the operation $\ulcorner q \urcorner; j$ *adds the atomic proposition* q *to the propositional availability set of the agent* j *in all worlds of the model.*

- *Take* $\sigma \in \mathcal{L}_r$ *and* $j \in \mathsf{A}$. *The* inference *operation yields the model* $M_{j;\sigma} = \langle W, R_i, V, \mathsf{PA}_i, \mathsf{AC}'_i, \mathsf{R}_i \rangle$, *differing from* M *just in the access set function of the agent* j, *which is given by*

$$\mathsf{AC}'_j(w) := \begin{cases} \mathsf{AC}_j(w) \cup \{\mathrm{conc}(\sigma)\} & \text{if } \sigma \in \mathsf{R}_j(w) \text{ and } \mathrm{prem}(\sigma) \subseteq \mathsf{AC}_j(w) \\ \mathsf{AC}_j(w) & \text{otherwise} \end{cases}$$

for every world $w \in W$. In words, the operation $j; \sigma$ adds the conclusion of σ to the access set of an agent j at a world w iff her rule and access sets at w contain σ and its premises, respectively.[6]

− *Take $\chi \in \mathcal{L}_f$ and $j \in A$. The announcement operation yields the model $M_{j;\chi!} = \langle W', R'_i, V', \mathsf{PA}'_i, \mathsf{AC}'_i, \mathsf{R}_i \rangle$, which is given by*

$$W' := \{ w \in W \mid (M, w) \models \chi \}, \qquad R' := R \cap (W' \times W')$$

and, for all $w \in W'$ and $i \in A$,

$$V'(w) := V(w), \qquad \mathsf{PA}'_i(w) := \mathsf{PA}_i(w) \cup \mathrm{atom}(\chi), \qquad \mathsf{AC}'_i(w) := \mathsf{AC}_i(w) \cup \{\chi\}$$

where $\mathrm{atom}(\chi)$ is the set of atomic propositions occurring in χ. In words, the operation $j; \chi!$ removes worlds where χ does not hold, restricting the accessibility relation and the valuation to the new domain. It also extends propositional availability sets with the atoms occurring in χ and extends access sets with χ itself, preserving rule sets as in the original model.

Note how while the first two operations affect the model components of just one agent (the one that extends her available atomic propositions and the one that applies the inference, resp.), the third one affects those of all agents.

It can be easily proved that the three operations preserve models in $\mathbf{M}_K$.

Proposition 4. *If M is a $\mathbf{M}_K$-model, so are $M_{\ulcorner q \urcorner; j}$, $M_{j; \sigma}$ and $M_{j; \chi!}$.*

In order to express the effect of this operations over the agent's knowledge, we extend our language $\mathcal{L}$ with three new modalities, $\langle \ulcorner q \urcorner; j \rangle$, $\langle j; \sigma \rangle$ and $\langle j; \chi! \rangle$ (their "boxed" versions are defined as their correspondent dual, as usual). The semantic interpretation of this *extended* $\mathcal{L}$ is as follows.

Definition 10 (Semantic interpretation). *Let $M = \langle W, R_i, V, \mathsf{PA}_i, \mathsf{AC}_i, \mathsf{R}_i \rangle$ be a semantic model, and take a world $w \in W$. Define the following formulas:*

$$\mathrm{Pre}(j; \sigma) := \mathrm{Ex}_j(\sigma) \wedge \bigwedge_{\varphi \in \mathrm{prem}(\sigma)} \mathrm{Ex}_j(\varphi) \qquad\qquad \mathrm{Pre}(j; \chi!) := \mathrm{Ex}_j(\chi)$$

Then,

$$
\begin{aligned}
(M, w) &\models \langle \ulcorner q \urcorner; j \rangle \, \varphi &\quad \textit{iff} \quad& (M_{\ulcorner q \urcorner; j}, w) \models \varphi \\
(M, w) &\models \langle j; \sigma \rangle \, \varphi &\quad \textit{iff} \quad& (M, w) \models \mathrm{Pre}(j; \sigma) \text{ and } (M_{j; \sigma}, w) \models \varphi \\
(M, w) &\models \langle j; \chi! \rangle \, \varphi &\quad \textit{iff} \quad& (M, w) \models \mathrm{Pre}(j; \chi!) \text{ and } (M_{j; \chi!}, w) \models \varphi
\end{aligned}
$$

The semantic definitions rely on Proposition 4: the given operations preserve models in the relevant class, so we can evaluate formulas in them. Moreover, the precondition of each operation reflects its intuitive meaning: for the awareness operation, the agent can consider new possibilities or even remember those she knew and forgot at any point. For performing an inference with a rule σ, the

---

[6] For the sake of simplicity, here we have considered only a *modus ponens* application of a rule, but other ways of using them are also possible.

Table 5. Extra axioms for extended $\mathcal{L}$ w.r.t. $\mathbf{M}_K$

$\vdash \langle \ulcorner q \urcorner; j \rangle\, p \leftrightarrow p$	$\vdash \langle \ulcorner q \urcorner; j \rangle\, \mathrm{R}_i\, \rho \leftrightarrow \mathrm{R}_i\, \rho$
$\vdash \langle \ulcorner q \urcorner; j \rangle\, {}^{[i]}p \leftrightarrow {}^{[i]}p$ for $i \neq j$	$\vdash \langle \ulcorner q \urcorner; j \rangle\, \neg \varphi \leftrightarrow \neg \langle \ulcorner q \urcorner; j \rangle\, \varphi$
$\vdash \langle \ulcorner q \urcorner; j \rangle\, {}^{[j]}p \leftrightarrow {}^{[j]}p$ for $p \neq q$	$\vdash \langle \ulcorner q \urcorner; j \rangle\, (\varphi \vee \psi) \leftrightarrow ((\langle \ulcorner q \urcorner; j \rangle\, \varphi \vee \langle \ulcorner q \urcorner; j \rangle\, \psi)$
$\vdash \langle \ulcorner q \urcorner; j \rangle\, {}^{[j]}q \leftrightarrow \top$	$\vdash \langle \ulcorner q \urcorner; j \rangle\, \Box_i \varphi \leftrightarrow \Box_i \langle \ulcorner q \urcorner; j \rangle\, \varphi$
$\vdash \langle \ulcorner q \urcorner; j \rangle\, \mathrm{Ac}_i\, \varphi \leftrightarrow \mathrm{Ac}_i\, \varphi$	If $\vdash \varphi$, then $\vdash [\ulcorner q \urcorner; j]\, \varphi$
$\vdash \langle j; \sigma \rangle\, p \leftrightarrow \mathrm{Pre}(j; \sigma) \wedge p$	$\vdash \langle j; \sigma \rangle\, \mathrm{R}_i\, \rho \leftrightarrow \mathrm{Pre}(j; \sigma) \wedge \mathrm{R}_i\, \rho$
$\vdash \langle j; \sigma \rangle\, {}^{[i]}p \leftrightarrow \mathrm{Pre}(j; \sigma) \wedge {}^{[i]}p$	$\vdash \langle j; \sigma \rangle\, \neg \varphi \leftrightarrow (\mathrm{Pre}(j; \sigma) \wedge \neg \langle j; \sigma \rangle\, \varphi)$
$\vdash \langle j; \sigma \rangle\, \mathrm{Ac}_i\, \varphi \leftrightarrow \mathrm{Pre}(j; \sigma) \wedge \mathrm{Ac}_i\, \varphi$ for $i \neq j$	$\vdash \langle j; \sigma \rangle\, (\varphi \vee \psi) \leftrightarrow ((\langle j; \sigma \rangle\, \varphi \vee \langle j; \sigma \rangle\, \psi)$
$\vdash \langle j; \sigma \rangle\, \mathrm{Ac}_j\, \varphi \leftrightarrow \mathrm{Pre}(j; \sigma) \wedge \mathrm{Ac}_j\, \varphi$ for $\varphi \neq \mathrm{conc}(\sigma)$	$\vdash \langle j; \sigma \rangle\, \Box_i \varphi \leftrightarrow (\mathrm{Pre}(j; \sigma) \wedge \Box_i [j; \sigma]\, \varphi)$
$\vdash \langle j; \sigma \rangle\, \mathrm{Ac}_j\, \mathrm{conc}(\sigma) \leftrightarrow \mathrm{Pre}(j; \sigma)$	If $\vdash \varphi$, then $\vdash [j; \sigma]\, \varphi$
$\vdash \langle j; \chi! \rangle\, p \leftrightarrow \mathrm{Pre}(j; \chi!) \wedge p$	$\vdash \langle j; \chi! \rangle\, \mathrm{R}_i\, \varphi \leftrightarrow \mathrm{Pre}(j; \chi!) \wedge \mathrm{R}_i\, \varphi$
$\vdash \langle j; \chi! \rangle\, {}^{[i]}p \leftrightarrow \mathrm{Pre}(j; \chi!) \wedge {}^{[i]}p$ for $p \notin \mathrm{atom}(\chi)$	$\vdash \langle j; \chi! \rangle\, \neg \varphi \leftrightarrow (\mathrm{Pre}(j; \chi!) \wedge \neg \langle j; \chi! \rangle\, \varphi)$
$\vdash \langle j; \chi! \rangle\, {}^{[i]}p \leftrightarrow \mathrm{Pre}(j; \chi!)$ for $p \in \mathrm{atom}(\chi)$	$\vdash \langle j; \chi! \rangle\, (\varphi \vee \psi) \leftrightarrow ((\langle j; \chi! \rangle\, \varphi \vee \langle j; \chi! \rangle\, \psi)$
$\vdash \langle j; \chi! \rangle\, \mathrm{Ac}_i\, \varphi \leftrightarrow \mathrm{Pre}(j; \chi!) \wedge \mathrm{Ac}_i\, \varphi$ for $\varphi \neq \chi$	$\vdash \langle j; \chi! \rangle\, \Box_i \varphi \leftrightarrow (\mathrm{Pre}(j; \chi!) \wedge \Box_i [j; \chi!]\, \varphi)$
$\vdash \langle j; \chi! \rangle\, \mathrm{Ac}_i\, \chi \leftrightarrow \mathrm{Pre}(j; \chi!)$	If $\vdash \varphi$, then $\vdash [j; \chi!]\, \varphi$

agent needs to know explicitly σ and all its premises. For announcing χ, the announcing agent needs to *know* it explicitly, which implies that χ is true.

In order to provide a sound and complete axiom system for the extended language, we use a standard DEL technique. We extend our previous "static" system (Tables 1 and 3) with *reduction axioms* indicating how to translate a formula with the new modalities to a provably equivalent one without them. Then, completeness follows from the completeness of the basic system. We refer to [13] for an extensive explanation.

Theorem 3 (Sound and complete axiom system for extended $\mathcal{L}$ w.r.t. $\mathbf{M}_K$). *The axioms and rules of Tables 1 and 3 plus those of Table 5 form a sound and strongly complete axiom system for formulas in extended $\mathcal{L}$ with respect to models in $\mathbf{M}_K$.*

The reduction axioms for atomic propositions, negation, disjunction and the box modal operator are standard. The interesting ones are those expressing how propositional availability, access and rule sets are affected. For the $\ulcorner q \urcorner; j$ operation, the axioms indicate that only change is the addition of p to the propositional availability sets of j. For the $j; \sigma$ operation, the axioms indicate that only the access sets of agent j are modified, and the modification consist in adding the conclusion of the applied rule. Finally, axioms of the $j; \chi!$ operation indicate that while rule sets are not affected, propositional availability sets are extended with atoms of χ and access sets are extended with χ itself.

Though the *awareness* and the *inference* operations affect only the model components of the agent who performs them, they are in some sense *public*. In the *awareness* case, $\ulcorner q \urcorner; j$ makes ${}^{[j]}p$ true in all worlds any agent i considers possible ($\Box_i\, {}^{[j]}p$). Still, the operation is not *explicit*: ${}^{[j]}p$ will become i's explicit knowledge only after she becomes aware of p and has access to ${}^{[j]}p$. The *inference* operation behaves in a similar way. A further refinement of these operations, reflecting better the private character of such actions, can be found in [14].

It is not complicated to see that the following proposition holds.

Proposition 5

- *The formula $[\ulcorner q \urcorner; j] \, \mathrm{Aw}_j(q)$ is valid: after $\ulcorner q \urcorner; j$ the agent j is aware of q.*
- *The formula $[j; \sigma] \, \mathrm{Ex}_j(\mathrm{conc}(\sigma))$ is valid: after $j; \sigma$ the agent j is explicitly informed about $\mathrm{conc}(\sigma)$.*
- *For χ propositional, $[j; \chi!] \, \mathrm{Ex}_i(\chi)$ is valid, that is, after $j; \chi!$ all agents i are explicitly informed about χ.*

In the case of announcements, the proposition cannot be extended to arbitrary χ. This is because the well-known Moore-type formulas $p \wedge \neg \mathrm{Ex}_i(p)$, which become false after being announced.

3.1   The Example

Let us go back to the discussion room of Example 1. In Section 2.4, the static part of our framework allowed us to present a still image describing the agents' information before the discussion (Table 4). Here, the dynamic part allow us to "press play", and see a video describing how the agents interact and how their information evolves.

Stage 1. Agent D's action of scratching his nose makes A aware of both mkns and gls. Moreover, he becomes aware of the three relevant rules, since he was already questioning the eyesight of the woman (esq).

$$\langle \ulcorner \text{mkns} \urcorner; A \rangle \, \langle \ulcorner \text{gls} \urcorner; A \rangle \, \Big(\mathrm{Aw}_A(\text{mkns}) \wedge \mathrm{Aw}_A(\text{gls}) \wedge \mathrm{Aw}_A(\text{mkns} \Rightarrow \text{gls}) \wedge$$
$$\mathrm{Aw}_A(\text{gls} \Rightarrow \text{esq}) \wedge \mathrm{Aw}_A(\text{esq} \Rightarrow \neg \text{glt}) \Big)$$

Stage 2. By becoming aware of mkns, A can introduce it into the discussion. Moreover, mkns becomes part of his explicit knowledge, and he announces it.

$$\langle A; \mathrm{Aw}_A(\text{mkns})! \rangle \, \Big(\mathrm{Aw}_{JURY}(\text{mkns}) \wedge \mathrm{Ex}_A(\text{mkns}) \wedge \langle A; \text{mkns}! \rangle \mathrm{Ex}_{JURY}(\text{mkns}) \Big)$$

Stage 3. In particular, the simple introduction of mkns to the discussion makes it part of G's explicit knowledge, since he was just unaware of it.

$$\Box_G(\text{mkns} \wedge \mathrm{Ac}_G \text{mkns}) \wedge \neg \mathrm{Aw}_G(\text{mkns}) \wedge \langle A; \mathrm{Aw}_A(\text{mkns})! \rangle \mathrm{Ex}_G(\text{mkns})$$

Stage 4. Now, A can apply the rule sgns $\Rightarrow$ gls and, after doing it, he announces the conclusion gls.

$$\langle A; \text{mkns} \Rightarrow \text{gls} \rangle \, \Big(\mathrm{Ex}_A(\text{gls}) \wedge \langle A; \text{gls}! \rangle \mathrm{Ex}_{JURY}(\text{gls}) \Big)$$

Stage 5. With gls in his explicit knowledge (from A's announcement), C can apply gls $\Rightarrow$ esq, announcing esq after it.

$$\langle C; \text{gls} \Rightarrow \text{esq} \rangle \, \Big(\mathrm{Ex}_C(\text{esq}) \wedge \langle C; \text{esq}! \rangle \mathrm{Ex}_{JURY}(\text{esq}) \Big)$$

Stage 6. Finally, B draws the last inference and announces the conclusion.

$$\langle B; \text{esq} \Rightarrow \neg \text{glt} \rangle \, \Big(\mathrm{Ex}_B(\neg \text{glt}) \wedge \langle B; \neg \text{glt}! \rangle \mathrm{Ex}_{JURY}(\neg \text{glt}) \Big)$$

Obviously, Stages 1-6 could be written in one formula, and given Proposition 5, it is not difficult to check that such formula is a logical consequence of the information state of Table 4.

4   Conclusions and Further Work

We have defined a framework to represent not only different notions of agents' information (awareness of, implicit information and explicit information), but also the way they evolve through certain epistemic actions. The framework is expressive enough to deal with situations like our running example, an excerpt of Sydney Lumet's 1957 movie "12 Angry Men".

Among the questions that arise from the present work, we mention three that we consider interesting. (1) We have discussed individual notions of knowledge, but there is also the important notion of *common knowledge*. It will be interesting to look at implicit and explicit versions of the concept, as well as how it is affected by epistemic actions. (2) We have focused on the notion of knowledge, but there are several other notions, like *belief* and *safe belief* that are worthwhile to investigate from our fine-grained perspective. (3) We have provided a *dynamic logic* approach. A future research line consists in looking at correspondences between our proposal and work on dialogues in argumentation theory [15].

Acknowledgments. We gratefully thank two anonymous LORI 2009 referees for their comments. Davide Grossi is supported by **Nederlandse Organisatie voor Wetenschappelijk Onderzoek** (VENI grant Nr. 639.021.816). Fernando R. Velázquez-Quesada is supported by **Consejo Nacional de Ciencia y Tecnología (CONACyT)**, México (scholarship # 167693).

References

1. Kornhauser, L.A., Sager, L.G.: Unpacking the court. Yale Law Journal (1986)
2. List, C., Puppe, C.: Judgment aggregation: A survey. In: Oxford Handbook of Rational and Social Choice. Oxford University Press, Oxford (2009)
3. Dietrich, F., List, C.: Strategy-proof judgment aggregation. Economics and Philosophy 23, 269–300 (2007)
4. van Ditmarsch, H., van der Hoek, W., Kooi, B.: Dynamic Epistemic Logic. Synthese Library Series, vol. 337. Springer, Heidelberg (2007)
5. van Ditmarsch, H., French, T.: Awareness and forgetting of facts and agents. In: Proceedings of the 2009 IEEE/WIC/ACM International Joint Conference on Web Intelligence and Intelligent Agent Technologies, WI-IAT 2009 (to appear, 2009)
6. Fagin, R., Halpern, J.Y.: Belief, awareness, and limited reasoning. Artificial Intelligence 34(1), 39–76 (1988)
7. Duc, H.N.: Reasoning about rational, but not logically omniscient, agents. Journal of Logic and Computation 7(5), 633–648 (1997)
8. Jago, M.: Rule-based and resource-bounded: A new look at epistemic logic. In: Ågotnes, T., Alechina, N. (eds.) Proceedings of the Workshop on Logics for Resource-Bounded Agents, ESSLLI 2009, Malaga, Spain, August 2006, pp. 63–77 (2006)
9. van Benthem, J.: Merging observation and access in dynamic logic. Journal of Logic Studies 1(1), 1–17 (2008)
10. Velázquez-Quesada, F.R.: Inference and update. Synthese (Knowledge, Rationality and Action) 169(2), 283–300 (2009)

11. Velázquez-Quesada, F.R.: Dynamic logics for explicit and implicit information. Working paper. Current version (2009), http://staff.science.uva.nl/~fvelazqu/docs/InfBeliefs-06-25.pdf
12. Grossi, D.: A note on brute vs. institutional facts: Modal logic of equivalence up to a signature. In: Dagstuhl Seminar Proceedings 09121: Normative Multi-Agent Systems (2009)
13. van Benthem, J., Kooi, B.: Reduction axioms for epistemic actions. In: Schmidt, R., Pratt-Hartmann, I., Reynolds, M., Wansing, H. (eds.) Advances in Modal Logic (Technical Report UMCS-04-09-01), pp. 197–211. University of Manchester (2004)
14. van Benthem, J., Velázquez-Quesada, F.R.: Inference, promotion, and the dynamics of awareness. Submitted to Synthese (KRA) (2009)
15. Prakken, H., Vreeswijk, G.: Logics for defeasible argumentation. In: Gabbay, D., Guenthner, F. (eds.) Handbook of Philosophical Logic, 2nd edn., vol. 4, pp. 218–319. Kluwer Academic Publishers, Dordrecht (2002)

Dynamic Testimonial Logic

Wesley H. Holliday

Department of Philosophy, Stanford University
wesholliday@stanford.edu

Abstract. We introduce a *dynamic testimonial logic* (DTL) to model
belief change over sequences of multi-agent testimony. Our static base
logic is Baltag and Smets' [1] *conditional doxastic logic* (CDL). Our
dynamic base logic is van Benthem's [12] *dynamic logic of belief upgrade*,
which we extend with a "belief suspension" operator. After showing how
to extract from CDL models agents' beliefs about the *doxastic reliability*
of other agents, we add "authority graphs" to DTL models to capture
agents' *epistemic trust* in other agents' testimony. For DTL's dynamic
testimony operator, we give complete reduction axioms. Finally, we
describe an application of DTL in modeling *epistemic bandwagon effects*.

Keywords: dynamic logic, modal logic, belief revision, testimony, trust.

1 Introduction

As it is modeled formally, judgment aggregation is an instantaneous process:
given a group of agents, each with an opinion on some proposition, an aggregation
function takes their (possibly conflicting) individual opinions and returns a group
opinion, *all at once*. Yet in many contexts—from courtrooms to committees—the
protocol is to solicit individual opinions sequentially, not simultaneously. For one
example of how the temporal dimension matters in expert testimony, consider
what Sorensen [11] calls the *epistemic bandwagon effect*:

> An expert's epistemic preferences can be justifiably influenced by his
> knowledge of another expert's preferences. Yet this provides the basis
> for an epistemic bandwagon effect. For the sake of simplicity, suppose
> there are three highly respectful experts, 1, 2, and 3, who prior to the
> roll-call vote are respectively in favour, indifferent, and opposed to a
> proposition. However, they only learn the others' preferences by their
> votes. If the roll-call vote is taken in order 1, 2, 3, expert 1 votes in
> favour. Having learned that another expert favours the proposition, the
> opinion of 2 is swayed and he too votes in favour. Having learned that
> two experts favour the proposition, 3 reverses his opinion (since he is

X. He, J. Horty, and E. Pacuit (Eds.): LORI 2009, LNAI 5834, pp. 161–179, 2009.

highly respectful) and the proposition is unanimously favoured. However, if the roll-call vote is taken in order 3, 2, 1, incremental disclosure of preferences and high respect results in the proposition being unanimously opposed.... [T]he general point holds for more complicated cases involving larger groups with different degrees of respect and irregular preference revelation. Disclosure order bias indicates that epistemic respect is trickier than has been supposed. (49-50)

There is nothing essential about voting in this scenario. The experts could express their views as statements in a public forum or as posts on a blog. What is essential is the sequential nature of disclosure, together with the agents' attitudes of respect for one another.

In this paper we introduce a *dynamic testimonial logic* (DTL) to model belief change over sequences of multi-agent testimony. We use the term 'testimony' in the broad sense common in the philosophical literature, of a statement offered as evidence for a proposition, whether or not it occurs in a courtroom or committee (for further details, see [6]). For applications of DTL, one of our goals is to formalize solutions to the epistemic bandwagon problem, in terms of both policies for individual belief revision in response to testimony and protocols governing testimonial sequences. Here we present only a base system capable of modeling the bandwagon effect itself.

While Sorensen uses the notion of *respect* among experts, in DTL we use the notion of *trust*. Modal logics have recently been developed to represent several types of trust. To mention only two, an agent may (i) trust another agent to perform an action or (ii) trust another agent's judgment on the truth value of a proposition. For examples in the literature, Broersen et al. [3] and Herzig et al. [9] discuss logics of *practical* trust (i), while Demolombe [4,5] and Liau [10] discuss logics of *epistemic* trust (ii). In this paper we deal exclusively with epistemic trust. The logics developed so far to model epistemic trust are static logics without dynamic operators, which provides another motivation for DTL as a *dynamic logic of trust*. For discussion of the importance of epistemic trust among experts and between laypeople and experts, we refer the reader to the philosophical literature [7,8].

1.1   Modeling Trust and Authority

Consider those experts on whose authority you would be willing to believe a proposition φ. We will say that you "trust the judgment" of these experts on φ. Among your trusted experts, some may be more authoritative for you than others; if expert 1 testifies that φ, expert 2 testifies that $\neg\varphi$, and you come to believe φ, then 1 is more authoritative for you than 2. If 2 were more authoritative, then you should have come to believe $\neg\varphi$. And if 1 and 2 were equally authoritative, you should not have changed your mind on φ either way, or you should have suspended belief on φ altogether. The same points apply if 1 and 2 are groups of experts, rather than individuals.

Supposing that 1 is more authoritative for you than 2, then after 1 testifies that φ, you no longer "trust" 2 on $\neg\varphi$, in the sense that you will no longer believe $\neg\varphi$ on the authority of 2 alone, something you might have done before 1 testified. However, if another expert 3 joins 2 in testifying that $\neg\varphi$, you may believe $\neg\varphi$ on the authority of 2 *together with* 3, though perhaps not on the authority of either of them individually. This will be the intuitive picture motivating our formal representations of trust and authority.

1.2 Assumptions of the Model

The version of DTL given here makes a number of assumptions about testimony. First, testimony is always *public*: in every case of testimony, the identity of the testifier and the content of the testimony is information available to all agents. Second, testimony need not be true, but it is always *sincere*: an agent i testifies that φ only if i believes φ. Third, testimony is heard under the *assumption of sincerity*: if an agent i testifies that φ, all other agents come to believe that i believes φ. Finally, testimony is in the form of *assertion*: agents testify *that* φ, but do not provide further arguments for φ.[1]

1.3 Testimony vs. Public Announcement

To identify the information provided by testimony, it is useful to compare testimony with *public announcement*, the classic case of an informational event in dynamic epistemic logic (see [13]). For our purposes, the crucial difference between testimony and announcement is that unlike testimony, announcements are not typically thought to come from one of the agents within the model, but from some anonymous external source.

What difference does the individual source of testimony make? When an agent j testifies that φ, an agent i in j's audience will acquire the information *that j believes φ*, given our assumptions about sincerity. But then what is the difference between a *truthful public announcement* that j believes φ and j's *own* (public) *testimony* that φ, if both provide the information that j believes φ?

The difference is that j's testimony provides *more* information[2]: it provides the information *that j is willing to publicly assert* φ. As we might say, j is willing to "go on the record" for φ. If j is the kind of agent who only publicly asserts

---

[1] Each of the first three assumptions could be dropped or modified, leading to a different version of DTL. The fourth assumption, that we are dealing with testimony in the form of assertion rather than argument, is more important. Given this assumption, agents hearing testimony on a proposition φ must decide whether to change their minds on φ purely on the basis of whether they take the testifier to be an authority on the proposition and whether they trust the judgment of the testifier.

[2] Here we are not considering the difference that public announcement is usually conceived as a source of "hard information" that *eliminates* possibilities, while testimony is better conceived as a source of "soft information" that reorders the relative plausibility of possibilities (see [12] for this distinction).

a proposition if he has conducted a thorough inquiry into its truth, then the information that j is willing to publicly assert φ is vital information. A truthful public announcement (from no particular agent) that j believes φ does not provide this vital information. For it may be that j believes many propositions, while he only has the time and resources to investigate some few of them in such a way that he would be willing to make public assertions about them.

We can now make a distinction between two ways in which one agent might judge another agent to be "reliable" about the truth of a proposition. If according to i, if j believes φ, j's belief is likely to be true, then i considers j *doxastically reliable* on φ. If according to i, if j testifies that φ, j's testimony is likely to be true, then i considers j to be *testimonially reliable* on φ. The point is that judgments of doxastic reliability and testimonial reliability may come apart. Suppose that j has expressed a general lack of understanding of economics. Then i might judge j's doxastic reliability on each economic proposition to be low. But suppose, as above, that i knows that j would only publicly assert a proposition if he had conducted a thorough inquiry into its truth. Then i might judge j's testimonial reliability on each economic proposition to be high; if j were to ever make a public assertion about economics, i would take it seriously. As we will see, this distinction can be represented formally in DTL.

2   Conditional Doxastic Logic

In this section we define the static base logic for DTL, the *conditional doxastic logic* (CDL) of Baltag and Smets [1].

2.1   Language

Definition 1. *Let* At *be a set of atomic sentences and* Agt *a set of agent-symbols. The language of CDL is defined by:*

$$\varphi := p \mid \neg\varphi \mid \varphi \wedge \varphi \mid B_i^\varphi \varphi$$

where $p \in$ At, $i \in$ Agt.

The intended reading of $B_i^\varphi \psi$ is "i believes that ψ conditional on φ."

2.2   Semantics

Definition 2. *Let $\leq$ be a binary relation on a set W. A comparability class is a set $C = \{w \in W \mid w \leq v \text{ or } v \leq w\}$ for some $v \in W$.*

Definition 3. *$\leq$ is a well-preorder on W if it is reflexive, transitive and every non-empty subset of W has a $\leq$-minimal element. $\leq$ is locally well-preordered* *on W if the restriction of $\leq$ to each comparability class C in W is a well-preorder on C.*

Definition 4. *A multi-agent plausibility model is a triple $\mathcal{M} = \left\langle W, \{\leq_i\}_{i \in \mathsf{Agt}}, V \right\rangle$ where (i) W is a non-empty set (of "possible worlds"), (ii) $\leq_i$ (agent i's plausibility ordering) is a locally well-preordered relation on W, and (iii) $V : \mathsf{At} \to \mathcal{P}(W)$ (a valuation).*

Following convention, we read $w \leq_i v$ as "i considers world w at least as plausible as world v," so the *minimal* worlds in the ordering $\leq_i$ are the *most* plausible worlds for i. We also use the notation $w \sim_i v$ (comparability) for ($w \leq_i v$ or $v \leq_i w$) and $\sim_i(w) = \{v \in W \mid w \sim_i v\}$ for the comparability class of world w for i.

Definition 5. *For boolean formulas, the truth definitions are those of classical modal logic. For conditional belief, the truth definition is:*

○ $\mathcal{M}, w \vDash B_i^\varphi \psi$ iff for all $v \in \min_{\leq_i}(\llbracket \varphi \rrbracket \cap \sim_i(w)) : \mathcal{M}, v \vDash \psi$

where we denote the set of *most plausible worlds* for i in $P \subseteq W$ by $\min_{\leq_i} P = \{v \in P \mid v \leq_i u \text{ for all } u \in P\}$ and the *truth set* of φ by $\llbracket \varphi \rrbracket = \{u \in W \mid \mathcal{M}, u \vDash \varphi\}$.

In CDL, (unconditional) belief and knowledge are derived operators. For belief, we define $B_i \varphi := B_i^\top \varphi$, read "$i$ believes that φ," and $\hat{B}_i \varphi := \neg B_i \neg \varphi$, read "$i$ considers it plausible that φ." For knowledge, we define $K_i \varphi := B_i^{\neg \varphi} \bot$, read "$i$ knows that φ," and $\hat{K}_i \varphi := \neg K_i \neg \varphi$, read "$i$ considers it possible that φ."

For complete axiomatizations of CDL, see [1] and [2].

2.3 Doxastic Reliability Information in CDL Models

CDL models contain information about what agents would believe upon learning various facts. They also contain information about what agents would believe upon learning about *other agents' beliefs*. Consider a formula such as $B_i^{B_j p} p \wedge B_i^{B_j \neg p} \neg p$, which is true if and only if in all the $B_j p$ worlds that i considers most plausible, j's belief is true, and likewise for the most plausible $B_j \neg p$ worlds. If this formula is true, then after a public announcement that $B_j p$, i will believe p. Intuitively, the formula expresses that i takes j to be *doxastically reliable* on p, in the sense of Section 1.3. Various judgments of doxastic unreliability can be expressed in a similar way. We can even extend this observation to agents' beliefs about the relative doxastic reliability of other agents. For example, a formula such as $B_i^{B_j p \wedge B_k \neg p} p \wedge B_i^{B_j \neg p \wedge B_k p} \neg p$ $(*)$ expresses i's belief in the superior doxastic reliability of j relative to k on p.

Figure 1 shows a CDL model in which $B_i^{B_k p} p \wedge B_i^{B_k \neg p} \neg p$ is true (at every world), so i considers k doxastically reliable on p, but $(*)$ is also true (at every world), so when j and k disagree, i considers j more reliable. The lines represent plausibility orderings, with arrows pointing toward more plausible worlds and arrowless lines indicating equi-plausibility. The solid lines are for i, while the

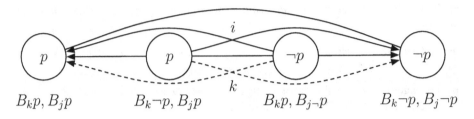

$$B_k p, B_j p \qquad B_k \neg p, B_j p \qquad B_k p, B_j \neg p \qquad B_k \neg p, B_j \neg p$$

Fig. 1.

dashed lines are for k. No lines are shown for j, indicating that at each world, j considers no other worlds possible. Note that every world is equi-plausible with itself for each agent, but we omit the reflexive loops.

If we view CDL models as containing information about agents' beliefs about the doxastic reliability of others, we can extract a relation $k <_{i,w}^{\varphi} j$ that holds just in case, e.g., $\mathcal{M}, w \vDash B_i^{B_j \varphi \wedge B_k \neg \varphi} \varphi \wedge B_i^{B_j \neg \varphi \wedge B_k \varphi} \neg \varphi$. Other definitions are also possible. Note that the relation $<_{i,w}^{\varphi}$ will not in general be an ordering, unless we make further assumptions about the relations and valuation in the model.

Although CDL models contain information about agents' views of the doxastic reliability of other agents, they do not contain information about agents' views of the *testimonial* reliability of other agents. If we were to assimilate j's testimony that φ to a public announcement of $B_j \varphi$, then CDL models would contain sufficient information to determine how agents' beliefs change in response to this "testimony."[3] However, as discussed in Section 1.3, j's testimony that φ is not equivalent to a public announcement of $B_j \varphi$, in terms of the information provided. For this reason, we will need to add additional structure to models for DTL in Section 4.

3   Dynamic Belief Revision

Having defined the static base logic for DTL, we turn to the dynamics. To model belief revision, we use van Benthem's [12] *dynamic logic of belief upgrade*. Two dynamic logics are presented in [12], one for *lexicographic upgrade* and one for *conservative upgrade*. Technically, either could serve as a dynamic base for DTL, but here we use the logic of conservative upgrade. It represents a "softer" way in which agents can revise their beliefs in response to incoming information, appropriate to settings in which authoritative sources conflict.

---

[3] Moreover, if we understand a judgment of doxastic reliability to be a kind of *trust*, then CDL provides a kind of logic of trust. An open problem, which we leave for future work, is to identify—and if possible, axiomatize—a fragment of CDL sufficient to capture the notion of trust given by judgments of doxastic reliability.

3.1   Language

Definition 6. *Let* At *be a set of atomic sentences and* Agt *a set of agent-symbols. The language of conservative upgrade is defined by:*

$$\varphi := p \mid \neg\varphi \mid \varphi \wedge \varphi \mid B_i^\varphi \varphi \mid [\uparrow_i \varphi]\varphi$$

where $p \in$ At, $i \in$ Agt.

The intended reading of $[\uparrow_i \varphi]\psi$ is "after i upgrades (or, revises his beliefs) with φ, ψ is the case." As an abbreviation, we write $[\uparrow_S \varphi]$ for $[\uparrow_{s_1} \varphi][\uparrow_{s_2} \varphi] \dots [\uparrow_{s_n} \varphi]$ where $S = \{s_1, ..., s_n\} \subseteq$ Agt.

3.2   Semantics

Models for the logic of conservative upgrade are the same multi-agent plausibility models as before.

Definition 7. *Given a model* $\mathcal{M} = \left\langle W, \{\leq_j\}_{j \in \mathsf{Agt}}, V \right\rangle$, *the updated model* $\mathcal{M} \uparrow_i \theta = \left\langle W, \{\leq_j\}_{j \in \mathsf{Agt}}^{\uparrow_i \theta}, V \right\rangle$ *is obtained by changing the plausibility ordering* $\leq_i$ *to* $\leq_i^{\uparrow \theta}$ *as follows: in each* $\leq_i$-*comparability class, the most plausible* θ *worlds become most plausible overall, but otherwise the old ordering remains.*

Definition 8. *The truth definitions for static formulas are those of CDL. The truth definition for belief upgrade is:*

○ $\mathcal{M}, w \vDash [\uparrow_i \theta]\varphi$ iff $\mathcal{M} \uparrow_i \theta, w \vDash \varphi$

3.3   Axiomatization

In [12] van Benthem provides *reduction axioms* that allow the rewriting of any formula with the dynamic operator $[\uparrow_i \varphi]$ as an equivalent formula in the static base logic of CDL. Given a complete axiomatization for CDL, these reduction axioms provide a complete axiomatization for the dynamic logic of conservative upgrade. Here we give only the crucial reduction axioms for conditional belief:

$$[\uparrow_i \theta] B_i^\varphi \psi \leftrightarrow \left(\left(\hat{B}^\theta [\uparrow_i \theta]\varphi \wedge B_i^{\theta \wedge [\uparrow_i \theta]\varphi} [\uparrow_i \theta]\psi \right) \vee \right.$$
$$\left. \left(B^\theta \neg [\uparrow_i \theta]\varphi \wedge B^{[\uparrow_i \theta]\varphi} [\uparrow_i \theta]\psi \right) \right)$$

$$[\uparrow_i \theta] B_j^\varphi \psi \leftrightarrow B_j^{[\uparrow_i \theta]\varphi} [\uparrow_i \theta]\psi \text{ for } j \neq i$$

The first axiom captures how i's upgrade affects i's conditional beliefs. The left side expresses that after i upgrades with θ, the best (most plausible) φ worlds for i are ψ worlds. The right side expresses an equivalent condition in terms of what is true *before* the upgrade, given a case distinction. Case 1: some of the best θ worlds in the original model become φ after the upgrade, indicated by

$\hat{B}^{\theta}\left[\uparrow_i\theta\right]\varphi$. Since the upgrade makes the best θ worlds best overall, in this case the best φ worlds after the upgrade are the best worlds satisfying $\theta\wedge\left[\uparrow_i\theta\right]\varphi$ in the original model. Case 2: none of the best θ worlds in the original model become φ after the upgrade, indicated by $B^{\theta}\neg\left[\uparrow_i\theta\right]\varphi$. In this case the best φ worlds after the upgrade are simply the best worlds satisfying $\left[\uparrow_i\theta\right]\varphi$ in the original model.

The second axiom captures how i's upgrade affects the conditional beliefs of other agents.[4] Since i's upgrade may change j's higher-order beliefs (about i's beliefs), to determine whether j believes ψ conditional on φ in the new model, we check whether j believes $\left[\uparrow_i\theta\right]\psi$ conditional on $\left[\uparrow_i\theta\right]\varphi$ in the original model.

3.4   Suspension of Belief

In modeling belief change due to testimony, we wish to model not only how agents form new beliefs about propositions, but also how agents *suspend belief* about propositions. Suppose that agents j and k are equally authoritative in the eyes of i. If j testifies that φ and then k testifies that $\neg\varphi$, one policy for i would be to believe whoever testified first—in this case, agent j. A more sensible policy, in a situation where equally authoritative sources conflict, would be to suspend belief about φ. Alternatively, i might not revise his beliefs at all, ignoring the conflicting testimony of j and k. Yet conflicting testimony from *authoritative* sources does not seem to "cancel out" to provide i with no information. Something informative has occurred—two authoritative sources have testified for φ and $\neg\varphi$ respectively—and the manner in which i suspends belief on φ should reflect this.

To extend the logic to a *dynamic logic of conservative upgrade and suspension*, we add a suspension operator $\left[\downarrow_i\varphi\right]$ to the language defined in the previous section. The intended reading of $\left[\downarrow_i\varphi\right]\psi$ is "after i suspends belief on φ, ψ is the case." As an abbreviation, we write $\left[\downarrow_S\varphi\right]$ for $\left[\downarrow_{s_1}\varphi\right]\left[\downarrow_{s_2}\varphi\right]\cdots\left[\downarrow_{s_n}\varphi\right]$ where $S=\{s_1,...,s_n\}\subseteq\mathsf{Agt}$.

Definition 9. *Given a model* $\mathcal{M}=\left\langle W,\{\leq_j\}_{j\in\mathsf{Agt}},V\right\rangle$, *the updated model* $\mathcal{M}\downarrow_i$ $\theta=\left\langle W,\{\leq_j\}_{j\in\mathsf{Agt}}^{\downarrow_i\theta},V\right\rangle$ *is obtained by changing the plausibility ordering* $\leq_i$ *to* $\leq_i^{\downarrow\theta}$ *as follows: in each* $\leq_i$-*comparability class, the most plausible* $\neg\theta$ *worlds and the most plausible* θ *worlds becomes equally plausible and most plausible overall, but otherwise the old ordering remains.*

---

[4] In the multi-agent setting, belief revision via relation change is a kind of *public* belief revision; when one agent's plausibility ordering changes, other agents may "notice" the change. This does not mean that if i upgrades with φ, all other agents necessarily come to believe that i believes φ. The effect can be more subtle, changing only the conditional beliefs of other agents. While it is natural to understand belief revision as a private mental change on the part of an agent, which other agents can at best infer or learn of through communication, in the version of DTL introduced below it makes sense for belief revision to be public in the sense that relation change is public, because information about whom agents trust will also be public. Hence if j testifies that φ and all agents know that i trusts j on φ, it is reasonable that agents' beliefs about i's beliefs change after j's testimony.

Definition 10. *The truth definition for belief suspension is:*

○ $\mathcal{M}, w \vDash [\downarrow_i \theta] \varphi$ iff $\mathcal{M} \downarrow_i \theta, w \vDash \varphi$

Proposition 1. *Together with a complete axiomatization of CDL and the reduction axioms for conservative upgrade, the following reduction axiom for suspension and belief (the other axioms are of the same form as those for upgrade) provides a complete axiomatization for the dynamic logic of conservative upgrade and suspension:*

$$[\downarrow_i \theta] B_i^\varphi \psi \leftrightarrow$$

$$\left(\left(B_i^\theta \left(\neg [\downarrow_i \theta] \varphi \right) \wedge B_i^{\neg\theta} \left(\neg [\downarrow_i \theta] \varphi \right) \wedge B_i^{[\downarrow_i\theta]\varphi} [\downarrow_i \theta] \psi \right) \vee \right.$$
$$\left(\hat{B}_i^\theta [\downarrow_i \theta] \varphi \wedge B_i^{\neg\theta} \left(\neg [\downarrow_i \theta] \varphi \right) \wedge B_i^{\theta\wedge[\downarrow_i\theta]\varphi} [\downarrow_i \theta] \psi \right) \vee$$
$$\left(B_i^\theta \left(\neg [\downarrow_i \theta] \varphi \right) \wedge \hat{B}_i^{\neg\theta} [\downarrow_i \theta] \varphi \wedge B_i^{\neg\theta\wedge[\downarrow_i\theta]\varphi} [\downarrow_i \theta] \psi \right) \vee$$
$$\left. \left(\hat{B}_i^\theta [\downarrow_i \theta] \varphi \wedge \hat{B}_i^{\neg\theta} [\downarrow_i \theta] \varphi \wedge B_i^{\theta\wedge[\downarrow_i\theta]\varphi} [\downarrow_i \theta] \psi \wedge B_i^{\neg\theta\wedge[\downarrow_i\theta]\varphi} [\downarrow_i \theta] \psi \right) \right)$$

The same style of analysis that we gave for the reduction axiom for upgrade and belief can be used here for suspension and belief. We leave it to the reader to confirm the soundness of the above axiom.

4   Basic DTL

In this section we develop basic DTL. For the design of the system, we make two simplifying assumptions about testimony. First, we assume that it is publicly known how agents trust the testimony of others. Second, we assume that the testimonial sequences of interest are on a single issue; as in Sorensen's bandwagon example, agents testifying in a given sequence either testify for a single proposition or against it (or pass). Here we cannot treat the many questions raised by private relations of trust and multi-issue testimony on related propositions. We plan to address both in future work.

4.1   Language

Definition 11. *Let* At *be a set of atomic sentences and* Agt *a finite set of agent-symbols. The language of basic DTL is defined by:*

$$\varphi_0 := p \mid \neg\varphi_0 \mid \varphi_0 \wedge \varphi_0$$
$$\varphi_1 := U\varphi_0$$
$$\varphi := \varphi_0 \mid \varphi_1 \mid \neg\varphi \mid \varphi \wedge \varphi \mid B_i^\varphi \varphi \mid [\uparrow_i \varphi] \varphi \mid [\downarrow_i \varphi] \varphi$$
$$Rec\,(i, \varphi_0) \mid S \preceq_i^{\varphi_0} S' \mid \langle i, !\varphi_0 \rangle \varphi$$

where $p \in$ At, $i \in$ Agt, and $S, S' \subseteq$ Agt.

The language of basic DTL includes four types of formula not in CDL. We have added the universal modality U for φ_0 formulas for technical reasons—in particular, so we can express in our axioms that two φ_0 formulas are equivalent in a model (see the Appendix).

The intended reading of $Rec\,(i, \varphi)$ is "i is (most recently) on the record as testifying that φ." The intended reading of $S \preceq_i^\varphi S'$ is "S' is as (testimonially) authoritative as S on φ for i." Intuitively, we have in mind that S's authoritativeness on φ for i is a matter of the extent to which i judges S to be testimonially reliable on φ, in the sense of Section 1.3. One may also wish to think of non-cognitive or non-rational factors that contribute to authoritativeness.

We use the abbreviations $S \prec_i^\varphi S' := S \preceq_i^\varphi S' \wedge S' \npreceq_i^\varphi S$, read "$S'$ is more authoritative than S on φ for i," and $S \approx_i^\varphi S' := S \preceq_i^\varphi S' \wedge S' \preceq_i^\varphi S$, read "$S$ and S' are equally authoritative on φ for i." In Section 4.3, we will use formulas of these kinds to define *trust formulas* of the form $T_i\,(j, \varphi)$, read "i trusts j's testimony on φ."

We also allow $\emptyset$ to occur in formulas: we read $\emptyset \prec_i^\varphi S$ as "S's testimony on φ is *authoritative* for i," $\emptyset \approx_i^\varphi S$ as "S's testimony on φ is *neutral* for i," and $S \prec_i^\varphi \emptyset$ as "S's testimony on φ is *anti-authoritative* for i." We can think of the formula $\emptyset \approx_i^\varphi S$ as expressing that i has no information about the testimonial (as opposed to doxastic) reliability of S. The formula $S \prec_i^\varphi \emptyset$ expresses that according to i, S's public testimony that φ (taken by itself) would actually support $\neg\varphi$.

The intended reading of $\langle i, !\varphi \rangle \, \psi$ is "(i can sincerely testify that φ and) after i testifies that φ, ψ is the case," the parenthetical phrase reflecting the precondition for testimony that i *believes* φ. We may also want to express what is the case after i testifies with his actual belief on φ, whatever that may be. In Section 4.3, we will define a formula, $\langle i, ?\varphi \rangle \, \psi$, read "after i testifies *with his opinion on* φ, ψ is the case." Note that in basic DTL, for simplicity we consider only testimony on factual formulas φ_0, so agents do not testify about the beliefs of others. It is for this reason that only φ_0 formulas can appear inside testimony operators and in authority and record formulas.

4.2   Semantics

Definition 12. *A testimonial model* $\mathcal{M} = \left(W, \{\leq_i\}_{i \in \mathsf{Agt}}, V, \mathsf{rec}, E \right)$ *is a multi-agent plausibility model together with two new functions,* (i) $\mathsf{rec} : \mathcal{P}\,(W) \to \mathcal{P}\,(\mathsf{Agt})$ *(a public record) such that if* $j \in \mathsf{rec}\,(P)$ *then* $j \notin \mathsf{rec}\,(W - P)$ *and* (ii) E *(an expertise function), which sends each triple* (i, w, P) *of an agent* $i \in \mathsf{Agt}$, *a world* $w \in W$, *and a proposition* $P \subseteq W$ *to an authority graph* $\left(\mathcal{P}\,(\mathsf{Agt}), \preceq_{i,w}^P \right)$. *The authority relation* $\preceq_{i,w}^P$ *is a total preorder on* $\mathcal{P}\,(\mathsf{Agt})$, *satisfying the condition:* $S \preceq_{i,w}^P S' \Leftrightarrow S \preceq_{i,w}^{W-P} S'$.

The function of the public record is to record which agents have testified on which propositions, where a proposition is now understood as a *set of worlds*, not a formula. For reasons discussed below, we require that an agent cannot be on the record for both a proposition and its complement at the same time. The function of the authority graphs is to encode agents' judgments of the relative testimonial authority of other (sets of) agents. For simplicity, we require that an authority relation be a total preorder, but for specific applications this could be modified. The condition that $S \preceq_{i,w}^{P} S' \Leftrightarrow S \preceq_{i,w}^{W-P} S'$ could also be dropped, at the expense of complicating the system.[5]

Definition 13. *A pointed testimonial model* $(\mathcal{M}, w)$ *is a model* $\mathcal{M} = \left(W, \{\leq_i\}_{i \in \mathsf{Agt}}, V, \mathsf{rec}, E \right)$ *together with a distinguished world* $w \in W$.

Definition 14. *A testimonial model* $\mathcal{M}$ *is* introspective *iff for all* $i \in \mathsf{Agt}$, $w, v \in W$, $P \subseteq W$: *if* $w \sim_i v$, *then* $E(i, w, P) = E(i, v, P)$.

In an introspective model, agents have knowledge of their own authority relations, reflecting the assumption that agents have introspective access to their views about other agents.

Definition 15. *A testimonial model is* fully public *iff for all* $i \in \mathsf{Agt}$, $w, v \in W$, $P \subseteq W$: $E(i, w, P) = E(i, v, P)$.

In such a model, each agent's authority relations are commonly known. Hence fully public models are introspective. In this paper we consider only fully public models, so we write $\preceq_i^{\varphi}$ instead of $\preceq_{i,w}^{\varphi}$.

Definition 16. *Given a pointed model* $(\mathcal{M}, w)$, *the updated pointed model* $(\mathcal{M}, w)^{\langle j, !\varphi \rangle} = \left(W, \{\leq_i\}_{i \in \mathsf{Agt}}^{\langle j, !\varphi \rangle}, V, \mathsf{rec}^{\langle j, !\varphi \rangle}, E, w \right)$ *is defined as follows (taking* $[\![\alpha]\!]$ *to be the truth set of* α *in* $(\mathcal{M}, w)$):

1. For $P \neq [\![\varphi]\!]$ and $P \neq W - [\![\varphi]\!]$, let $\mathsf{rec}^{\langle j, !\varphi \rangle}(P) = \mathsf{rec}(P)$.
   Let $\mathsf{rec}^{\langle j, !\varphi \rangle}([\![\varphi]\!]) = \mathsf{rec}([\![\varphi]\!]) \cup \{i\}$.
   Let $\mathsf{rec}^{\langle j, !\varphi \rangle}(W - [\![\varphi]\!]) = \mathsf{rec}(W - [\![\varphi]\!]) - \{i\}$.
2. Let $\leq_j^{\langle j, !\varphi \rangle} = \leq_j$. For $i \in \mathsf{Agt}$, $i \neq j$:
   (a) If $\mathsf{rec}^{\langle j, !\varphi \rangle}(W - [\![\varphi]\!]) \prec_i^{[\![\varphi]\!]} \mathsf{rec}^{\langle j, !\varphi \rangle}([\![\varphi]\!])$ and
       $[\![B_j \varphi \wedge \varphi]\!] \cap \sim_i (w) \neq \emptyset$, then let $\leq_i^{\langle j, !\varphi \rangle} = \leq_i^{\uparrow B_j \varphi \wedge \varphi}$.
   (b) If $\mathsf{rec}^{\langle j, !\varphi \rangle}([\![\varphi]\!]) \prec_i^{[\![\varphi]\!]} \mathsf{rec}^{\langle j, !\varphi \rangle}(W - [\![\varphi]\!])$ and
       $[\![B_j \varphi \wedge \neg\varphi]\!] \cap \sim_i (w) \neq \emptyset$, then let $\leq_i^{\langle j, !\varphi \rangle} = \leq_i^{\uparrow B_j \varphi \wedge \neg\varphi}$.

---

[5] Without the condition, we would have to give $S \preceq_i^{\varphi} S'$ a more complicated reading than "S' is as authoritative as S on φ for i," since both $S \preceq_i^{\varphi} S'$ and $S' \preceq_i^{\neg\varphi} S$ could be true. A little reflection also shows that transitivity for $\preceq_{i,w}^{P}$ would no longer hold. One reason to drop the condition that $S \preceq_{i,w}^{P} S' \Leftrightarrow S \preceq_{i,w}^{W-P} S'$ is that it implies that j is authoritative on φ for i if and only if j is authoritative on $\neg\varphi$ for i. Yet this principle is subject to counterexamples. (We leave it to the reader to think of some.) In cases where such counterexamples arise, the condition may be dropped. In the cases we have in mind, those of testimony among scientific experts, it seems more plausible to require that a scientist j is authoritative on a scientific proposition φ if and only if j is authoritative on $\neg\varphi$. Since counterexamples can be produced for other conditions that one might consider imposing on authority relations, we do not build other "principles of authority" or "principles of trust" into the logic.

(c) If $\mathrm{rec}^{\langle j,!\varphi\rangle}\left(W - [\![\varphi]\!]\right) \approx_i^{[\![\varphi]\!]} \mathrm{rec}^{\langle j,!\varphi\rangle}\left([\![\varphi]\!]\right) \not\approx_i^{[\![\varphi]\!]} \emptyset,$
$[\![B_j\varphi \wedge \varphi]\!] \cap \sim_i(w) \neq \emptyset$ and $[\![B_j\varphi \wedge \neg\varphi]\!] \cap \sim_i(w) \neq \emptyset,$
then let $\leq_i^{\langle j,!\varphi\rangle} = \leq_i^{\uparrow B_j\varphi,\downarrow\varphi}.$

(d) Otherwise let $\leq_i^{\langle j,!\varphi\rangle} = \leq_i^{\uparrow B_j\varphi}.$

Note that if $(\mathcal{M}, w)$ is introspective/fully public, then $(\mathcal{M}, w)^{\langle j,!\varphi\rangle}$ is introspective/fully public, since upgrade and suspension do not change the comparability relations $\sim_i$ or the expertise function E.

The definition states when j testifies that φ, j goes on the record for $[\![\varphi]\!]$ and comes off the record for $W - [\![\varphi]\!] = [\![\neg\varphi]\!]$. For another agent i, after j testifies that φ, if i judges those on the record for $[\![\varphi]\!]$ to be more authoritative than those on the record for $[\![\neg\varphi]\!]$, and i considers it possible for j to truly believe φ, then i upgrades with $B_j\varphi \wedge \varphi$. If, on the other hand, i judges those on the record for $[\![\neg\varphi]\!]$ to be more authoritative, and i considers it possible for j to falsely believe φ, then i upgrades with $B_j\varphi \wedge \neg\varphi$. If i judges both groups to be authoritative and equally so, and i considers it possible both that j truly believes φ and that j falsely believes φ, then i upgrades with $B_j\varphi$ but suspends belief on φ itself. Otherwise i simply upgrades with $B_j\varphi$, and whether i comes to believe φ in this case depends on i's beliefs about j's doxastic reliability.[6]

We should note several features of the testimony update. First, we do not allow an agent to be on the record for a proposition and its complement at the same time, because if this were possible, then i could count the same agent's authority both in favor of and against a proposition.[7] Second, note that after j testifies that φ, while i compares the authority of those on the record for $[\![\varphi]\!]$ with the authority of those on the record for $[\![\neg\varphi]\!]$, i does not consider those who have testified that $\psi \wedge \varphi$ or those who have testified that ψ and $\psi \rightarrow \neg\varphi$ (assuming these are not equivalent to φ or $\neg\varphi$ in the model). This is unproblematic if we wish to model single-issue testimonial sequences, but if we wish to model multi-issue sequences in which agents testify on related formulas, the definition of update for testimony must be more complex. Finally, note that i considers all and only those on the record for $[\![\varphi]\!]$ and those on the record for $[\![\neg\varphi]\!]$. Other policies are possible. For example, i may choose not to consider k's authority in favor of $[\![\varphi]\!]$ if although k testified that φ, i believes that k no longer believes φ. By counting the authority of those agents who have testified that φ but who (i believes) no longer believe φ, we assume that i judges the testimonial reliability

---

[6] Hence when i does not have any view of the relative testimonial reliability of the $[\![\varphi]\!]$-testifiers versus the $[\![\neg\varphi]\!]$-testifiers (or he does, but he considers it impossible for j to believe φ and for φ to have the truth value claimed by the more authoritative group of agents), i "falls back" on his beliefs (if he has any) about j's doxastic reliability. Note that our semantics for the testimony operator gives a kind of priority to the assumption of sincerity. Whatever else i believes as a result of j's testimony that φ, i will believe that j believes φ.

[7] We do, however, allow an agent to count his own past testimony in favor of accepting a proposition. We do not rule this out because the modeler can do so when specifying the authority graphs: simply let $S \approx_i^P S \cup \{i\}$ for every $S \subseteq \mathsf{Agt}$ and $P \subseteq W$.

of k in terms of how reliable k's past *testimony* has been, even in cases where (i believes) k later gave up the belief that the testimony expressed.[8]

Two other model transformations are useful to include in DTL, but we do not define them formally here. The first is a simpler testimony transformation that updates only agents' beliefs about *factual* formulas, ignoring how agents' higher-order beliefs (about others' beliefs) change in response to testimony. Such a transformation is sufficient to model the bandwagon effect of Section 1, where higher-order beliefs are not essential, and using an associated operator in formulas reduces the complexity of deriving what agents' factual beliefs will be after a testimonial sequence. The second transformation is *public retraction*. So far, the only way for an agent to get off the record for $[\![\varphi]\!]$ is to go on the record for $[\![\neg\varphi]\!]$. Yet we can easily define a model transformation that allows an agent j to publicly retract his past testimony so that he will no longer be on the record for $[\![\varphi]\!]$, and other agents will upgrade with $\neg B_j\varphi$.

Definition 17. *The truth definitions for the new formulas are:*

o $\mathcal{M}, w \vDash U\varphi$ iff for all $v \in W$, $\mathcal{M}, v \vDash \varphi$
o $\mathcal{M}, w \vDash Rec\,(i, \varphi)$ iff $i \in \mathsf{rec}\,([\![\varphi]\!])$
o $\mathcal{M}, w \vDash S \leq_i^\varphi S'$ iff $S \leq_i^{[\![\varphi]\!]} S'$
o $\mathcal{M}, w \vDash \langle j, !\varphi \rangle\,\psi$ iff $\mathcal{M}, w \vDash B_j\varphi$ and $(\mathcal{M}, w)^{\langle j, !\varphi \rangle}, w \vDash \psi$

In the Appendix, we give a sound and complete axiomatization for basic DTL over the class of fully public models.

4.3   Reliability, Opinion, and Trust

We can now express formally the difference between i judging j doxastically reliable and i judging j testimonially reliable. For example, the satisfiable formula $\neg B_i^{B_j\varphi}\varphi \wedge \langle j, !\varphi \rangle\,B_i\varphi$ expresses that if i were to learn that $B_j\varphi$ (and nothing stronger), i would not come to believe φ, but if j were to publicly testify that φ, i would come to believe φ, reflecting the kind of situation described in Section 1.3.

Given the precondition of $B_j\varphi$ for $\langle j, !\varphi \rangle$, we can express "after j testifies with his opinion on φ, ψ is the case" as follows[9]:

$$\langle j, ?\varphi \rangle\,\psi := (\neg B_j\varphi \wedge \neg B_j\neg\varphi \wedge \langle j, !\top \rangle\,\psi) \vee \langle j, !\varphi \rangle\,\psi \vee \langle j, !\neg\varphi \rangle\,\psi$$

We can also define the notion of trust we have adopted. First, we need two abbreviations for $S \subseteq \mathsf{Agt}$:

---

[8] We could distinguish such *pure* testimonial reliability from a *hybrid* testimonial-doxastic reliability, judged in terms of the reliability of an agent's testimony in just those cases in which the agent retained the belief that his testimony expressed.

[9] One defect of this definition is that if j has no opinion on φ, then j testifies to $\top$, which provides no information to other agents. Yet intuitively, the fact that j has no opinion is itself informative. We leave it to future work to capture these subtleties.

○ $Rec\,(S,\varphi) := \bigwedge_{s \in S} Rec\,(s,\varphi)$

○ $Abs\,(S,\varphi) := \bigwedge_{s \in S} (\neg Rec\,(s,\varphi) \wedge \neg Rec\,(s,\neg\varphi))$

We read $Abs\,(S,\varphi)$ as "the agents in S have (so far) *abstained* from testifying on φ."

Let $Part_X$ be the set of *partitions* of X. Then the trust formula $T_i\,(j,\varphi)$ is:

$$\alpha \wedge \left(\bigvee_{\{X,Y,Z\} \in Part_{\mathsf{Agt}}} (Rec\,(X,\varphi) \wedge Rec\,(Y,\neg\varphi) \wedge Abs\,(Z,\varphi) \wedge \beta) \right)$$

where $\alpha := \hat{K}_i\,(B_j\varphi \wedge \varphi)$ and $\beta := Y - \{j\} \prec_i^{\varphi} X \cup \{j\}$. We read $T_i\,(j,\varphi)$ as "i trusts j's testimony on φ." As indicated in the semantics for $\langle j, !\varphi \rangle$, in order for i to trust j on φ, i must consider it possible for i to believe φ and for φ to be true at the same time.[10]

Given our definition of trust, the formulas $(B_j\varphi \wedge T_i\,(j,\varphi)) \rightarrow \langle j, !\varphi \rangle B_i\varphi$ and $T_i\,(j,\varphi) \rightarrow K_i T_i\,(j,\varphi)$ are basic DTL validities (and analogues of axioms C3 and C2 in Liau's [10] static logic of trust).

5   Application: Bandwagon Effects

Given the intended readings of DTL formulas, we can sketch how to model Sorensen's bandwagon scenario. Let the initial premises about the three agents in the example be:

1. $B_1\varphi$ $\emptyset \approx_1^{\varphi} \{2\} \approx_1^{\varphi} \{3\} \prec_1^{\varphi} \{2,3\}$
2. $\neg\,(B_2\varphi \vee B_2\neg\varphi)$ $\emptyset \prec_2^{\varphi} \{1\} \approx_2^{\varphi} \{3\} \prec_2^{\varphi} \{1,3\}$
3. $B_3\neg\varphi$ $\emptyset \approx_3^{\varphi} \{1\} \approx_3^{\varphi} \{2\} \prec_3^{\varphi} \{1,2\}$

Since the epistemic bandwagon effect need not involve agents' higher-order beliefs (but only beliefs about who is on the record), we can model it using the simpler kind of testimony operator alluded to in Section 4.2. If we denote that

---

[10] Note that whether i trusts j's testimony on φ depends on who else is on the record for $[\![\varphi]\!]$ and who else is on for $[\![\neg\varphi]\!]$. This explains why $T_i\,(j,\varphi)$ is consistent with $\emptyset \not\prec_i^{\varphi} \{j\}$ and even $\{j\} \prec_i^{\varphi} \emptyset$. If we wished to define "individual trust" or "isolated trust," we could let $IT_i\,(j,\varphi) := \emptyset \prec_i^{\varphi} \{j\}$. Here a peculiarity arises if we consider *anti-authorities*, those agents j such that $\{j\} \prec_i^{[\![\varphi]\!]} \emptyset$. Agent i can "trust" the anti-authority j, even though $\{j\} \cup \mathsf{rec}\,([\![\varphi]\!]) \prec_i^{[\![\varphi]\!]} \emptyset$. This occurs in a situation in which although $\{j\} \cup \mathsf{rec}\,([\![\varphi]\!]) \prec_i^{[\![\varphi]\!]} \emptyset$, we also have $\mathsf{rec}\,([\![\neg\varphi]\!]) \prec_i^{[\![\varphi]\!]} \{j\} \cup \mathsf{rec}\,([\![\varphi]\!])$. Suppose $\mathsf{rec}\,([\![\varphi]\!])$ and $\mathsf{rec}\,([\![\neg\varphi]\!])$ are both anti-authoritative for i. If $\mathsf{rec}\,([\![\neg\varphi]\!])$ is *more* anti-authoritative than $\mathsf{rec}\,([\![\varphi]\!]) \cup \{j\}$ for i, then when j testifies that φ, i will actually come to believe φ. It is certainly a stretch to say that i "trusts" j in this case, but it would unduly complicate our terminology to introduce distinctions for this case.

operator by $\langle i, \varphi! \rangle$ instead of $\langle i, !\varphi \rangle$, then the following formulas, which represent the outcomes of the two testimonial sequences, are easily derivable from the premises above[11]:

$$\langle 1, \varphi? \rangle \langle 2, \varphi? \rangle \langle 3, \varphi? \rangle B_1\varphi \wedge B_2\varphi \wedge B_3\varphi$$
$$\langle 3, \varphi? \rangle \langle 2, \varphi? \rangle \langle 1, \varphi? \rangle B_1\neg\varphi \wedge B_2\neg\varphi \wedge B_3\neg\varphi$$

A list of DTL validities facilitates the derivation of these formulas, but for reasons of space we do not give the list or the derivation here.

6 Conclusion

We have proposed a logic to model belief change over sequences of multi-agent testimony. In Section 1 we began by distinguishing *testimony* from *public announcement*, considering the informational difference that the individual source of testimony makes. We also distinguished judgments of *doxastic* reliability from those of *testimonial* reliability, showing in Section 2 how multi-agent plausibility models capture the former and in Section 4 how testimonial models capture both. In Section 3 we showed how to represent *suspension of belief*, providing the key reduction axiom for the associated dynamic operator. We introduced *dynamic testimonial logic* (DTL) in Section 4, and we defined a dynamic testimony operator for one policy of belief revision in response to testimony. In Section 5, we showed how DTL can represent agents' judgments of the two kinds of reliability, as well as agents *trusting* the testimony of others. Finally, in Section 6 we sketched how to model *epistemic bandwagon effects* in DTL.

In future work we plan to extend basic DTL to represent further aspects of testimony, including testimonial sequences on multiple issues, authority graphs that are not fully public, operations that change authority graphs, trust about doxastic as well as factual formulas, and agents' reasoning about how testimony has influenced the beliefs of others. In a full version of DTL, we hope to model not only how bandwagons start, but also how to stop them.

Acknowledgements. I wish to thank Johan van Benthem, Tomohiro Hoshi, Thomas Icard, and Eric Pacuit for generous and helpful feedback on this paper. I also benefited from discussion of this work in the 2009 Logical Dynamics Workshop at Stanford University.

References

1. Baltag, A., Smets, S.: A qualitative theory of dynamic belief revision. In: Bonanno, G., van der Hoek, W., Wooldridge, M. (eds.) Logic and the Foundations of Game and Decision Theory (LOFT7). Texts in Logic and Games, vol. 3, pp. 13–60. Amsterdam University Press (2008)

---

[11] If we use the $\langle i, !\varphi \rangle$ operator defined formally in Section 4.2, then in order to derive the desired formulas we must add additional premises (not mentioned in Sorensen's description) about the three experts.

2. Board, O.: Dynamic interactive epistemology. Games and Economic Behavior 49, 49–80 (2004)
3. Broersen, J., Dastani, M.M., Huang, Z., van der Torre, L.W.N.: Trust and commitment in dynamic logic. In: Shafazand, M.H., Tjoa, A.M. (eds.) EurAsia-ICT 2002. LNCS, vol. 2510, pp. 677–684. Springer, Heidelberg (2002)
4. Demolombe, R.: To trust information sources: a proposal for a modal logical framework. In: Castelfranchi, C., Tan, Y.-H. (eds.) Trust and deception in virtual societies, pp. 111–124. Kluwer Academic Publishers, Dordrecht (2001)
5. Demolombe, R.: Reasoning about trust: A formal logical framework. In: Jensen, C., Poslad, S., Dimitrakos, T. (eds.) iTrust 2004. LNCS, vol. 2995, pp. 291–303. Springer, Heidelberg (2004)
6. Graham, P.: What is testimony? The Philosophical Quarterly 47(187), 227–232 (1997)
7. Hardwig, J.: Epistemic dependence. The Journal of Philosophy 82(7), 335–349 (1985)
8. Hardwig, J.: The role of trust in knowledge. The Journal of Philosophy 88(12), 693–708 (1991)
9. Herzig, A., Lorini, E., Hübner, J.F., Ben-Naim, J., Boissier, O., Castelfranchi, C., Demolombe, R., Longin, D., Perrussel, L., Vercouter, L.: Prolegomena for a logic of trust and reputation. In: Boella, G., Pigozzi, G., Singh, M.P., Verhagen, H. (eds.) NorMAS 2008, pp. 143–157. University of Luxembourg Press, Luxembourg (2008)
10. Liau, C.-J.: Belief, information acquisition, and trust in multi-agent systems–a modal logic formulation. Artificial Intelligence 149, 31–60 (2003)
11. Sorensen, R.: Problems with electoral evaluation of expert opinions. The British Journal for the Philosophy of Science 35(1), 47–53 (1984)
12. van Benthem, J.: Dynamic logic of belief revision. Journal of Applied Non-Classical Logics 17(2), 129–155 (2007)
13. van Ditmarsch, H., van der Hoek, W., Kooi, B.: Dynamic Epistemic Logic. Springer, Heidelberg (2008)

Appendix

Theorem 1. *Together with the reduction axioms for belief upgrade and suspension and with an axiomatization for CDL plus the universal modality, the following axiom system is sound and complete for the class of* fully public testimonial models. The static axioms are:

(R1) $Rec\,(j,\varphi) \rightarrow K_i Rec\,(j,\varphi)$ (R2) $(Rec\,(j,\varphi) \wedge U\,(\varphi \leftrightarrow \psi))$
(R3) $Rec\,(j,\varphi) \rightarrow \neg Rec\,(j,\neg\varphi)$ $\rightarrow Rec\,(j,\psi)$
(A1) $S \preceq_i^\varphi S' \rightarrow K_j S \preceq_i^\varphi S'$ (A2) $(S \preceq_i^\varphi S' \wedge U\,(\varphi \leftrightarrow \psi))$
(A3) $S \preceq_i^\varphi S' \vee S' \preceq_i^\varphi S$ $\rightarrow S \preceq_i^\psi S'$
(A4) $S \preceq_i^\varphi S$ (A5) $(S \preceq_i^\varphi S' \wedge S' \preceq_i^\varphi S'')$
(A6) $S \preceq_i^\varphi S' \leftrightarrow S \preceq_i^{\neg\varphi} S'$ $\rightarrow S \preceq_i^\varphi S''$

The new reduction axioms for conservative upgrade/suspension are:

(C1) $[\uparrow_i \varphi]\,\alpha \leftrightarrow \alpha$ for α of the form $U\psi, Rec\,(j,\psi)$ or $S \preceq_j^\psi S'$
(S1) Same as (C1) but for $[\downarrow_i \varphi]$

The reduction axioms for the DTL testimony operator are:

(T1) $\langle j, !\varphi \rangle \, Rec\,(j, \psi) \leftrightarrow \big(B_j\varphi \wedge$
$\quad ((Rec\,(j, \psi) \wedge \neg U\,(\psi \leftrightarrow \neg\varphi)) \vee U\,(\psi \leftrightarrow \varphi)))$

(T2) For $i \neq j$, $\langle j, !\varphi \rangle \, Rec\,(i, \psi) \leftrightarrow (B_j\varphi \wedge Rec\,(i, \psi))$

(T3) $\langle j, !\varphi \rangle \, \alpha \leftrightarrow (B_j\varphi \wedge \alpha)$ for α of the form $p, U\psi$ or $S \preceq_i^{\psi} S'$

(T4) $\langle j, !\varphi \rangle \, \neg\psi \leftrightarrow (B_j\varphi \wedge \neg \langle j, !\varphi \rangle \, \psi)$

(T5) $\langle j, !\varphi \rangle \, (\psi \wedge \psi') \leftrightarrow (\langle j, !\varphi \rangle \, \psi \wedge \langle j, !\varphi \rangle \, \psi')$

(T6) For θ and ψ that do not contain testimony operators,

$$\langle j, !\varphi \rangle \, B_i^{\theta} \psi \leftrightarrow \Bigg(B_j\varphi \wedge$$

$$\left(\bigvee_{\{X,Y,Z,V\} \in Part_{\mathsf{Agt}\left(B_i^{\theta}\psi\right) - \{j\}}} \Big(T_X\,(j, \varphi) \wedge T_Y^-\,(j, \varphi) \wedge T_Z^{\approx}\,(j, \varphi) \wedge U_V\,(j, \varphi)\right.$$

$$\left.\wedge [\uparrow_X B_j\varphi \wedge \varphi] [\uparrow_Y B_j\varphi \wedge \neg\varphi] [\uparrow_Z B_j\varphi] [\downarrow_Z \varphi] [\uparrow_V B_j\varphi] B_i^{\theta^*} \psi^* \Big)\right)\Bigg)$$

where we use the following notation: $\mathsf{Agt}\,(B_i^{\theta}\psi)$ is the set of agent-symbols occurring in $B_i^{\theta}\psi$; $T_i^-\,(j, \varphi)$ and $T_i^{\approx}\,(j, \varphi)$ are obtained by changing α and β in the definition of $T_i\,(j, \varphi)$ in Section 4.3. For $T_i^-\,(j, \varphi)$ set $\alpha := \hat{K}_i\,(B_j\varphi \wedge \neg\varphi)$, $\beta := X \cup \{j\} \prec_i^{\varphi} Y - \{j\}$; for $T_i^{\approx}\,(j, \varphi)$ set $\alpha := \hat{K}_i\,(B_j\varphi \wedge \varphi) \wedge \hat{K}_i\,(B_j\varphi \wedge \neg\varphi)$, $\beta := Y - \{j\} \approx_i^{\varphi} X \cup \{j\} \wedge \neg(X \cup \{j\} \approx_i^{\varphi} \emptyset)$. Given $S \subseteq \mathsf{Agt}$, we also define:

○ $T_S\,(j, \varphi) := \bigwedge_{s \in S} T_s\,(j, \varphi)$, and similarly for $T_S^-\,(j, \varphi)$ and $T_S^{\approx}\,(j, \varphi)$

○ $U_S\,(j, \varphi) := \bigwedge_{s \in S} (\neg T_s\,(j, \varphi) \wedge \neg T_s^-\,(j, \varphi) \wedge \neg T_s^{\approx}\,(j, \varphi))$

Finally, the formulas θ^* and ψ^* are obtained from θ and ψ respectively by replacing, for all α, each occurrence of $Rec\,(j, \alpha)$ in θ and ψ by $((Rec\,(j, \alpha) \wedge \neg U\,(\alpha \leftrightarrow \neg\varphi)) \vee U\,(\alpha \leftrightarrow \varphi)))$.

Proof. (rough sketch) **Soundness.** Axiom (R1) reflects the global nature of the rec function, while (A1) reflects the *full publicity* of models. (R2) and (A2) hold because the $\preceq_i^P$ relations and rec function deal with propositions as sets of worlds. (R3) (with (R2)) corresponds to the condition that an agent cannot be on the record for both a proposition and its complement at the same time. (A3)−(A5) express the basic properties of the $\preceq_i^P$ relations, while (A6) (with (A2)) expresses the condition that authority relations must agree for a proposition and its complement. (C1) and (S1) indicate that belief upgrade and suspension do not change universal propositional facts, the rec function, or the $\preceq_i^P$ relations.

(T1) says that after j testifies that φ, j is on the record for ψ iff j was already on the record for ψ and ψ is not equivalent in the model to $\neg\varphi$ (for if it were, j would have been taken off the record for ψ when he testified that φ) or ψ is equivalent in the model to φ (in which case j was added to the record for ψ when he testified that φ). (T2) says that for agents other than j, the record does not change after j's testimony. (T3) reflects the fact that testimony does not change local or universal propositional facts or the $\preceq_i^P$ relations. (T4) − (T5) give standard properties of dynamic operators.

(T6) captures the effect of j's testimony that φ on agents' beliefs, which follows exactly the semantic definition of model transformation due to testimony. Note that we only consider what happens to the beliefs of those agents whose symbols appear in $B_i^\theta \psi$, since others do not matter for evaluating the formula. Following the semantics for the testimony operator, we do not update j's beliefs. The reason for the change from θ and ψ to θ^* and ψ^* is that θ and ψ may contain a formula of the form $Rec(j,\alpha)$, the truth value of which may change after j's testimony. Hence we use the same idea as in (T1) and express what must be true in the original model in order for $Rec(j,\alpha)$ to be true in the updated model.[12]

Completeness. Using the reduction axioms (C1) − (T6), every formula of basic DTL with dynamic operators is reducible to an equivalent formula in the static part of the language.[13] It therefore suffices to show completeness for the static part of DTL. We use the standard canonical model construction, only we must show how to construct the expertise function and the public record for the canonical model.

Suppose Γ is a consistent set of CDL formulas plus $\preceq_i^\varphi$ and Rec formulas. Extend Γ to a maximally consistent set Γ^+ in the usual way. Let $\Gamma_{\preceq}^+$ be the set of authority formulas $S \preceq_i^\varphi S'$ in Γ^+ and Γ_{rec}^+ the set of record formulas $Rec(i,\varphi)$ in Γ^+. The *canonical testimonial model* based on $\Gamma_{\preceq}^+$ and Γ_{rec}^+ has the domain:

$$W^C = \left\{ \Delta \mid \Delta \text{ is an MCS and } \Delta \cup \Gamma_{\preceq}^+ \cup \Gamma_{\text{rec}}^+ \text{ is consistent} \right\}$$

We extract a canonical set of authority relations from $\Gamma_{\preceq}^+$ as follows: if $S \preceq_i^\varphi S' \in \Gamma_{\preceq}^+$, then for every $\Delta \in W^C$, construct the relation $\preceq_{i,\Delta}^{[\![\varphi]\!]}$ such that $S \preceq_{i,\Delta}^{[\![\varphi]\!]} S'$. This construction clearly produces a (fully public) authority relation. For suppose $\preceq_{i,\Delta}^{[\![\varphi]\!]}$ is not reflexive, because $S \not\preceq_{i,\Delta}^{[\![\varphi]\!]} S$. Then by construction, $S \preceq_i^\varphi S \notin \Gamma_{\preceq}^+$. Since $\Gamma_{\preceq}^+$ is a maximally consistent and hence complete set of authority formulas, we have $S \not\preceq_i^\varphi S \in \Gamma_{\preceq}^+$. But then $\Gamma_{\preceq}^+$ is inconsistent by (A4). The arguments for the other properties of authority relations follow similarly. We define the canonical expertise function E^C as the function that maps each triple (i, Δ, P) of $i \in \mathsf{Agt}$, $\Delta \in W^C$, and $P \subseteq W^C$ to the relation $\preceq_{i,\Delta}^P$ constructed as above (except for those $P \subseteq W^C$ that are not

---

[12] The replacement of θ and ψ by θ^* and ψ^* does not achieve the correct result if either θ or ψ contains additional testimony operators, hence the restriction on θ and ψ.

[13] We reduce DTL formulas by applying the reduction axioms from the "inside out," eliminating dynamic operators from subformulas first. This explains why the restriction on θ and ψ in (T6) is acceptable; since θ and ψ are subformulas of $B_i^\theta \psi$, any testimony operators will be eliminated from them by the time we get to $\langle j, !\varphi \rangle B_i^\theta \psi$.

definable by φ_0 formulas, for which we let $\preceq^P_{i,\Delta}$ and $\preceq^{W-P}_{i,\Delta}$ be the same, arbitrary total preorder on $\mathcal{P}(\mathsf{Agt})$). We define the canonical rec^C function by setting $\mathsf{rec}^C(P) = \{i \in \mathsf{Agt} \mid Rec(i,\varphi) \in \Gamma^+_{\mathsf{rec}}$ and $[\![\varphi]\!] = P\}$ for all $i \in \mathsf{Agt}$. It is straightforward to check that this construction produces a proper public record.

The rest of the completeness proof for DTL uses the standard methods, following the completeness proof for CDL (see [2]).

From the Logical Point of View: The Chain Store Paradox Revisited

Li Li[1], Robert C. Koons[2], and Jianjun Zhang[1]

[1] Department of Philosophy, Nanjing University
[2] Department of Philosophy, University of Texas at Austin

Abstract. The standard approach to a rational action paradox in game theory (namely, the chain store paradox) has presupposed that the player's beliefs are probabilities represented by functions with values between 0 and 1. However, a general solution must include the possibility that the subjective probabilities take only the values 1 and 0, requiring a non-Bayesian account of belief revision. In this paper, we propose a situation-theoretic diagnosis and solution to the paradox, based on the conception of Austinian propositions relativized to particular situations, as developed by Barwise and Etchemendy.

Keywords: Selten chainstore paradox, situation theory, belief revision.

1 Introduction

Reinhard Selten has raised a rational choice paradox in game theory: the chain store paradox. Its extensive form produces inconsistency between game theory and plausible human behavior.

To explain in detail, imagine a market situation: a monopolist A has twenty chain stores in different cities of a region. Assume he faces some fixed number of potential competitors. Each potential competitor has only a single opportunity to enter the market. When the opportunity comes, each competitor has to decide whether to enter or not. If he does enter, the monopolist has to choose whether to accommodate or to deter (by aggressive pricing). If the monopolist accommodates, it will cost him less and yield a profit for the competitor.

According to standard game theory, the monopolist deduces the following by backward induction: after the last competitor, there is no point to aggressive pricing, since there are no more competitors to attempt to deter. So, to accommodate in the last round is the rational strategy for the monopolist to take, since it costs less. Thus, if the last competitor is rational, knowing the monopolist will cooperate, he will choose to enter the market, and the monopolist will accommodate.

Consequently, the next to the last competitor will enter, too, knowing the rationality of the monopolist, who will certainly not act aggressively in the next to last round, since attempting to deter the last competitor would be futile. Thus, the induction argument reaches the conclusion that each of the players should

X. He, J. Horty, and E. Pacuit (Eds.): LORI 2009, LNAI 5834, pp. 180–188, 2009.

choose to enter and the monopolist should always react with his accommodating response [6].

However, intuitively, if the monopolist were to choose to deter at an early stage, it is probable that the next competitor would not enter the market, since the monopolist's deviation from the predicted strategy would force him to revise his estimate of the monopolist's utility structure. Given the likely success of deterrence, the monopolist should choose to deter in the first round. Selten calls this the conclusion of "deterrence theory" [6, pg. 131 - 132].

Thus, there is a contradiction between the conclusion of the induction argument and a plausible deterrence theory. Selten named it "the chain store paradox."

2   Game Theory Solution

Philip Reny [4] argued that game theory assumes that both the monopolist and the competitor are rational agents and that the knowledge of the structure of the game and of the principles of game theory is a matter of mutual belief between them. So they both know (via the induction argument) that the monopolist will not deter. However, if we add the assumption that the monopolist does deter in any round, a contradiction is generated, since we are assuming both that the monopolist is rational and that he is irrational. So, the natural way to avoid the paradox is simply to avoid making assumptions that contradict the theory of the game.

However, Reny's assumption that deterrence behavior by the monopolist is irrational is itself subject to a paradox. If we modify the game theory by assuming that the monopolist acts "irrationally" by attempting to deter in the first round, we must conclude that this attempted deterrence would be successful, which makes his choice rational after all. The paradox still exists.

Christina Bicchieri [2] defended the claim that the monopolist's efforts to deter wouldn't succeed, for the competitor's belief will not be updated in the appropriate way, since their prior probability of the monopolist's irrationality is zero. If the retaliation did emerge, it would just be treated by the succeeding competitors as a freak accident, without implications for the monopolist's future behavior. Thus, even if the assumption is added, it will not affect the results, and the backward induction argument succeeds.

Reny and Bicchieri's arguments are based on the work of David M. Kreps and Robert Wilson.[5] They argued that the assumption of complete information (with certainty) in the Chain-Store Paradox is more a modeling artifact than a good representation of real situations. It is easy to think that potential entrants are in fact uncertain about the monopolist's payoffs.

Their solution depends on the use of mixed strategy equilibrium , and in the general case, on concentroid mixed strategies. Here, the mixed strategy equilibrium is defined as a Nash or correlated equilibrium in which one or more players play a mixed strategy. A Nash equilibrium contains a noncentroid mixed strategy if it contains a mixed strategy in which some player do not play these equally valued pure strategies with equal probabilities.[3, pg. 30] A correlated equilibrium assigns a probability to each combination of strategies, and thus assigns to each strategy of each player a probabilistic distribution over the strategies of the other players.

Kreps et al. interpret the mixed-strategy equilibrium as representing the players' collective uncertainty about each others' "type", where each type assigns a different utility to the outcomes of various joint pure strategies. The competitors are uncertain about whether the monopolist enjoys aggressive behavior as an end in itself, and the monopolist is uncertain about how sensitive the competitors are to the threat of future aggression. The players update their subjective probabilities using Bayes's theorem. At the equilibrium point, the monopolist's acting aggressively in the first round results in belief updates that induce a state of near indifference in the competitors between entering and not entering: each competitor's entrance depends on his particular type. Similarly, the equilibrium point induces near indifference in the monopolist between responding aggressively and accommodating. What had been a contradiction (when uncertainty was overlooked) resolves itself into an equilibrium point of mutually supporting states of uncertainty.

3   Situation Theory Approach

Selten believed that the paradox was an unexpected, surprising result in game theory, but we will show: "it is a paradoxical in the strong sense: a logical antinomy of rational belief." [3, pg. 26]

The solution of Kreps at al. presupposes that the monopolist's and competitors' beliefs are represented by subjective probabilities with values between 0 and 1 (exclusive).But what if all the relevant probabilities took the values of 0 and 1 only? This possibility cannot be excluded by mere stipulation. If we just replace the probability with a model that includes only believing or disbelieving, we must address explicitly the problem of belief revision, since Bayesian conditioning will no longer be well-defined (since the prior probability of unexpected behavior by the monopolist would be equal to zero).

Let Jip represent the rational belief of player $i(i = m, c)$ in proposition p, and let K represent the subjunctive conditional proposition that if the monopolist were to retaliate against the first competitor, the second competitor would stay out. The truth of K depends on how the competitors revise their beliefs. If the competitor does not think that the monopolist believes that if he retaliates, the second competitor will stay out, he will be sure that the monopolist would not act aggressively, since the backward induction argument would in that case be dispositive. Given this conviction on the part of the second competitor, if the monopolist were, contrary to expectation, to act aggressively in the first round, the second competitor would have to revise his beliefs about the monopolist's utilities. Let's assume that, in such a case, the second competitor would come to believe that the monopolist is an inveterate retaliator. On this assumption, retaliation by the monopolist in the first round would in fact successfully deter the second competitor, and so would be an optimal strategy for the monopolist, given his actual, non-aggressive utility function. Thus, if the competitor believes that the monopolist does not believe in K, then K would in fact be true.

Conversely, assume that the second competitor does believe that the monopolist believes that if he were to act aggressively in the first round, then the second competitor would stay out. The second competitor will then expect the monopolist to act aggressively in the first round (given the opportunity). Consequently, such aggressive action will result in no revision of the second competitor's beliefs. Since the game ends in the second round, and since the monopolist has no possible reason to act aggressively in that round, the second competitor would not be deterred. Thus, if the second competitor believes that the monopolist does believe in K, then K would in fact be false.

Since the monopolist has available all of the information that we used in reaching this conclusion, we should assume that the monopolist himself believes $K \leftrightarrow \neg JcJmK$, that is, the second competitor would be deterred from entering if and only if he does not believe that the monopolist believes in K. If the monopolist also believes that the competitor believes JmK if and only if the monopolist himself believes it (a reasonable assumption, given their mutual rationality and their common knowledge of the game), then the monopolist will infer.

(J1) and (J5) are axioms and rules on a very plausible of rational belief, which is meant to capture the properties of justifiable or ideally rational belief (as opposed to knowledge):

(J1) $\neg J\bot$
(J2) $J\varphi$,where φ is a logical axiom
(J3) $J(\varphi \rightarrow \phi) \rightarrow (J\varphi \rightarrow J\phi)$
(J4) $J\varphi \rightarrow JJ\varphi$
(J5)   from φ infer $J\varphi$.

(J1) is a schema which express the consistency of rational justified belief. Schema (J4) is too strong, in general, but it is also dispensable. [3, pg. 15] We employ it here simply for simplicity's sake. Schemata (J2) and (J3) simply ensure that the property of being rationally justifiable in a situation is closed under logical entailment. (J5) guarantees that certain obviously true axioms of the logic of rationally justifiable belief are themselves rationally justifiable in the situation under consideration. We will need the following lemma for the following proof:

Lemma 1. $J\neg J\varphi \rightarrow \neg J\varphi$.

Lemma 1 follows very quickly from (J1)–(J4). The inconsistency can be proved as follows:

(1) $J(K \leftrightarrow \neg JJK)$ (assumption for reductio)
(2) $JK,$ (assumption for conditional proof)
(3) $J\neg JJK,$ (1) (2) (J2) (J3) (1) ,(2)
(4) $\neg JJK,$ (3),(lemma 1) (1),(2)
(5) $JK \rightarrow \neg JJK,$ conditional proof ,(2)-(4), (1)

(6) $J(K \leftrightarrow \neg JJK) \to [JK \to \neg JJK]$, (conditional proof,(1)-(5))
(7) $J[J(K \leftrightarrow \neg JJK)] \to [JK \to \neg JJK]$, (6),(J5)
(8) $JJ(K \leftrightarrow \neg JJK)$, (1),(J4)(1)
(9) $J(JK \to \neg JJK)$, (7),(8),(J3)(1)
(10) JJK, (assumption for reducito)
(11) $J\neg JJK$, (9),(10), (J3) (1)(10)
(12) $\neg JJK$, (11),(lemma 1)(1)(10)
(13) $\neg JJK$, (reductio (10)-(12) (1)
(14) $J(K \leftrightarrow \neg JJK)$, (conditional proof (1)-(13))
(15) $J[J(K \leftrightarrow \neg JJK) \to \neg JJK]$, (14),(J5)
(16) $J\neg JJK$, (8),(15),(J3),(1)
(17) JK, (1),(16),(J2),(J3)(1)
(18) JJK, (17),(J4) ,(1)
(19) $\neg J(K \leftrightarrow \neg JJK)$. Reductio ,(1),(12),(18)

By diagnosing the paradox, let us look into the propositions that represent the objects of belief. By treating truth as a property of propositions, not sentences, Jon Barwise and John Etchemendy have modeled two distinct conceptions of propostions: one is the Russelian proposition and the other is the Austinian proposition. [1] In Russell's view, a sentence alone expresses a proposition, which is true or not. However, for Austin, there is always a contextual parameter — the situation the statement is about — that comes between the sentence and a proposition. "All propositions contain an additional contextually determined feature, namely, the situation they are about." [1, pg. 1]

"According to Austin, a legitimate statement A provides two things: a historical (or an actual) situation Sa, and a type of situation Ta ... the statement A is true if Sa is of type Ta; otherwise it is false."[1, pg. 29]

"Sentences express Austinian propositions, which consist of three elements: a sentence type of English assignment of extensions to the indexical and demonstrative elements of that type, and a partial model of the world that the propositions is about."[3, pg. 100]

Based on the Austin's view on propositions, Jon Barwise and John Etchemendy provided a solution to the liar's paradox. Here is a typical liar propositions (ls is a proposition that Barwise and Etchemendy call a "denial liar") [3, pg. 100]:

$$ls = \neg[ls; true(ls)]$$

S must be an actual situation for ls to be expressible. For any actual S, ls is true relative to any sufficiently large part of the world, but not relative to situation S. No actual situation S can contain the information the denial liar proposition ls is true relative to it. Hence ls will be true relativized to some larger situation S'.

In the case of the chain store paradox, the biconditional proposition $K \leftrightarrow \neg JJK$ is like a liar proposition. If we treated the proposition as Russellian, that is, both the monopolist and the competitor are in the same situation S (representing the whole world), a paradox would be generated as we discussed above. In contrast, by using Austinian propositions, situation theory can solve the paradox by treating the biconditional $K \leftrightarrow \neg JJK$ as relativized to one or more situation parameters. The biconditional itself concludes two propositional parts each of which is the argument of the rational justifiability of operator J. On the Austinlian account, each proposition must have its own situation parameter. Let us assume, to begin with that all the proposition are in the same situation, call it S. So the biconditional proposition can be represented as $K \leftrightarrow \neg JJSKS$.we argued above this biconditional is obviously true, it should be rationally believed. However, if we suppose it can be rationally believed in situation S, then the contradiction we derived above will still be derivable, the proof we give above could be used to prove the following in situation theory:

$$\neg J(K \leftrightarrow S \neg JJSKS).$$

In other words, the biconditional can not be rationally believed by competitors trapped in situation S, but could be rationally believed by outsider observers, as long as these observers are in a larger situation S', which contains more information than the situation S. Thus, no paradox is generated. In this way, without modifying the general belief axioms and rules, the paradox can be solved.

This situation solution given above is one of the phenomenons similar to that Roy Sorensen calls a "blindspot". Some fact is a blindspot for a person if there is some reason in principle why that person couldn't know it . Sorensen reach the conclusion about the chain store paradox. "Thus we have another illustration of the fact that one cannot foresee the choices of an equally well-informed ideal agent if that agent has the means and the motive to undermine the prediction."[7, pg. 359]

In fact we have shown the biconditional $K \leftrightarrow S \neg JJSKS$ is a blindspot for the players in the game but not for the rest of us. Situation theory provides some explanation of why these occurs by hypothesis some propositions are not universally accessible, it is not possible for some persons to think about the very situation in which they are embedded, only the outsider can obtain the right "perspective" on the game. The players have suffered from intellectual tunnel vision.

More details about how the situations are different from each other, and how the reasoning works in the situation remain open questions in this field.

4   Situation Theory Approach to Infinite Chain-Store-Like Paradox

We can see from the chain store paradox that a finite game which involves self-reference can generate paradox. Someone may argue that if we just imagine a infinite game, there would be many equilibria, which corresponds to many different behavioral patterns. In the infinite case, without the backward induction argument, the paradox will not appear. However, by giving the example of promise keeping, which comes from the work of D. H. Hodgson, we can show that the infinite game can generate paradox as well.

To generate the Liar-like paradox, three conditions are required.

1. The promise-receivers must expect that promise-maker to conform to one specific behavioral pattern, corresponding to one of the many Nash equilibria.
2. The promise-receivers would revise their beliefs about the promise-maker in a non-Bayesian way, should they observe off-equilibrium behavor.
3. The way in which they change their beliefs depends on the beliefs they have about the rationality (given the promise-maker's utilities) of the observed deviation. More specifically: if they believe that the deviation is rational, then they will assume that the deviation was a simple mistake (the result of what game-theorists call a "trembling hand"), and they will expect the promisor to return to the equilibrium path in the future. Alternatively, if they believe that the deviation was irrational, then they will completely abandon their expectation of the original equilibrium pattern.

Let's compare the familiar Prisoner's Dilemma with that of Hodgson's Promise Keeping Dilemma:

In PD, the payoffs would be

	DC	C
$Dont'\ confess$	$3,3$	$0,4$
$Confess$	$4,0$	$1,1$

In Hodgson's game, the payoffs would be:

	$Promise$	$Don't\ trust$
$Keep$	$3,3$	$0,1$
$Break$	$4,0$	$1,1$

In the promise keeping game, the action of Breaking the promise dominates the action of Keeping it. Since this is common knowledge, in a single-play game, the promisee can be certain that the promisor will Break, and so he will choose Don't trust.

To move from a finite game to an infinite game, let's assume that the payoffs in each successive round shrink at an exponential rate, in such a way that keeping a promise in one round has a higher expected utility than breaking the promise if and only if keeping the promise induces the promisee in the next round to play Trust. Such an infinite game has infinitely many Nash equilibria. Keeping and trusting the promise in every round is one such equilibrium. However, there is also an equilibrium in which the promisor breaks his promise in the first n round and keeps the promise in all succeeding rounds (with the promises playing Don't trust in the first n rounds and Trust thereafter), and one additional equilibrium in which the promisor breaks his promise in every round and every promisee plays Don't trust.

Let's suppose that the players start the game in a state of expectation corresponding to the equilibrium of Break/Don't trust in the first n rounds, and

Keep/trust thereafter. Let's also suppose that n is a relatively large number. Now, let's consider what happens if, contrary to the certain expectations of the promisees, the promisor keeps his promise in the first round. We have to make some assumptions about the upshot of whatever non-Bayesian belief-revision mechanism is at work.

Let's suppose that, if the promisees believe that the promisor believes that, by playing off the equilibrium in round 1, he can induce the second promisee to deviate from the equilibrium by playing Trust in round 2. In this case, the promisor's deviation is consistent with the basic assumptions about his rationality and utility function, and we can assume that the second promisee will make a minimal change in his beliefs about the promisor's future play. Such minimal change would consist in believing that the observed deviation was a mere fluke, to be followed by a return by the promisor to the originally expected pattern (of promise-breaking in the rest of the first n rounds). Alternatively if the promisees believe that the promisor does not believe that deviating by keeping the promise in the first round would induce trust in the second round, then the promisees are forced to make much more radical changes in their beliefs about the rationality and motivation of the promisor. Let's suppose that they will come to believe that the monopolist enjoys keeping his promise so much that it strictly dominates promise-breaking in each round. In that case, they would expect the promisor to keep his promises in all rounds, and the second promisee would be induced to play Trust.

Thus, if the promisor is believed to believe that promise-keeping in the first round would create expectations of future promise-keeping, then doing so would in fact fail to create those expectations. Conversely, if the promisor is not believed to believe that keeping the promise would create such expectations, then keeping the promise would in fact create those expectations. Thus, we have a paradoxical blindspot of precisely the same kind as the chain store paradox.

Hodgson believed that his argument demonstrated that the moral theory of act-utilitarianism is self-refuting, since it cannot sustain the obviously advantageous practice of promising. We believe that the argument shows, at the very least, that utilitarian moral theory is seriously incomplete. Because of the pervasiveness of liar-like blindspots, utilitarian moral theory fails to provide the participants in social situations with any concrete guidance with respect to such important institutions as promising. An adequate moral and legal theory must supplement the principle of utility with respect for established social rules.

References

1. Barwise, J., Etchemendy, J.: The Liar. Oxford University Press, Oxford (1987)
2. Bicchieri, C.: Common knowledge and backward induction:A solution to the paradox. In: Varid, M. (ed.) Proceedings of the second conference on theoretical aspects of reasoning about knowledge (1988)
3. Koons, R.C.: Paradox(es) of belief and strategic rationality. Cambridge University Press, Cambridge (1992)

4. Reny, P.: Rationality, common knowledge and the theory of games. Unpblished Manuscript, University of Western Ontario, Dept of Economics (1988)
5. Kreps, D.M., Wilson, R.: Reputation and imperfect information. Journal of Economic Theory 27, 253–279 (1982)
6. Selten, R.: The chainstore paradox. Theory and decision 9(2), 127–159
7. Sorensen, R.A.: Blindspots. Clarendon Press, Oxford (1988)

A Cooperation Logic for Declaration Structures

Hu Liu*

Institute of Logic and Cognition,
Sun Yat-sen University, China
liuhu2@mail.sysu.edu.cn

Abstract. Two types of game structures for logics of ability have been
proposed: Concurrent game structures and models of propositional control. The former takes an abstract view and can be used for general
purposes; the latter is a restriction but is much easier for implementation. We present a game structure in between namely a declaration
structure. A cooperation logic based on declaration structures is given.
We present its deductive system and algorithms for model checking and
satisfiability checking.

Keywords: cooperation logics, propositional control, declarations.

1 Introduction

Cooperation logics are relatively new area of interest in game logics. They provide
formalisms to analyze ability of agents in game like systems. Modal operators
are used to express properties that a group of agents can cooperate by choosing
their game strategies to bring about certain situations. The best known examples
of cooperation logics are alternating-time temporal logic (ATL) [1] and *coalition
logic* (CL) [6]. The two are closely related and their semantics are essentially
equivalent [4]. Their models reflect an abstract view of a strategic game in which
whenever all agent have chosen their strategies, the system will come up with an
unique outcome by a *prescribed* outcome function. They do not explain where
the outcome function comes from.

From the point of view of a software agent designer, ATL and CL are not easy
to handle in the sense that they presume a detailed system specification, where all
possible system transitions are explicitly presented. Such a specification would be
impractical if the system contains a large set of variables, and therefore contains
an exponentially larger set of states. *Coalition logic of propositional control* (CL-
PC) [9] is a recently proposed logical framework that presumes a much simpler
specification: A game structure is defined by assigning to each variable an agent
that controls it. Outcomes are determined by agents choosing valuations over
variables they control. There are several succedent works along this way, such as
embedding delegation [10], action [7], or preference [8] in CL-PC, geralizations
of CL-PC [3], and applications of CL-PC [5].

---

* This research was supported by A Foundation for the Author of National Excellent
Doctoral Dissertation of PR China (FANEDD 2007B01).

X. He, J. Horty, and E. Pacuit (Eds.): LORI 2009, LNAI 5834, pp. 189–197, 2009.

Knowing who controls what is enough for a designer to build models of CL-PC. It is a natural way to specify game like structures, and is easy to use. However, arguably due to its static allocation function, CL-PC has received much less attention than ATL and CL. In CL-PC, each proposition is controlled by an appointed agent throughout the whole model, whereas applications may require loosing or regaining controls. It is meaningless to define temporal operators on a CL-PC model since it consists of a product of a family of equivalent relations. Any temporal operator would collapse to the "next" operator. [3] presented a generalization of CL-PC's model such that a proposition may be controlled by more than one agent, and may not be controlled by any agent. However, an outcome function was introduced to ensure an unique outcome under each situation. According to the above observation, such generalizations sacrifice the main advantage of propositional control. Also, a dynamic extension of CL-PC was suggested in [10], where dynamic operators were constructed by actions $a_1 \rightsquigarrow a_2$, reading as that agent a_1 gives proposition p to agent a_2. We take a different approach in this paper. First, we deal with the issue within a cooperation logic, that is, we keep the simplicity of cooperation logics where only coalition operators are presented. Second, losing or gaining control over propositions is not a result of agents' actions of transferring control; it is programmed within a system: agents losing or gaining control according certain conditions.

We define a generalization of CL-PC's model called *declaration structures* (DS). DS does not require a prescribed outcome function so that it is indeed in the school of propositional control. The basic idea is to replace allocation functions by declarations that says "who controls what in what conditions". An agent gains control of a proposition if certain condition holds, and loses control if the condition fails. We distinguishes two types of declarations: Agent declaration is the condition for an agent to control a proposition; environment declaration is the condition for no agent to control a proposition, and its value is determined by system background.

2   Declaration Structures

The set of formulas is constructed from a given set of propositions At and a given set of agents Ag, both finite and non-empty. Where $p \in At$ and $A \subseteq Ag$,

$$\varphi =:: p \mid \neg\varphi \mid \varphi_1 \vee \varphi_2 \mid \diamond_A \varphi.$$

$\diamond_A \varphi$ is a cooperation modality, and is read as that agents in A can make sure the system reaches a next state at which φ is true. Other Boolean connectives are defined as usual. Let $\square_A \varphi =_{df} \neg \diamond_A \neg\varphi$. A formula that contains no modality is called an *objective formula*.

By $l(p)$, we denote a literal p or $\neg p$. An *environment declaration* is an objective formula denoted by $\psi_{l(p)}$ with intended meaning that $l(p)$ is bound to be true if $\psi_{l(p)}$ holds. An *agent declaration* is an objective formula denoted by ψ_p^i with intended meaning that agent i controls p if ψ_p^i holds.

Definition 21. *A declaration structure is a tuple* $DS = (Ag, At, \mathcal{D}_e, \mathcal{D}_{Ag})$, *where*

(1) $Ag = \{1, \ldots, n\}$ is a finite non-empty set of agents;
(2) $At = \{p_1, \ldots, p_k\}$ is a finite non-empty set of propositional variables; a Boolean valuation $s : At \to \{0, 1\}$ is called a state.
(3) $\mathcal{D}_e$ is a set of environment declarations such that for each literal $l(p)$, there is a single declaration $\psi_{l(p)}$ in $\mathcal{D}_e$.
(4) $\mathcal{D}_{Ag}$ is a set of agent declarations such that for each agent i and each proposition p, there is a single declaration ψ_p^i in $\mathcal{D}_{Ag}$.
  We impose the following constraints on declarations: For any $p \in At$,
  (i) $\psi_p^1 \vee \ldots \vee \psi_p^n \vee \psi_p \vee \psi_{\neg p}$ is a tautology, and
  (ii) for any two different declarations ψ and ψ' in $\{\psi_p^1, \ldots, \psi_p^n, \psi_p, \psi_{\neg p}\}$, $\psi \to \neg\psi'$ is a tautology.

Remark. The constraints guarantee that at each state each proposition is controlled by a single agent or determined by the environment. Without the constraints, there will be undetermined situations in the system, and will require an outcome function to solve the problem, which should be avoided in structures based on propositional control.

Let $A = \{i_1 \ldots i_j\} \subseteq Ag$. We write ψ_p^A for declarations $\psi_p^{i_1}, \ldots, \psi_p^{i_j}$, where $\psi_p^A = \psi_p^{i_1} \vee \ldots \vee \psi_p^{i_j}$. We write $\overline{A}$ for Ag/A, and i for a singleton $\{i\}$.

Let i be an agent, s a state. $Control(i, s)$ is the set of propositions controlled by agent i at state s, i.e. $Control(i, s) = \{p | s \models \psi_p^i\}$. For any $A \subseteq Ag$, let $Control(A, s) = \bigcup\{Control(i, s) | i \in A\}$. For each state s, $Control(+, s)$ $(Control(-, s))$ is the set of propositions whose values are positively (negatively) determined by the environment, i.e. $Control(+, s) = \{p | s \models \psi_p\}$ $(Control(-, s) = \{p | s \models \psi_{\neg p}\})$.

Given a state s and $A \subseteq Ag$, s_A is the set of next states to s that can be guaranteed by agents A. Formally, $s' \in s_A$ if the followings hold for any proposition p:

(i) $s'(p) = 1$ if $p \in Control(+, s)$.
(ii) $s'(p) = 0$ if $p \in Control(-, s)$.
(iii) $s'(p) = s(p)$ if $p \in Control(Ag/A, s)$.

Definition 22. *We write $s \models_{DS} \varphi$ to indicate that φ is satisfied at state s in declaration structure DS. $\models_{DS} \varphi$ is to denote that φ is satisfied at every state in DS. The semantic rules for $\models_{DS}$ are as follows:*

$s \models_{DS} p$ *iff $s(p) = 1$, where p is a proposition.*
$s \models_{DS} \neg\varphi$ *iff $s \not\models_{DS} \varphi$.*
$s \models_{DS} \varphi_1 \vee \varphi_2$ *iff $s \models_{DS} \varphi_1$ or $s \models_{DS} \varphi_2$.*
$s \models_{DS} \Diamond_A\varphi$ *iff $s' \models_{DS} \varphi$ for some $s' \in s_A$.*

It can be checked that a declaration structure DS is semantically equivalent to a standard Kripke model that consists of all states in DS with obvious valuations, and with binary relations R_A such that sR_At iff $t \in s_A$. We will freely use DS as a Kripke model.

3   Axiom Systems for Declaration Structures

Fix a declaration structure $DS = (Ag, At, \mathcal{D}_e, \mathcal{D}_{Ag})$, a deductive system LDS for DS (*logic for DS*) consists of axioms and rules

(*Prop*) propositional tautologies,
 (K_A) $\square_A(\varphi_1 \to \varphi_2) \to (\square_A\varphi_1 \to \square_A\varphi_2)$,
 (D_A) $\square_A\varphi \to \diamond_A\varphi$,
 (*Envi*) $\psi_{l(p)} \to \square_A l(p)$,
 (*effect*) $\psi_p^A \to \diamond_A l(p)$,
(*non-effect*) $l(p) \wedge \neg\psi_p^A \wedge \neg\psi_{\neg l(p)} \to \square_A l(p)$,
 (*Comb*) $\diamond_A l(p_1) \wedge \ldots \wedge \diamond_A l(p_j) \to \diamond_A(l(p_1) \wedge \ldots \wedge l(p_j))$, where $p_1, \ldots, p_j$ are different propositions.
 (*Incl*) $\diamond_A\varphi \to \diamond_B\varphi$, where $A \subseteq B$,
 (*MP*) $\varphi, \varphi \to \psi/\psi$, and
 (N_A) $\varphi/\square_A\varphi$.

Notions of a deduction ($\Gamma \vdash_{LDS} \varphi$), a proof ($\vdash_{LDS} \varphi$) are defined as usual.

Theorem 31. *(soundness) For any formula φ, $\vdash_{LDS} \varphi$ implies $\models_{DS} \varphi$.*

By a similar argument to CL-PC's in [9], it can be shown that every formula is equivalent to an objective one.

Lemma 32. *Let $A \subseteq Ag$, $p_1, \ldots, p_j$ be different propositions. Then*

$$\vdash_{LDS} \diamond_A(l(p_1) \wedge \ldots \wedge l(p_j)) \leftrightarrow \bigwedge_{1 \leq i \leq j}(\psi_{l(p_i)} \vee \psi_{p_i}^A \vee (l(p_i) \wedge \psi_{p_i}^{\overline{A}})).$$

Proof. First, by *(Comb)* and properties of modal diamond, we have that

$$\vdash_{LDS} \diamond_A(l(p_1) \wedge \ldots \wedge l(p_j)) \leftrightarrow (\diamond_A l(p_1) \wedge \ldots \wedge \diamond_A l(p_j)).$$

It is sufficient to show that for each $1 \leq i \leq j$,

$$\vdash_{LDS} \diamond_A l(p_i) \leftrightarrow (\psi_{l(p_i)} \vee \psi_{p_i}^A \vee (l(p_i) \wedge \psi_{p_i}^{\overline{A}}))$$

For the right to left direction, by *(Envi)*, and D_A, (1) $\vdash_{LDS} \psi_{l(p_i)} \to \diamond_A l(p_i)$. By *(non-effect)*, (D_A) and definition 21-4, (2) $\vdash_{LDS} (l(p_i) \wedge \psi_{p_i}^A) \to \diamond_A l(p_i)$. From (1), (2) and *(effect)*, it is a propositional consequence that $\vdash_{LDS} (\psi_{l(p_i)} \vee \psi_{p_i}^A \vee (l(p_i) \wedge \psi_{p_i}^{\overline{A}})) \to \diamond_A l(p_i)$.

For the other direction, by *(Envi)*, (3) $\vdash_{LDS} \diamond_A l(p_i) \to \neg\psi_{\neg l(p_i)}$. By (3) and definition 21-4, (4) $\vdash_{LDS} \diamond_A l(p_i) \to (\psi_{l(p_i)} \vee \psi_{p_i}^A \vee \psi_{p_i}^{\overline{A}})$. By *(non-effect)*, (5) $\vdash_{LDS} \diamond_A l(p_i) \to (\psi_{l(p_i)} \vee \psi_{p_i}^A \vee l(p_i))$. From (4) and (5), it is a propositional consequence that $\vdash_{LDS} \diamond_A l(p_i) \to (\psi_{l(p_i)} \vee \psi_{p_i}^A \vee (l(p_i) \wedge \psi_{p_i}^{\overline{A}}))$. □

Theorem 33. *Any formula φ is equivalent in LDS to an objective formula.*

Proof. The proof is by induction on the structure of φ. Let $\varphi = \diamond_A \chi$. Other cases are straightforward. By inductive hypothesis, χ is equivalent to an objective formula χ'. We write χ' in its disjunctive normal form as $\pi^1 \vee \ldots \vee \pi^d$, where each π^i is a conjunction of different literals. Then $\vdash_{LDS} \diamond_A \chi \leftrightarrow (\diamond_A \pi^1 \vee \ldots \vee \diamond_A \pi^d)$. By lemma 32, each $\diamond_A \pi^i$ is equivalent to an objective formula. $\qquad\square$

As having been noted in [9], the translation does not imply that the cooperation logic is redundant, since it involves an exponential blow-up in the size of a formula. Theorem 33 gives a procedure of computing an objective formula for any given formula φ. We will use $\mathcal{T}(\varphi)$ to denote the objective formula.

Definition 34. *(Canonical model) Given a LDS system, we define its canonical model as a tuple $M_{LDS}^{can} = (S^{can}, R^{can}, \theta^{can})$, where S^{can} is the set of LDS maximal consistent sets; $R^{can} = \{R_A^{can} | A \subseteq Ag\}$ and $\Gamma R_A^{can} \Delta$ iff for all $\varphi \in \Delta, \diamond_A \varphi \in \Gamma$, or equivalently, iff for all $\square_A \varphi \in \Gamma, \varphi \in \Delta; \theta^{can}(\Gamma, p) = 1$ iff $p \in \Gamma$.*

Lemma 35. *(Truth lemma) $\Gamma \models_{M_{LDS}^{can}} \varphi$ iff $\varphi \in \Gamma$.*

Lemma 36. *M_{LDS}^{can} is isomorphic to DS.*

Proof. We define a function f from M_{LDS}^{can} to DS such that for any proposition p, $f(\Gamma)(p) = 1$ iff $p \in \Gamma$. Clearly f is indeed a function. Suppose that $f(\Gamma) = f(\Delta)$. Then Γ and Δ contains exactly the same literals, and therefore contains the same objective formulas. By theorem 33, $\Gamma = \Delta$. Thus f is an injection. Given a state s, $\{p|s(p) = 1\}$ is consistent and can be extended to a maximal consistent set Γ. Since $f(\Gamma) = s$, f is a surjection. clearly Γ and $f(\Gamma)$ satisfy the same propositions.

It is left to prove that $\Gamma R_A^{can} \Delta$ iff $f(\Gamma) R_A f(\Delta)$. For the left to right direction, assume that $f(\Gamma) R_A f(\Delta)$ does not hold. Then we have two cases:

Case 1: For some proposition p such that $f(\Gamma) \models_{DS} \psi_{l(p)}$, $f(\Delta) \not\models_{DS} l(p)$. Then $l(p) \notin \Delta$. Since $\psi_{l(p)}$ is objective, and Γ and $f(\Gamma)$ satisfy the same propositions, we have $\Gamma \models_{M_{LDS}^{can}} \psi_{l(p)}$. By truth lemma, $\psi_{l(p)} \in \Gamma$. By *(Envi)*, $\square_A l(p) \in \Gamma$. Then for all Δ' such that $\Gamma R_A^{can} \Delta'$, $l(p) \in \Delta'$. Thus $\Gamma R_A^{can} \Delta$ does not hold.

Case 2: $f(\Gamma)(p) \neq f(\Delta)(p)$ for some $p \in Control(Ag/A, f(\Gamma))$. By definition 21-4, $f(\Gamma) \models_{DS} \neg \psi_p^A$ and $f(\Gamma) \models_{DS} \neg \psi_{\neg p}$. Since $\neg \psi_p^A$ and $\neg \psi_{\neg p}$ are objective, similarly we have $\neg \psi_p^A \in \Gamma$ and $\neg \psi_{\neg p} \in \Gamma$. Without losing generality, we assume that $f(\Gamma)(p) = 1$ and $f(\Delta)(p) = 0$. Then $p \in \Gamma$ and $p \notin \Delta$. By *(non-effect)*, $\square_A p \in \Gamma$. Then for all Δ' such that $\Gamma R_A^{can} \Delta'$, $p \in \Delta'$. We conclude that $\Gamma R_A^{can} \Delta$ does not hold.

For the other direction, assume that $f(\Gamma) R_A f(\Delta)$. Let φ be any formula in Δ. Let π be the conjunction of all literals in Δ. By theorem 33, $\vdash_{LDS} \pi \to \varphi$. Then $\vdash_{LDS} \diamond_A \pi \to \diamond_A \varphi$. Let $l(p)$ be a literal such that $l(p) \in \Delta$. Then $f(\Delta) \models_{DS} l(p)$. By the semantical definition, $f(\Gamma) \models_{DS} \psi_{l(p)} \vee \psi_p^A \vee (l(p) \wedge \psi_p^{\overline{A}})$. Since the formula is objective, we have $\psi_{l(p)} \vee \psi_p^A \vee (l(p) \wedge \psi_p^{\overline{A}}) \in \Gamma$. By lemma 32, $\diamond_A \pi \in \Gamma$. We conclude that $\diamond_A \varphi \in \Gamma$. $\qquad\square$

Theorem 37. *(completeness) LDS is complete with respect to the semantics on DS, i.e. for any formula φ, $\models_{DS} \varphi$ implies $\vdash_{LDS} \varphi$.*

Theorem 38. *Let DS and DS' be different declaration structures based on the same propositions and agents. Then there exists a formula φ such that $\vdash_{LDS} \varphi$ and $\nvdash_{LDS'} \varphi$.*

Proof. We use ψ to denote declarations in DS, and ψ' to denote declarations in DS'. Let

$$\varphi_p^A = \psi_p^A \leftrightarrow (\Diamond_A p \wedge \Diamond_A \neg p), \text{ where } A \subseteq Ag,$$
$$\varphi_p^+ = \psi_p \leftrightarrow \Diamond_\emptyset p \wedge \neg(\Diamond_{Ag} p \wedge \Diamond_{Ag} \neg p),$$
$$\varphi_p^- = \psi_{\neg p} \leftrightarrow \Diamond_\emptyset \neg p \wedge \neg(\Diamond_{Ag} p \wedge \Diamond_{Ag} \neg p).$$

$\varphi_p'^A, \varphi_p'^+, \varphi_p'^-$ are similarly defined. It is easy to check that $\models_{M_{DS}} \varphi_p^A \wedge \varphi_p^+ \wedge \varphi_p^-$ and $\models_{M_{DS'}} \varphi_p'^A \wedge \varphi_p'^+ \wedge \varphi_p'^-$.

Since declarations in DS and DS' are different, there exists a proposition p such that one of the following pairs are inequivalent: (ψ_p, ψ_p'), $(\psi_{\neg p}, \psi_{\neg p}')$, $(\psi_p^A, \psi_p'^A)$ for $A \subseteq Ag$. Consider the case where $\nvdash \psi_p^A \leftrightarrow \psi_p'^A$. Other cases are similar.

Suppose that $\models_{M_{DS'}} \psi_p^A \leftrightarrow (\Diamond_A p \wedge \Diamond_A \neg p)$. Since $\models_{M_{DS'}} \psi_p'^A \leftrightarrow (\Diamond_A p \wedge \Diamond_A \neg p)$, we have that $\models_{M_{DS'}} \psi_p^A \leftrightarrow \psi_p'^A$. Since ψ_p^A and $\psi_p'^A$ are objective, and every propositional evaluation is included in $M_{DS'}$, we have that $\models \psi_p^A \leftrightarrow \psi_p'^A$. Contradict to $\nvdash \psi_p^A \leftrightarrow \psi_p'^A$. Thus, $\nvDash_{M_{DS'}} \psi_p^A \leftrightarrow (\Diamond_A p \wedge \Diamond_A \neg p)$. By coherence and completeness theorem, we conclude that $\vdash_{LDS} \psi_p^A \leftrightarrow (\Diamond_A p \wedge \Diamond_A \neg p)$ and $\nvdash_{LDS'} \psi_p^A \leftrightarrow (\Diamond_A p \wedge \Diamond_A \neg p)$. $\square$

LDS is a deductive system for individual declaration structures. A deductive system LDS^g for general declaration structures can be constructed from each LDS due to the facts that any formula is equivalent to an objective one, and the alphabet is finite. LDS^g will be such that $\vdash_{LDS^g} \varphi$ iff $\models_{DS} \varphi$ for every declaration structure DS. We leave details of the construction.

4   More about Declarations

Fixing a set of agents and a set of propositions, behaviors of cooperation modalities are totally determined by declarations. In order to obtain desired properties of coalitions, we can impose more constraints on declarations besides the two in definition 21-4

Let $\Psi = \{\psi_{p_i}^j | 1 \leq i \leq k, 1 \leq j \leq n\} \cup \{\psi_{l(p_i)} | 1 \leq i \leq k\} \cup At$. A *constraint* (on declarations) is a Boolean combinations of elements in Ψ. Note that declarations ψ are prime in a constraint, though they are objective formulas in LDS. Each constraint defines a set of declaration structures. A simple example is a constraint $\psi_p \vee \psi_{\neg p}$ that defines the set of declaration structures in which proposition p is always not controlled by any agent. We say that a declaration structure *satisfies* a constraint if the constraint is a tautology in the structure.

First, we consider for each given formula its corresponding constraint. It is by theorem 33 that we can construct an objective formula $T(\varphi)$ that is equivalent to any given φ. It follows that any property of coalitions (any formula) can be fulfilled by specifying certain declarations.

Theorem 41. *For any formula φ, $\vdash_{LDS} \varphi$ if and only if DS satisfies $T(\varphi)$ (where $T(\varphi)$ is understood as a constraint, not a formula).*

The property that p is not controlled by any agent is formalized as $\neg(\diamond_{Ag}p \wedge \diamond_{Ag}\neg p)$. The formula is translated by T into $\neg((\psi_p \vee \psi_p^{Ag} \vee (p \wedge \psi_p^{\emptyset}) \wedge (\psi_{\neg p} \vee \psi_p^{Ag} \vee (\neg p \wedge \psi_p^{\emptyset}))$, which can be reduced to $\psi_p \vee \psi_{\neg p}$.

The reverse is also true, that is, each constraint corresponds to a modal formula. Let $[\chi]$ be a modal formulas that is obtained from a constraint χ by respectively substituting $(\diamond_A p \wedge \diamond_A \neg p)$ and $\square_{Ag} l(p)$ for each occurrence of ψ_p^A and $\psi_{l(p)}$.

Theorem 42. *For each constraint χ, DS satisfies χ if and only if $\vdash_{LDS} [\chi]$.*

Now we consider the relationship between constraints and property of Kripke models. For example, it is easy to see that for each R_A in DS, R_A is reflexive if and only if DS satisfies $\bigwedge_{1 \leq i \leq k}((\psi_{p_i} \to p_i) \wedge (\psi_{\neg p_i} \to \neg p_i))$. For a general result, we use the standard translation ST_x defined in [2] (page 84). For each modal formula φ, $ST_x(\varphi)$ is a first order formula with a single free variable x. Without introducing confusion, $DS \models \forall x ST_x(\varphi)$ means that sentence $\forall x ST_x(\varphi)$ is true in the first order model DS. Readers shall consult [2] for details.

Theorem 43. *For each constraint χ, $DS \models \forall x ST_x([\chi])$ if and only if DS satisfies χ.*

It can be verified that $\forall x ST_x[\bigwedge_{1 \leq i \leq k}((\psi_{p_i} \to p_i) \wedge (\psi_{\neg p_i} \to \neg p_i))]$ is equivalent to that $\forall x(x R_A x)$ for all $A \subseteq Ag$ (or for a single $A \subseteq Ag$).

5   Complexity

LDS model checking is the problem of finding whether or not a state in a declaration structure satisfies a given formula. Our algorithm is much similar to the one for CL-PC presented in [9] (figure 4, page 108):

```
function eval(φ, (Ag, At, De, DAg), s) returns 0 or 1
if φ ∈ At then
    return s(φ)
else if φ = ¬φ1 then
    return 1 − eval(φ1, (Ag, At, De, DAg), s)
else if φ = φ1 ∨ φ2 then
    return max(eval(φ1, (Ag, At, De, DAg), s), eval(φ2, (Ag, At, De, DAg), s))
else if φ = ◇Aφ1 then
    for each s′ ∈ sA do
        if eval(φ1, (Ag, At, De, DAg), s′) then
            return 1
```

```
        end if
      end for
    end if
    end function
```

The only extra cost of our algorithm in contrast to CL-PC's is in the case of $\varphi = \diamond_A \varphi_1$, where we have to calculate s_A. This can be done by a pre-procedure that determines for each declarations a set of valuations that satisfy it, which is a Boolean satisfiability problem and is NP-complete. Since CL-PC model checking problem is PSPACE-complete. We conclude that model checking problem of LDS is as tractable as CL-PC's.

Theorem 51. *The model checking problem of LDS is PSPACE-complete.*

LDS satisfiability checking is the problem that given a formula φ and a declaration structure DS, whether or not φ is true at some state of DS. The present version of satisfiability checking is different from the one in [9], where the only argument is a formula φ, and it is to check whether or not φ is true at some state of some structure. We argue that the present version is a more important problem to be checked, because properties of a concrete declaration structure and its deductive system LDS are usually in question. Properties of general declaration structures and its system LDS^g is less important.

Theorem 52. *The satisfiability checking problem of LDS is PSPACE-complete.*

Proof. Given a declaration structure DS and a formula φ, the problem is checked with a non-deterministic algorithm as follows: First guess a state s in DS; then determine whether or not φ is true at s. By theorem 51, the problem is in NPSPACE, therefore in PSPACE.

To see PSPACE hardness, we reduce QSAT to the LDS satisfiability problem in the same way as in the proof of theorem 5.1 in [9], where a CL-PC model is created for each given quantified Boolean formula $\alpha = \exists x_1 \forall x_2 \exists x_3 \ldots Q x_m \varphi(x_1, x_2, x_3, \ldots, x_m)$. Since the valuation in the CL-PC model is irrelevant to the proof, we can regard it as a declaration structure that consists of agents $1, 2, 3, \ldots, m$ and propositions $x_1, x_2, x_3, \ldots, x_m$, and declarations $\psi_{l(x_i)} = False$ and $\psi^i_{x_i} = True$ for each $i \in \{1, \ldots, m\}$. We conclude that α is true iff $\diamond_1 \square_2 \diamond_3 \ldots G_m \varphi(x_1, x_2, x_3, \ldots, x_m)$ is satisfiable in the declaration structure, where G is $\diamond$ if Q is $\exists$, $\square$ if Q is $\forall$. □

6 Conclusion

We have defined a notion of declaration structures as a general framework of propositional control. We used the same type of cooperation operators as CL-PC's to express properties of the so-called *contingent ability*. It was shown in [9] that this seemingly weaker notion of ability is enough for defining α-ability and β-ability. The former is captured in ATL and CL. An obvious challenge for future research is to answer the question: How far can we go along this road? Models of CL-PC is a restricted version of declaration structures, and declaration structures is a restricted version of models of ATL and CL. Is there a further generalization of declaration structures that is equivalent to models of ATL and CL, but keep the key idea of propositional control of no prescribed outcome functions? A generalization of allowing disjunctive choices of agents would be the first step and is not discussed in the paper. If we have had a full general structure in propositional control, we could investigate the possibility of a logic of propositional control that has the full power of ATL and CL. That certainly requires temporal operators in the logic.

References

1. Alur, R., Henzinger, T.A., Kupferman, O.: Alternating-time temporal logic. Journal of ACM 49(5), 672–713 (2002)
2. Blackburn, P., Rijke, M., Venema, Y.: Modal Logic. Cambridge University Press, Cambridge (2001)
3. Gerbrandy, J.: Logics of propositional control. In: Proceedings of the 5th international conference on Autonomous agents and multiagent systems (2006)
4. Goranko, V., Jamroga, W.: Comparing semantics of logics for multi-agent systems. Synthese 139(2), 241–280 (2004)
5. Oravec, V., Fogel, J.: Coalition logic of propositional control based multi agent system modeling. In: Proceedings of 3rd IEEE International Conference on Mechatronics (2006)
6. Pauly, M.: Logic for social software, Ph.D. Thesis, University of Amsterdam (2001)
7. Sauro, L., Gerbrandy, J., van der Hoek, W., Wooldridge, M.: Reasoning about action and cooperation. In: Proceedings of the 5th international conference on Autonomous agents and multiagent systems (2006)
8. Troquard, N., van der Hoek, w., Wooldridge, M.: A logic of games and propositional control. In: Proceedings of the 8rd international conference on Autonomous agents and multiagent systems (2009)
9. van der Hoek, W., Wooldridge, M.: On the logic of cooperation and propositional control. Artificial Intelligence 164, 81–119 (2005)
10. van der Hoek, W., Wooldridge, M.: On the synamics of delegation, cooperation, and control: a logical account. In: Proceedings of the 4th international conference on Autonomous agents and multiagent systems (2005)

Intentions and Assignments

Emiliano Lorini[1], Mehdi Dastani[2], Hans van Ditmarsch[3,*],
Andreas Herzig[1], and John-Jules Meyer[2]

[1] Université de Toulouse, IRIT-CNRS, France
[2] Utrecht University, The Netherlands
[3] University of Sevilla, Spain

Abstract. The aim of this work is propose a logical approach to intention dynamics based on the notion of *assignment* [3, 7]. The function of an assignment is to associate the truth value of a certain formula φ to a propositional atom p. We combine a static modal logic of belief and choice with three kinds of dynamic modalities and corresponding three kinds of assignments: assignments operating on an agent's beliefs, assignments operating on the agent's choices and assignments operating on the objective world. An agent's intention is defined in our approach as the agent's choice to perform a given action and two basic operations on intentions called intention generation and intention reconsideration are defined as specific kinds of assignments on choices.

1 Introduction

Since the seminal work of Cohen & Levesque [6] aimed at implementing Bratman's philosophical theory of intention [5], many formal logics for reasoning about intentions, their dynamics and their relationships with beliefs have been developed (see, e.g., [12, 13, 14, 16, 17, 19]). Most of them are frameworks based on a blend of dynamic logic with doxastic logic, enriched with modal operators for motivational attitudes such as preferences, goals and intentions. But, although logical analysis of intention dynamics are available in the literature, the issue of a *formal semantics* for the dynamics of intentions has received much less attention. Indeed, all previous approaches are mostly interested in characterizing in the object language the epistemic conditions under which an agent's intention persists over time and the epistemic conditions under which an an agent's intention is generated, but they do not provide a semantic characterization of the process of generating an intention and of the process of reconsidering an intention.

The aim of this work is to shed light on this unexplored area by proposing a formal semantics of intention dynamics based on the notion of *assignment*. The function of an assignment is to associate the truth value of a certain formula φ to a propositional atom p. We combine a static modal logic including modal operators for belief and choice with three kinds of dynamic modalities and corresponding three kinds of assignments: assignments operating on an agent's beliefs, assignments operating on the agent's choices and assignments operating on the objective world. An agent's intention is defined in our approach as the agent's choice to perform a given action and two basic

---

* Hans van Ditmarsch is also affiliated to the University of Otago.

X. He, J. Horty, and E. Pacuit (Eds.): LORI 2009, LNAI 5834, pp. 198–211, 2009.

operations on intentions called *intention generation* and *intention reconsideration* are defined as specific kinds of assignments on choices.

Assignments were studied before in the literature on modal logic for information dynamics. However they were only applied to the dynamics of belief and knowledge [3, 7], and there is still no application of this notion to the theory of intention. In this paper we show that assignments are well-suited to model intention dynamics. Indeed, assignments capture the *locality* of intention dynamics better than other operations like announcements [8] and upgrades [2, 11], where locality means that the process of reconsidering (or generating) an agent's intention does not affect the other intentions of the agent.

The rest of the paper is organized as follows. The first part (Section 2) introduces a static logic of belief, choice and intention. In the second part (Section 3) we move from a static perspective on agents' attitudes to a dynamic perspective. We first present the syntax and semantics of three kinds of assignments on beliefs, on choices and on the objective world. Then in Section 4, we analyze the notion of *executability preconditions* for assignments. We devote special attention to *executability preconditions* of assignments which are responsible for the generation (resp. reconsideration) of an intention. In Section 5 we compare our approach with existing logical approaches to belief and preference dynamics.

2 A Modal Logic of Beliefs, Choices and Intentions

We introduce a modal logic called **L** which supports reasoning about three different kinds of mental attitudes: beliefs, choices (or chosen goals), and intentions.

2.1 Syntax

Let $ATM^{Fact} = \{f_1, f_2, \ldots\}$ be a nonempty finite set of atoms denoting facts (or state of affairs), and let $ATM^{Act} = \{\alpha, \beta, \ldots\}$ be a nonempty finite set of atoms denoting actions. The atom α is meant to stand for 'the agent performs a certain action α'. We also have special atoms of type $good_\alpha$ expressing that 'performing action α is good for the agent'. ATM^{Good} is the corresponding set, that is, $ATM^{Good} = \{good_\alpha | \alpha \in ATM^{Act}\}$. We define $ATM = ATM^{Fact} \cup ATM^{Act} \cup ATM^{Good}$ to be the set of atomic formulas. We note $p, q, \ldots$ the elements in ATM.

The language $\mathcal{L}$ of the logic **L** is the set of formulas defined by the following BNF:

$$\varphi ::= p \mid \neg\varphi \mid \varphi \wedge \varphi \mid [B]\varphi \mid [C]\varphi$$

where p ranges over ATM. The other Boolean constructions $\top$, $\bot$, $\vee$, $\rightarrow$ and $\leftrightarrow$ are defined from $\neg$ and $\wedge$ in the standard way.

The two modal operators of our logic have the following reading: $[B]\varphi$ means 'the agent believes φ' and $[C]\varphi$ means 'the agent has chosen φ' (or 'the agent wants φ to be true'). Operators $[C]$ are used to denote the agent's choices, that is, the state of affairs that the agent has decided to pursue. Similar operators have been studied in [6, 12, 14].

The following abbreviation will also be convenient for every $\alpha \in ATM^{Act}$:

$$\mathtt{I}(\alpha) \stackrel{\text{def}}{=} [C]\alpha.$$

$\mathtt{I}(\alpha)$ is meant to stand for 'the agent intends to do action α'.

2.2 Semantics

Models of the logic **L** (**L**-models) are tuples $F = \langle W, \mathscr{B}, \mathscr{C}, \mathscr{V} \rangle$ defined as follows:

- W is a nonempty set of possible worlds or states;
- $\mathscr{B} \subseteq W \times W$ is a serial, transitive and Euclidean accessibility relation for belief;
- $\mathscr{C} \subseteq W \times W$ is a serial, transitive and Euclidean accessibility relation for choice;
- $\mathscr{V} : ATM \longrightarrow 2^W$ is a valuation function.

Accessibility relations on W can be viewed as functions from W to 2^W. Therefore, we write $\mathscr{B}(w) = \{v | (w, v) \in \mathscr{B}\}$ and $\mathscr{C}(w) = \{v | (w, v) \in \mathscr{C}\}$. $\mathscr{B}(w)$ is the set of worlds that are compatible with the agent's beliefs at w (or belief accessible worlds at w), $\mathscr{C}(w)$ is the set of worlds that are compatible with the agent's choices at w (or choice accessible worlds at w).

The accessibility relations $\mathscr{B}$ and $\mathscr{C}$ satisfy the following additional constraints for every $w \in W$:

S1   if $v \in \mathscr{B}(w)$ then $\mathscr{C}(v) = \mathscr{C}(w)$;
S2   if $v \in \mathscr{C}(w)$ and $u \in \mathscr{B}(v)$ then $u \in \mathscr{C}(w)$;
S3   if $v \in \mathscr{C}(w)$ then $v \in \mathscr{B}(v)$.

Constraint S1 expresses that the agent's choices are positively and negatively introspective (i.e. if v is compatible with the agent's beliefs at w then the set of worlds which are compatible with the agent's choices at w is identical to the set of worlds which are compatible with the agent's choices at v). According to constraint S2, the agent always chooses the states that he considers possible from the states that he chooses. According to constraint S3, if at state w the agent chooses state v then at v the agent considers v a possible state. In other words, the agent always chooses that the current state belongs to the set of states that he considers possible.

Given a model M, a world w and a formula φ, we write $M, w \models \varphi$ to mean that φ is true at world w in M. The rules defining the truth conditions of formulas are just standard for atomic formulas, negation and disjunction. The following are the remaining truth conditions for $[B]\varphi$, $[C]\varphi$:

- $M, w \models [B]\varphi$ iff $M, v \models \varphi$ for all w' such that $v \in \mathscr{B}(w)$;
- $M, w \models [C]\varphi$ iff $M, v \models \varphi$ for all w' such that $v \in \mathscr{C}(w)$.

We write $\models_L \varphi$ if φ is *valid* (i.e. φ is true in all **L**-models). We say that φ is *satisfiable* if $\neg\varphi$ is not valid.

2.3 Axiomatization

Fig. 1 contains the axiomatization of the logic **L**. We adopt a standard KD45 logic for beliefs [9] and a standard KD45 logic for choices [6]. Thus, we have positive and negative introspection for beliefs (Axioms 4 and 5 for $[B]$), and we assume that an agent cannot have inconsistent beliefs (Axiom D for $[B]$). We assume that if the agent chooses something then he chooses to choose it and if the agent does not choose something then he chooses not to choose it (Axioms 4 and 5 for $[C]$), and we assume that the

(PC)	All principles of classical propositional calculus
(KD45$_{[B]}$)	All principles of modal logic KD45 for [B]
(KD45$_{[C]}$)	All principles of modal logic KD45 for [C]
(PIntr$_{[C]}$)	$[C]\varphi \rightarrow [B][C]\varphi$
(NIntr$_{[C]}$)	$\neg[C]\varphi \rightarrow [B]\neg[C]\varphi$
(AchieveAware)	$[C]\varphi \rightarrow [C][B]\varphi$
(NotIncorrectBel)	$[C](\varphi \rightarrow \neg[B]\neg\varphi)$

Fig. 1. Axiomatization of **L**

agent cannot have inconsistent choices (Axiom D for [C]). We have negative and positive introspection for choices (Axioms **PIntr**$_{[C]}$ and **NIntr**$_{[C]}$). Moreover, we suppose that if the agent chooses φ then he chooses to believe φ (Axiom **AchieveAware**). In other words, if the agent wants that φ will be true then he wants to achieve a state in which he believes that φ is true (i.e. he wants to achieve φ knowingly). Finally, we suppose that the agent always wants that if φ is true then he does not believe $\neg\varphi$ (Axiom **NotIncorrectBel**). In other words, the agent always wants not to have incorrect beliefs.

We call **L** the logic axiomatized by the principles given in Fig. 1. We write $\vdash_{\mathbf{L}}$ φ if φ is a **L**-theorem. For instance, the following theorem is provable by Axiom **NotIncorrectBel**, Axiom K and necessitation rule for [C]:

$$\vdash_{\mathbf{L}} [C][B]\varphi \rightarrow [C]\varphi.$$

The theorem just says that if the agent wants to acquire a certain belief then he wants the content of this belief to be true. Therefore, our logic does not allow self-deception. For example, it excludes the situation of a person who wants to believe that God exists in order to feel better when thinking about the afterlife (i.e. $[C][B]\,GodExists$) and, at the same time, she does not want that God exists (i.e. $\neg[C]\,GodExists$).

It is to be noted that $[B]\varphi \wedge [C]\neg\varphi$ are satisfiable in our logic. For example, our logic allows the situation of a person who smokes and believes this (i.e. $[B]\,smoke$) and, she decides to stop smoking (i.e. $[C]\neg smoke$).

Theorem 1. *The logic **L** is completely axiomatized by the principles in Fig. 1.*

Proof. It is a routine task to check that the axioms of the logic **L** correspond one-to-one to their semantic counterparts on the models. In particular, Axioms D, 4 and 5 for [B] correspond to the seriality, transitivity and Euclideanity of the accessibility relation $\mathscr{B}$. Axioms D, 4 and 5 for [C] correspond to the seriality, transitivity and Euclideanity of $\mathscr{C}$. Axioms **PIntr**$_{[C]}$ and **NIntr**$_{[C]}$ together correspond to the constraint S1. Finally, Axiom **AchieveAware** corresponds to S2 and Axiom **NotIncorrectBel** corresponds to S3. It is routine, too, to check that all of our axioms are in the Sahlqvist class. This means that the axioms are all expressible as first-order conditions on models and that they are complete with respect to the defined model classes, cf. [4, Th. 2.42].   □

3 From Static to Dynamic Mental States

In this section we extend the logic **L** of Section 2 by modal operators for objective world change and mental attitude change. We distinguish two kinds of mental attitude change: belief change and choice change. We call $\mathbf{L}^{dyn}$ the extended logic.

3.1 Syntax

Atomic events for world change (or *atomic world assignments*) are of the form $p \overset{W}{\rightsquigarrow} \varphi$ whereas atomic events for belief change (or *atomic belief assignments*) and for choice change (or *atomic choice assignments*) are of the form $p \overset{B}{\rightsquigarrow} \varphi$ and $p \overset{C}{\rightsquigarrow} \varphi$: $p \overset{B}{\rightsquigarrow} \varphi$ is the event 'the truth value of φ is assigned to p in the agent's beliefs'; $p \overset{C}{\rightsquigarrow} \varphi$ is the event 'the truth value of φ is assigned to p in the agents' choices'; $p \overset{W}{\rightsquigarrow} \varphi$ is the event 'the truth value of φ is assigned to p in the objective world'.

We respectively note $BASG_B = \{p \overset{B}{\rightsquigarrow} \varphi | p \in ATM \text{ and } \varphi \in \mathcal{L}\}$, $BASG_C = \{p \overset{C}{\rightsquigarrow} \varphi | p \in ATM \text{ and } \varphi \in \mathcal{L}\}$ and $BASG_W = \{p \overset{W}{\rightsquigarrow} \varphi | p \in ATM \text{ and } \varphi \in \mathcal{L}\}$ the set of atomic belief assignments, the set of atomic choice assignments and the set of atomic world assignments. We define $EVT = BASG_B \cup BASG_C \cup BASG_W$ to be the set of atomic assignments.

The following two types of atomic choice assignments characterize two basic operations on an agent's intentions:

$$gen(\alpha) \overset{def}{=} \alpha \overset{C}{\rightsquigarrow} \top;$$
$$rec(\alpha) \overset{def}{=} \alpha \overset{C}{\rightsquigarrow} \bot.$$

The event $gen(\alpha)$ is the agent's mental operation of *generating* the intention to do action α, whereas the event $rec(\alpha)$ is the agent's mental operation of *reconsidering* (or *erasing*) his intention to perform action α.[1]

Complex assignments are defined as *partial functions* from ATM to $\mathcal{L}$. We distinguish three kinds of complex assignments: *complex belief assignments*, *complex choice assignments* and *complex world assignments*. We note $\sigma_B, \sigma'_B, \ldots$ the complex belief assignments, $\sigma_C, \sigma'_C, \ldots$ the complex choice assignments and $\sigma_W, \sigma'_W, \ldots$ the complex world assignments. Moreover, we respectively note $CASG_B$, $CASG_C$ and $CASG_W$ the set of all complex belief assignments, the set of all complex choice assignments and the set of all complex world assignments. Given a complex belief assignment σ_B, $D(\sigma_B)$ is the *domain* of σ_B and $C(\sigma_B)$ is its *codomain*. Similarly, $D(\sigma_C)$ (resp. $D(\sigma_W)$) is the domain of the complex choice (resp. world) assignment σ_C (resp. σ_W)

---

[1] For some different approaches to intention generation and intention reconsideration, see e.g. [10, 13].

and $C(\sigma_C)$ (resp. $C(\sigma_W)$) is its codomain. We extend partial functions to total functions by stipulating that when $p \notin D(\sigma_B)$ then $\sigma_B(p) = p$. Similarly, we stipulate that $\sigma_C(p) = p$ when $p \notin D(\sigma_C)$, and $\sigma_W(p) = p$ when $p \notin D(\sigma_W)$.

For every complex belief assignment σ_B, we define

$$s(\sigma_B) = \{p \overset{B}{\rightsquigarrow} \sigma_B(p) | p \in D(\sigma_B)\}$$

to be the corresponding set of atomic belief assignments. Similarly, for every complex choice assignment σ_B and complex world assignments σ_W

$$s(\sigma_C) = \{p \overset{C}{\rightsquigarrow} \sigma_C(p) | p \in D(\sigma_C)\} \text{ and}$$
$$s(\sigma_W) = \{p \overset{W}{\rightsquigarrow} \sigma_W(p) | p \in D(\sigma_W)\}$$

are the corresponding sets of atomic choice and atomic world assignments.

The elements of ASG are all sets including a set of atomic belief assignments, a set of atomic choice assignments and a set of atomic world assignments, that is,

$$ASG = \{\{s(\sigma_B), s(\sigma_C), s(\sigma_W)\} | \sigma_B \in CASG_B, \sigma_C \in CASG_C, \sigma_W \in CASG_W\}.$$

The language $\mathcal{L}^{dyn}$ of the logic $\mathbf{L}^{dyn}$ is defined by the following BNF:

$$\varphi ::= p \mid \neg\varphi \mid \varphi \wedge \varphi \mid [\mathrm{B}]\varphi \mid [\mathrm{C}]\varphi \mid [\Sigma{:}w]\psi \mid [\Sigma{:}B]\psi \mid [\Sigma{:}c]\psi$$

where p ranges over ATM and Σ ranges over ASG.

The formula $[\Sigma{:}w]\psi$ is meant to stand for: ψ holds in the objective world after the occurrence of the event Σ. The formula $[\Sigma{:}B]\psi$ is meant to stand for: ψ holds in the context of the agent's beliefs after the occurrence of the event Σ. The formula $[\Sigma{:}c]\psi$ is meant to stand for: ψ holds in the context of the agent's choices after the occurrence of the event Σ. The duals of the operators $[\Sigma{:}w]$, $[\Sigma{:}B]$ and $[\Sigma{:}c]$ are defined as usual: $\langle\Sigma{:}w\rangle\psi \overset{def}{=} \neg[\Sigma{:}w]\neg\psi$, $\langle\Sigma{:}B\rangle\psi \overset{def}{=} \neg[\Sigma{:}B]\neg\psi$, and $\langle\Sigma{:}c\rangle\psi \overset{def}{=} \neg[\Sigma{:}c]\neg\psi$.

3.2   Semantics

We introduce a function Pre from EVT to $\mathcal{L}$ which returns the *executability preconditions* of every atomic belief assignment, of every atomic choice assignment and of every atomic world assignment. The function Pre is generalized to the events Σ in ASG in a straightforward manner. Suppose $\Sigma = \{s(\sigma_B), s(\sigma_C), s(\sigma_W)\}$. Then:

$$Pre(\Sigma) = \bigwedge_{p \overset{B}{\rightsquigarrow}\varphi \in s(\sigma_B)} Pre(p \overset{B}{\rightsquigarrow} \varphi) \wedge \bigwedge_{p \overset{C}{\rightsquigarrow}\varphi \in s(\sigma_C)} Pre(p \overset{C}{\rightsquigarrow} \varphi) \wedge \bigwedge_{p \overset{W}{\rightsquigarrow}\varphi \in s(\sigma_W)} Pre(p \overset{W}{\rightsquigarrow} \varphi).$$

Note that this formula is indeed in the language and is not infinitary. Indeed, the set of atoms ATM has been supposed to be finite.

For every $\Sigma \in ASG$, $Pre(\Sigma)$ denotes the executability preconditions of the event Σ, i.e. the conditions which are together necessary and sufficient to ensure that the event Σ will possibly *occur*.

Suppose that $\Sigma = \{s(\sigma_B), s(\sigma_C), s(\sigma_W)\}$. In order to give semantics to the operators $[\Sigma{:}w]$, $[\Sigma{:}B]$ and $[\Sigma{:}c]$ we define the model

$$M^{\Sigma} = \langle W^{\Sigma}, \mathscr{B}^{\Sigma}, \mathscr{C}^{\Sigma}, \mathscr{V}^{\Sigma} \rangle$$

which results from the occurrence of the event Σ in the model M. The elements of M^{Σ} are defined as follows:

$$
\begin{aligned}
W^{\Sigma} =&\{w_W | w \in W \text{ and } M, w \models Pre(\Sigma)\} \cup \\
&\{w_B | w \in W \text{ and } M, w \models Pre(\Sigma)\} \cup \\
&\{w_C | w \in W \text{ and } M, w \models Pre(\Sigma)\}; \\
\mathscr{B}^{\Sigma} =&\{(w_W, v_B) | v, w \in W \text{ and } (w, v) \in \mathscr{B}\} \cup \\
&\{(w_B, v_B) | v, w \in W \text{ and } (w, v) \in \mathscr{B}\} \cup \\
&\{(w_C, v_C) | v, w \in W \text{ and } (w, v) \in \mathscr{B}\}; \\
\mathscr{C}^{\Sigma} =&\{(w_W, v_C) | v, w \in W \text{ and } (w, v) \in \mathscr{C}\} \cup \\
&\{(w_B, v_C) | v, w \in W \text{ and } (w, v) \in \mathscr{C}\} \cup \\
&\{(w_C, v_C) | v, w \in W \text{ and } (w, v) \in \mathscr{C}\}; \\
\mathscr{V}^{\Sigma}(p) =&\{w_W | w \in W \text{ and } M, w \models \sigma_W(p)\} \cup \\
&\{w_B | w \in W \text{ and } M, w \models \sigma_B(p)\} \cup \\
&\{w_C | w \in W \text{ and } M, w \models \sigma_C(p)\}.
\end{aligned}
$$

M^{Σ} is obtained by creating three copies of each state of the original model M (a copy for the objective world, a copy for belief, a copy for choice), and by restricting the original model to the set of states in which the executability preconditions of Σ hold. Moreover, for every atom p, the effect of a model update by Σ is to assign the truth value of $\sigma_B(p)$ to the atom p in all belief copies of the original states, to assign the truth value of $\sigma_C(p)$ to the atom p in all choice copies, and to assign the truth value of $\sigma_W(p)$ to the atom p in all world copies.

For every world copy w_W, at w_W the agent considers possible all belief copies of those states that he considered possible before the event Σ, and he chooses all choice copies of those states that he chose before the event Σ.

For every belief copy w_B, at w_B the agent considers possible all belief copies of those states that he considered possible before the event Σ, and he chooses all choice copies of those states that he chose before the event Σ.

For every choice copy w_C, at w_C the agent considers possible all choice copies of those states that he considered possible before the event Σ, and he chooses all choice copies of those states that he chose before the event Σ.

This construction of the updated model M^{Σ} ensures that the agent is aware that his choices have been changed accordingly so that the properties of positive and introspection over the agent's choices (constraint S1) are preserved after the occurrence of the event Σ. Moreover, it ensures that the agent chooses that his beliefs change as his choices change so that the constraints S2 and S3 are preserved after the occurrence of the event Σ. On the contrary, the operation of world change is independent of the operations of belief and choice change, and the operations of belief and choice change are independent of the operation of world change.

Theorem 2. *If M is an **L**-model then M^{Σ} is an **L**-model.*

Proof. It is just trivial to prove that our operation of model update preserves the seriality, transitivity and Euclideanity of the accessibility relations $\mathscr{B}$ and $\mathscr{C}$.

We just prove that the constraints S1, S2 and S3 are also preserved.

We start with S1. Assume $v_B \in \mathscr{B}^{\Sigma}(w_W)$ and $u_C \in \mathscr{C}^{\Sigma}(v_B)$. It follows that $v \in \mathscr{B}(w)$ and $u \in \mathscr{C}(v)$ and $M, w \models Pre(\Sigma)$ and $M, u \models Pre(\Sigma)$. Then, by constraint S1, we have $u \in \mathscr{C}(w)$. Therefore, $u_C \in \mathscr{C}^{\Sigma}(w_W)$. In a similar way we can prove that if $v_B \in \mathscr{B}^{\Sigma}(w_B)$ and $u_C \in \mathscr{C}^{\Sigma}(v_B)$ then $u_C \in \mathscr{C}^{\Sigma}(w_B)$, and if $v_C \in \mathscr{B}^{\Sigma}(w_C)$ and $u_C \in \mathscr{C}^{\Sigma}(v_C)$ then $u_C \in \mathscr{C}^{\Sigma}(w_C)$.

Now, assume $v_C \in \mathscr{C}^{\Sigma}(w_W)$ and $u_B \in \mathscr{B}^{\Sigma}(w_W)$. It follows that $v \in \mathscr{C}(w)$ and $u \in \mathscr{B}(w)$ and $M, v \models Pre(\Sigma)$ and $M, u \models Pre(\Sigma)$. Then, by constraint S1, we have $v \in \mathscr{C}(u)$. Therefore, $v_C \in \mathscr{C}^{\Sigma}(u_B)$. In a similar way we can prove that if $v_C \in \mathscr{C}^{\Sigma}(w_B)$ and $u_B \in \mathscr{B}^{\Sigma}(w_B)$ then $v_C \in \mathscr{C}^{\Sigma}(u_B)$, and if $v_C \in \mathscr{C}^{\Sigma}(w_C)$ and $u_C \in \mathscr{B}^{\Sigma}(w_C)$ then $v_C \in \mathscr{C}^{\Sigma}(u_C)$.

Let us prove S2. Assume $v_C \in \mathscr{C}^{\Sigma}(w_W)$ and $u_C \in \mathscr{B}^{\Sigma}(v_C)$. It follows that $v \in \mathscr{C}(w)$ and $u \in \mathscr{B}(v)$ and $M, w \models Pre(\Sigma)$ and $M, u \models Pre(\Sigma)$. Then, by constraint S2, we have $u \in \mathscr{C}(w)$. We can conclude that $u_C \in \mathscr{C}^{\Sigma}(w_W)$. In a similar way we can prove that if $v_C \in \mathscr{C}^{\Sigma}(w_B)$ and $u_C \in \mathscr{B}^{\Sigma}(v_C)$ then $u_C \in \mathscr{C}^{\Sigma}(w_B)$; and if $v_C \in \mathscr{C}^{\Sigma}(w_C)$ and $u_C \in \mathscr{B}^{\Sigma}(v_C)$ then $u_C \in \mathscr{C}^{\Sigma}(w_C)$.

Let us prove S3. Assume $v_C \in \mathscr{C}^{\Sigma}(w_W)$. It follows that $v \in \mathscr{C}(w)$ and $M, v \models Pre(\Sigma)$. Then, by constraint S3, we have that $v \in \mathscr{B}(v)$. Therefore, $v_C \in \mathscr{B}^{\Sigma}(v_C)$.

In a similar way we can prove that if $v_C \in \mathscr{C}^{\Sigma}(w_B)$ then $v_C \in \mathscr{B}^{\Sigma}(v_C)$ and if $v_C \in \mathscr{C}^{\Sigma}(w_C)$ then $v_C \in \mathscr{B}^{\Sigma}(v_C)$.   □

The truth conditions are those of Section 2 plus the following:

- $M, w \models [\Sigma{:}W]\psi$ iff, if $M, w \models Pre(\Sigma)$ then $M^{\Sigma}, w_W \models \psi$;
- $M, w \models [\Sigma{:}B]\psi$ iff, if $M, w \models Pre(\Sigma)$ then $M^{\Sigma}, w_B \models \psi$;
- $M, w \models [\Sigma{:}C]\psi$ iff, if $M, w \models Pre(\Sigma)$ then $M^{\Sigma}, w_C \models \psi$;

Note that $\langle \Sigma{:}W \rangle \top$ and $\langle \Sigma{:}B \rangle \top$ and $\langle \Sigma{:}C \rangle \top$ are individually equivalent to the executability preconditions of Σ (i.e. $Pre(\Sigma)$). Therefore $\langle \Sigma{:}W \rangle \top$, $\langle \Sigma{:}B \rangle \top$ and $\langle \Sigma{:}C \rangle \top$ should be respectively read 'Σ will possibly occur in the objective world', 'Σ will possibly occur in the context of the agent's beliefs' and 'Σ will possibly occur in the context of the agent's choices'.

In Section 4 we will specify in detail four general kinds of executability preconditions: the executability preconditions of the atomic world assignments $p \overset{W}{\leadsto} \bot$ and $p \overset{W}{\leadsto} \top$ (with $p \in ATM^{Fact}$); the executability preconditions of the atomic world assignment $\alpha \overset{W}{\leadsto} \top$ (with $\alpha \in ATM^{Act}$); the executability preconditions of the atomic choice assignment $gen(\alpha)$; the executability preconditions of the atomic choice assignment $rec(\alpha)$. The first are the executability preconditions of the event which consists in making true (resp. false) a certain objective fact p; the second are executability preconditions of the intentional action α; the third are the executability preconditions of the process of generating the intention to do α; the fourth are the executability preconditions of the process reconsidering the intention to do α.

A discussion about the relationships between the present approach and the action/event model with assignments à la [3, 7] is given in Section 5.

3.3 Axiomatization

We have reduction axioms for the three operators $[\Sigma{:}W]$, $[\Sigma{:}B]$ and $[\Sigma{:}C]$.

Theorem 3. *Suppose* $\Sigma = \{s(\sigma_B), s(\sigma_C), s(\sigma_W)\}$. *Then, the following schemata are valid in* $\boldsymbol{L}^{dyn}$:

R1a.	$[\Sigma{:}W]p \leftrightarrow (Pre(\Sigma) \to \sigma_W(p))$
R1b.	$[\Sigma{:}B]p \leftrightarrow (Pre(\Sigma) \to \sigma_B(p))$
R1c.	$[\Sigma{:}C]p \leftrightarrow (Pre(\Sigma) \to \sigma_C(p))$
R2a.	$[\Sigma{:}W]\neg\varphi \leftrightarrow (Pre(\Sigma) \to \neg[\Sigma{:}W]\varphi)$
R2b.	$[\Sigma{:}B]\neg\varphi \leftrightarrow (Pre(\Sigma) \to \neg[\Sigma{:}B]\varphi)$
R2c.	$[\Sigma{:}C]\neg\varphi \leftrightarrow (Pre(\Sigma) \to \neg[\Sigma{:}C]\varphi)$
R3a.	$[\Sigma{:}W](\varphi \wedge \psi) \leftrightarrow ([\Sigma{:}W]\varphi \wedge [\Sigma{:}W]\psi)$
R3b.	$[\Sigma{:}B](\varphi \wedge \psi) \leftrightarrow ([\Sigma{:}B]\varphi \wedge [\Sigma{:}B]\psi)$
R3c.	$[\Sigma{:}C](\varphi \wedge \psi) \leftrightarrow ([\Sigma{:}C]\varphi \wedge [\Sigma{:}C]\psi)$
R4a.	$[\Sigma{:}W][\mathsf{B}]\varphi \leftrightarrow (Pre(\Sigma) \to [\mathsf{B}][\Sigma{:}B]\varphi)$
R4b.	$[\Sigma{:}B][\mathsf{B}]\varphi \leftrightarrow (Pre(\Sigma) \to [\mathsf{B}][\Sigma{:}B]\varphi)$
R4c.	$[\Sigma{:}C][\mathsf{B}]\varphi \leftrightarrow (Pre(\Sigma) \to [\mathsf{B}][\Sigma{:}C]\varphi)$
R5a.	$[\Sigma{:}W][\mathsf{C}]\varphi \leftrightarrow (Pre(\Sigma) \to [\mathsf{C}][\Sigma{:}C]\varphi)$
R5b.	$[\Sigma{:}B][\mathsf{C}]\varphi \leftrightarrow (Pre(\Sigma) \to [\mathsf{C}][\Sigma{:}C]\varphi)$
R5c.	$[\Sigma{:}C][\mathsf{C}]\varphi \leftrightarrow (Pre(\Sigma) \to [\mathsf{C}][\Sigma{:}C]\varphi)$

Proof. We just prove **R4a** as an example.

$M, w \models [\Sigma{:}W][\mathsf{B}]\varphi$
IFF if $M, w \models Pre(\Sigma)$ then $M^\Sigma, w_W \models [\mathsf{B}]\varphi$
IFF if $M, w \models Pre(\Sigma)$ then, if $v_B \in \mathscr{B}^\Sigma(w_W)$ then $M^\Sigma, v_B \models \varphi$
IFF if $M, w \models Pre(\Sigma)$ then, if $v \in \mathscr{B}(w)$ and $M, w \models Pre(\Sigma)$ and $M, v \models Pre(\Sigma)$
then $M^\Sigma, v_B \models \varphi$
IFF if $M, w \models Pre(\Sigma)$ then, if $v \in \mathscr{B}(w)$ and $M, v \models Pre(\Sigma)$ then $M^\Sigma, v_B \models \varphi$
IFF if $M, w \models Pre(\Sigma)$ then, if $v \in \mathscr{B}(w)$ then if $M, v \models Pre(\Sigma)$ then $M^\Sigma, v_B \models \varphi$
IFF if $M, w \models Pre(\Sigma)$ then, if $v \in \mathscr{B}(w)$ then $M, v \models [\Sigma{:}B]\varphi$
IFF if $M, w \models Pre(\Sigma)$ then $M, w \models [\mathsf{B}][\Sigma{:}B]\varphi$
IFF if $M, w \models Pre(\Sigma) \to [\mathsf{B}][\Sigma{:}B]\varphi$. □

We call $\boldsymbol{L}^{dyn}$ the logic axiomatized by the principles of the logic $\mathbf{L}$ *plus* the axiom schemata of Theorem 3 and the rule of replacement of proved equivalence. We write $\vdash_{\boldsymbol{L}^{dyn}} \varphi$ if φ is a $\boldsymbol{L}^{dyn}$-theorem. The following are examples of $\boldsymbol{L}^{dyn}$-theorems about intention generation and intention reconsideration.

Proposition 1

(1a) $\vdash_{L^{dyn}} [\{\emptyset,\{gen(\alpha)\},\emptyset\}{:}w]I(\alpha)$

(1b) $\vdash_{L^{dyn}} [\{\emptyset,\{rec(\alpha)\},\emptyset\}{:}w]\neg I(\alpha)$

(1c) $\vdash_{L^{dyn}} [\{\emptyset,\{gen(\alpha)\},\emptyset\}{:}c]\alpha$

(1d) $\vdash_{L^{dyn}} [\{\emptyset,\{rec(\alpha)\},\emptyset\}{:}c]\neg\alpha$

(1e) $\vdash_{L^{dyn}} \neg I(\beta) \rightarrow [\{\emptyset,\{gen(\alpha)\},\emptyset\}{:}w]\neg I(\beta)$ *if* $\alpha \neq \beta$

(1f) $\vdash_{L^{dyn}} I(\beta) \rightarrow [\{\emptyset,\{rec(\alpha)\},\emptyset\}{:}w]I(\beta)$ *if* $\alpha \neq \beta$

Proof. We prove Theorem (1a) as an example. $[\emptyset,\{gen(\alpha)\},\emptyset{:}w]I(\alpha)$ is equivalent to $Pre(\Sigma) \rightarrow [\text{C}][\emptyset,\{gen(\alpha)\},\emptyset{:}c]\alpha$ (by reduction axiom **R5a.** and rule of replacement of proved equivalences). The latter is equivalent to $Pre(\Sigma) \rightarrow [\text{C}](Pre(\Sigma) \rightarrow \top)$ (by reduction axiom **R1c.** and rule of replacement of proved equivalences) which in turn is equivalent to $\top$. $\square$

According to Theorem 1a, after generating the intention to do α, the agent intends to do α in the objective world. According to Theorem 1b, after reconsidering the intention to do α, the agent does not intend to do α in the objective world. Theorems 1c and 1d express the corresponding effects of the processes of intention generation and of intention reconsideration in the context of the agent's choices: after generating (resp. reconsidering) the intention to do α, the agent performs (resp. does not perform) action α in the context of his choices. Theorems 1e and 1f express that the operations of intention generation and of intention reconsideration are local operations, that is, the process of generating (resp. reconsidering) an intention does not affect the other intentions of the agent: if α and β are different actions and the agent intends (resp. does not intend) to do β then, after reconsidering (resp. generating) the intention do α, the agent will intend (resp. not intend) to do β.

Theorem 4. *The logic L^{dyn} is completely axiomatized by principles in Fig. 1 together with the schemata of Theorem 3 and the rule of replacement of proved equivalence.*

Proof. By means of the principles **R1a-R6** in Theorem 3, it is straightforward to prove that for every L^{dyn} formula there is an equivalent L formula. In fact, each reduction axiom **R1a-R5c**, when applied from the left to the right by means of the rule of replacement of proved equivalence **R6**, yields a simpler formula, where 'simpler' roughly speaking means that the dynamic operators are pushed inwards. Once the dynamic operators attain an atom they are eliminated by one of the equivalences **R1a-R1c**. Hence, the completeness of L^{dyn} is a straightforward consequence of Theorem 1. $\square$

4   Applications

It is now time to study in detail the notion of executability preconditions introduced in Section 3.2. We will study four general kinds of executability preconditions: the executability preconditions of the event which consists in making true (resp. false) a certain objective fact p; the executability preconditions of an intentional action; the executability preconditions of a process of intention generation; the executability preconditions of a process of intention reconsideration. Thus, we will be able to clarify why the distinction between atoms denoting facts and atoms denoting actions given in Section 2.1 is not merely a syntactic distinction. On the contrary, it has concrete effects on the dynamic level of our model. Indeed, we will show that the executability preconditions for assignments to atoms denoting facts are qualitatively different from the executability preconditions for assignments to atoms denoting actions.

Positive and negative effect preconditions. Executability preconditions can be used to describe how the world will change after the occurrence of a certain action, that is, how a fact p might become true (resp. false) when the agent acts in a certain way. In way similar to [15], we denote with $\gamma^+(\alpha, p)$ the positive effect preconditions of action α with respect to p (i.e. the conditions which ensure that p will be settled to be true when action α occurs) and with $\gamma^-(\alpha, p)$ the negative effect preconditions of action α with respect to p (i.e. the conditions which ensure that p will be settled to be false when action α occurs). For example, we might suppose that $\gamma^+(pullTrigger, scaredEnemy) = holdsGun \wedge loadedGun$, i.e. the positive effect preconditions of the action 'pull the trigger of the gun' with respect to the fact 'the enemy gets scared' consist in 'holding a loaded gun in a hand'. As the following definition highlights, we can say that a certain fact will possibly become true (resp. false) if and only if, there exists an action α performed by the agent, the positive (resp. negative) effect preconditions of α with respect to p hold and there is no action β performed by the agent such that the negative (resp. positive) effect preconditions of β with respect to p hold.

Definition 1. *For every $p \in ATM^{Fact}$ we define:*
$$Pre(p \overset{W}{\rightsquigarrow} \top) = \bigvee_{\alpha \in ATM^{Act}}(\alpha \wedge \gamma^+(\alpha, p)) \wedge \neg \bigvee_{\beta \in ATM^{Act}}(\beta \wedge \gamma^-(\alpha, p));$$
$$Pre(p \overset{W}{\rightsquigarrow} \bot) = \bigvee_{\alpha \in ATM^{Act}}(\alpha \wedge \gamma^-(\alpha, p)) \wedge \neg \bigvee_{\beta \in ATM^{Act}}(\beta \wedge \gamma^+(\alpha, p)).$$

Executability preconditions for action execution. The following general principle clarifies the connection between mental level and intentional action level (a similar principle is discussed in [12]).

(*) An action α will be possibly performed by an agent if and only if the agent has the intention to perform action α and he does not believe that doing α is something bad for him.

Therefore, if an agent intends to a certain action α and, before starting to perform the action, he learns that doing α is something bad for him, he will not start to execute the action and he will reconsider his corresponding intention (see Principle *** below). As the following definition highlights, the Principle * can be expressed in our logic.

Definition 2. *For every $\alpha \in ATM^{Act}$ we define:*[2]
$$Pre(\alpha \overset{W}{\rightsquigarrow} \top) = \mathtt{I}(\alpha) \wedge \neg[\mathtt{B}]\neg good_\alpha.$$

From Definition 2 it follows that all action occurrences of type α are occurrences of intentional actions.

---

[2] We here suppose that the agent believes that doing an action α is *bad* for him if and only if he believes doing α is *not good* for him.

Executability preconditions for intention generation and for intention reconsideration.
The executability preconditions of a process of intention generation correspond to general principles of instrumental rationality which specify the beliefs that an agent uses as premises of a practical argument (viz. the argument that concludes in an intention). Such beliefs are generally called *reasons for acting* or *reasons for intending* and have been extensively studied in the philosophical literature (see, e.g., [1, 18]).

We here suppose that an agent will possibly decide to perform a certain action (and will possibly form the corresponding intention) on the basis of the following general principle of instrumental rationality:

(**) An agent will possibly form the intention to perform action α if and only if, he does not have this intention and he believes that doing action α is something good for him.

The Principle ** can be expressed in our logic in terms of the executability preconditions of the event $gen(\alpha)$.

Definition 3. *For every* $\alpha \in ATM^{Act}$ *we define:*

$$Pre(gen(\alpha)) = \neg\mathtt{I}(\alpha) \wedge [\mathtt{B}]good_\alpha.$$

The following principle is about intention reconsideration.

(***) An agent will possibly reconsider his intention to perform action α if and only if, he intends to perform action α and he believes that performing action α is something bad for him.

The Principle *** can be expressed in our logic in terms of the executability preconditions of the event $rec(\alpha)$.

Definition 4. *For every* $\alpha \in ATM^{Act}$ *we define:*

$$Pre(rec(\alpha)) = \mathtt{I}(\alpha) \wedge [\mathtt{B}]\neg good_\alpha.$$

An example. We provide a general example in order to show how $\mathbf{L}^{dyn}$ can be concretely used to model intention dynamics. Consider the scenario represented in Fig. 2.

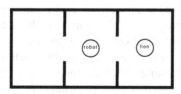

Fig. 2. Example

The agent is a robot moving in a space with three rooms (left room, middle room, right room). It can either move right or move left, that is, $ATM^{Act} = \{moveL, moveR\}$. Moreover, if the robot is in the middle room or in the left room and it moves to the left, it will be in the left room afterwards. If the robot is in the middle room or in the left room or in the right room and it moves to the right, it will not be in the left room afterwards. Finally, if the robot is in the right room and it moves left, it will not be in the right room afterwards. These three facts are formally expressed by the following three positive and negative effect preconditions: $\gamma^+(moveL, robotL) = robotL \lor robotM$, $\gamma^-(moveL, robotL) = robotR$ and $\gamma^-(moveR, robotL) = robotL \lor robotM \lor robotR$.

The robot just entered into the middle room from the left room by moving right, it has still the intention to move right and it does not have the intention to move left in order to escape from the middle room:

H1. $robotM \land \neg robotL \land \neg robotR \land \mathtt{I}(moveR) \land \neg \mathtt{I}(moveL)$.

From the middle room the robot can see that a ferocious lion is inside the right room. Hence, the robot has the following beliefs: it believes that moving right is something bad for him and it believes that moving left is something good for him. Indeed, if it moves right it will be eaten by the lion, and if it moves left it will escape from the lion:

H2. $[\mathtt{B}]\neg good_{moveR} \land [\mathtt{B}]good_{moveL}$.

As the following proposition highlights, the previous Hypothesis H1 and H2 ensure that the following three-event sequence will possibly occur: the robot will reconsider its intention to move right and will generate the intention to move left; the robot will perform the action of moving left; the robot will enter into the left room. Moreover, at the end of the three-event sequence, the robot will be in the left room.

Proposition 2. $\vdash_{L^{dyn}} (H1 \land H2) \rightarrow \langle \{\emptyset, \{rec(moveR), gen(moveL)\}, \emptyset\}{:}w\rangle$
$\langle \{\emptyset, \emptyset, \{moveL \overset{W}{\rightsquigarrow} \top\}\}{:}w\rangle \langle \{\emptyset, \emptyset, \{robotL \overset{W}{\rightsquigarrow} \top\}\}{:}w\rangle robotL$.

5   Related Works and Perspectives

Note that the three operators $[\Sigma{:}w]$, $[\Sigma{:}B]$ and $[\Sigma{:}c]$ can be seen as nothing but the three points e_W, e_B and e_C of an action/event model à la [3, 7], such that $Pre(e_W) = Pre(e_B) = Pre(e_C) = Pre(\Sigma)$ and such that $(e_W, e_B), (e_B, e_B), (e_C, e_C) \in \mathcal{B}$ and $(e_W, e_C), (e_B, e_C), (e_C, e_C) \in \mathcal{C}$, and such such that $Post(e_W)(p) = \sigma_W(p)$ for all atoms $p \in D(\sigma_W)$ and similarly for e_C and e_B, where $Post(e_W)(p)$ is the postcondition of the event e_W applied to the atom p. In the surprising variation on the event models with assignments proposed in this paper, we have defined preconditions *per atomic proposition* p and not all at once per event e_C, e_B, e_W. As only for a finite number of atoms such preconditions are given, the precondition à la DEL indeed corresponds to the conjunction defining $Pre(\Sigma)$ at the beginning of Section 4. So this is all neat and nice, and we have therefore more variation in finetuning preconditions than in standard DEL with assignments.

In [2, 11] a logic of knowledge and preference dynamics is provided. In van Benthem & Liu's approach knowledge dynamics are modeled by means of announcements (or updates), whereas preference dynamics are modeled by means of operations on accessibility relations called upgrades. We have provided here an approach to choice change and intention dynamics based on assignments. We think indeed that assignments, rather than announcements and upgrades, are more suited to model intention dynamics (intention generation and intention reconsideration). Indeed, intention dynamics are obtained by means of *local* operations on an agent's choices and assignments are a natural candidate to formalize these kinds of operations. This locality aspect of intention dynamics has been discussed in Section 3.3 in which we have shown that a process of generating (resp. reconsidering) an intention defined in terms of assignments does not affect the other intentions of the agent (Theorems 1e and 1f).

Directions for future research are manifold. For instance, the logic $\mathbf{L}^{dyn}$ does not allow to distinguish between *present-directed intention* and *future-directed intention*. According to [5], a future-directed intention is an intention to do some action later whereas a present-directed intention is an intention to do some action now. An interesting direction to be explored is an extension of $\mathbf{L}^{dyn}$ with temporal modalities in order to be able to express this distinction. Furthermore, in this paper we only considered the single-agent case. We plan to extend our approach to the multi-agent case.

References

1. Audi, R.: A theory of practical reasoning. American Philosophical Quarterly 19, 25–39 (1982)
2. van Benthem, J., Liu, F.: Dynamic logic of preference upgrade. Journal of Applied Non-Classical Logics 17(2), 157–182 (2007)
3. van Benthem, J., van Eijck, J., Kooi, B.: Logics of communication and change. Information and Computation 204(11), 1620–1662 (2006)
4. Blackburn, P., de Rijke, M., Venema, Y.: Modal Logic. Cambridge University Press, Cambridge (2001)
5. Bratman, M.: Intentions, plans, and practical reason. Harvard University Press (1987)
6. Cohen, P.R., Levesque, H.J.: Intention is choice with commitment. Artificial Intelligence 42, 213–261 (1990)
7. van Ditmarsch, H., Kooi, B.: Semantic results for ontic and epistemic change. In: Proceedings of LOFT 7, pp. 87–117 (2008)
8. van Ditmarsch, H., van der Hoek, W., Kooi, B.: Dynamic Epistemic Logic. Kluwer Academic Publishers, Dordrecht (2007)
9. Hintikka, J.: Knowledge and Belief. Cornell University Press (1962)
10. van der Hoek, W., Jamroga, W., Wooldridge, M.: Towards a theory of intention revision. Synthese 155(2), 265–290 (2007)
11. Liu, F.: Changing for the better: preference dynamics and agent diversity. PhD thesis, University of Amsterdam (2008)
12. Lorini, E., Herzig, A.: A logic of intention and attempt. Synthese 163(1), 45–77 (2008)
13. Meyer, J.J.C., van der Hoek, W., van Linder, B.: A logical approach to the dynamics of commitments. Artificial Intelligence 113(1-2), 1–40 (1999)
14. Rao, A.S., Georgeff, M.P.: Modelling rational agents within a BDI-architecture. In: Proceedings of KR 1991, pp. 473–484 (1991)
15. Reiter, R.: Knowledge in action: logical foundations for specifying and implementing dynamical systems. MIT Press, Cambridge (2001)
16. Shoham, Y.: Agent-oriented programming. Artificial Intelligence 60, 51–92 (1993)
17. Singh, M., Asher, N.: A logic of intentions and beliefs. Journal of Philosophical Logic 22, 513–544 (1993)
18. Von Wright, G.H.: On so-called practical inference. The Philosophical Review 15, 39–53 (1972)
19. Wooldridge, M.: Reasoning about rational agents. MIT Press, Cambridge (2000)

Epistemic Games in Modal Logic: Joint Actions, Knowledge and Preferences All Together

Emiliano Lorini, François Schwarzentruber, and Andreas Herzig

Université de Toulouse, CNRS,
Institut de Recherche en Informatique de Toulouse (IRIT)

Abstract. We propose a modal logic called $\mathcal{EDLA}$ (*Epistemic Dynamic Logic of Agency*) that allows to reason about epistemic games in strategic form. $\mathcal{EDLA}$ integrates the concepts of joint action, preference and knowledge. In the first part of the paper we introduce $\mathcal{EDLA}$ and provide soundness, completeness and complexity results. In the second part we study in $\mathcal{EDLA}$ the epistemic and rationality conditions of some classical solution concepts like Nash equilibrium and iterated strict dominance. In the last part of the paper we combine $\mathcal{EDLA}$ with Dynamic Epistemic Logic (DEL) in order to model epistemic game dynamics.

1 Introduction

We present a modal logic integrating the concepts of joint action, preference and knowledge. Our logic supports reasoning about epistemic games in strategic form in which agents decide what to do according to some general principles of rationality while being uncertain about several aspects of the interaction such as other agents' choices, other agents' preferences, etc.

While epistemic games have been extensively studied in economics (in the so-called interactive epistemology area, see e.g. [1,8,3,9]) and while there have been some analysis of epistemic games in modal logic (see, e.g., [5,10,8,20]), no modal approach to epistemic games in strategic form has been proposed up to now which addresses all the following issues at the same time: to provide a *formal language*, and a corresponding *formal semantics*, which is sufficiently general to express solution concepts like Nash Equilibrium or Iterated Deletion of Strictly Dominated Strategies (IDSDS) and to deduce formally the epistemic and rationality conditions on which such solution concepts are based; to prove its *soundness* and *completeness*; to study its computational properties like *decidability* and *complexity*. In this paper, we try to fill this gap by proposing a sound and complete modal logic for epistemic games interpreted on a Kripke-style semantics. We also provide complexity results for our logic.

The remainder of the paper is organized as follows. In Section 2 we present our modal logic of joint actions, preference and knowledge called $\mathcal{EDLA}$ (*Epistemic Dynamic Logic of Agency*). Section 3 is devoted to the analysis in $\mathcal{EDLA}$ of the epistemic conditions of Nash equilibrium and IDSDS. In Section 4 we make $\mathcal{EDLA}$ dynamic by extending it with constructions of Dynamic Epistemic Logic (DEL) [11], and we show that this dynamic version of $\mathcal{EDLA}$ allows to express IDSDS in a more compact way than in the static $\mathcal{EDLA}$. In Section 5 we show how our logical framework

X. He, J. Horty, and E. Pacuit (Eds.): LORI 2009, LNAI 5834, pp. 212–226, 2009.

can be applied to the analysis of strategic interaction with imperfect information about the game structure. Finally, in Section 6, we compare our approach with some existing approaches to epistemic games in modal logic.

2   A Logic of Joint Actions, Knowledge And Preferences

The logic $\mathcal{EDLA}$ (*Epistemic Dynamic Logic of Agency*) is an extension of the logic $\mathcal{DLA}$ (*Dynamic Logic of Agency*) with modal operators for preference and knowledge modalities. $\mathcal{DLA}$ itself, which was presented in [14,17], extends dynamic logic by a modal operator of historic possibility quantifying over possible joint actions of all agents. This operator is borrowed from STIT theory [4]. In [14,17] the relationships between $\mathcal{DLA}$ and Coalition Logic (CL), and $\mathcal{DLA}$ and STIT have been studied. We will come back to this point in Section 2.3.

2.1   Syntax

The syntactic primitives of $\mathcal{EDLA}$ are the finite set of agents Agt, the set of atomic formulas Atm and a nonempty finite set of atomic action names $Act = \{a_1, a_2, \ldots, a_{|Act|}\}$.

The language $\mathcal{L}_{\mathcal{EDLA}}$ of the logic $\mathcal{EDLA}$ is given by the following BNF:

$$\varphi ::= p \mid \perp \mid \neg\varphi \mid \varphi \vee \varphi \mid \langle i{:}a \rangle \varphi \mid \Diamond\varphi \mid \mathsf{K}_i\varphi \mid [\mathsf{good}]_i\,\varphi$$

where p ranges over Atm, a ranges over Act, and i ranges over Agt.

It is supposed that every agent performs exactly one action at a time, that actions performed by different agents are independent, that actions of different agents are performed in parallel and lead to a unique successor state. Therefore the formula $\langle i{:}a \rangle \varphi$ reads "i performs action a and φ holds afterwards", and $\langle i{:}a \rangle \top$ reads "i performs a". Note that this is slightly different from the standard PDL reading "*there is* a possible execution of action a after which φ holds", which takes into account that there could be different executions of the same action leading to different successor states. $\langle i{:}a \rangle \top \wedge \langle j{:}b \rangle \top$ means that i and j respectively perform a and b in parallel.

The operator $\Diamond$ is an operator of historic possibility, and quantifies over possible joint actions of all agents, that is, over the strategy profiles of the current game (the terms "joint actions of all agents" and "strategy profiles" are supposed here to be synonymous). $\Diamond\varphi$ reads "φ holds for some alternative strategy profile of the current game", or simply "φ is possibly true".

The classical Boolean connectives $\wedge$, $\rightarrow$, $\leftrightarrow$ and $\top$ (tautology) are defined from $\perp$, $\vee$ and $\neg$ in the usual manner. Moreover, $[i{:}a]\,\varphi$ abbreviates $\neg\langle i{:}a \rangle\neg\varphi$, $\Box\varphi$ abbreviates $\neg\Diamond\neg\varphi$ and $\widehat{\mathsf{K}}_i\varphi$ abbreviates $\neg\mathsf{K}_i\neg\varphi$. $\Box\,\varphi$ means "φ is necessarily true". Therefore $[i{:}a]\perp$ reads "i does not perform action a", and $[i{:}a]\varphi$ reads "if i performs a then φ holds afterwards". The following abbreviations are convenient to speak about joint actions. Sets of agents are called *coalitions*, noted $C_1, C_2, \ldots$ To every agent $i \in Agt$ we associate the set Act_i of all possible ordered pairs $i{:}a$, that is, $Act_i = \{i{:}a \mid a \in Act\}$. Besides, we note Δ the set of all joint actions of all agents (alias strategy profiles), that is, $\Delta = \prod_{i \in Agt} Act_i$.

Elements in Δ are Agt-tuples noted α, β, γ, δ, ... Given $\delta \in \Delta$, we note δ_i the element in δ corresponding to agent i. Finally, we note $\delta_C = (\delta_i)_{i \in C}$ the tuple which consists of the vector of all δ_i for $i \in C$. Therefore $\delta_{Agt} = \delta$. Moreover, we write $\delta_{-i} = \delta_{Agt \setminus \{i\}}$.

The following abbreviation will be useful to axiomatize $\mathcal{EDLA}$. For every $\delta \in \Delta$ and $C \subseteq Agt$: $\langle \delta_C \rangle \varphi \stackrel{\text{def}}{=} \bigwedge_{j \in C} \langle \delta_j \rangle \varphi$. $\langle \delta_C \rangle \varphi$ reads "the joint action δ_C is going to be performed by coalition C and φ will be true afterwards". For example, $\langle i{:}a, j{:}b \rangle \varphi$ abbreviates $\langle i{:}a \rangle \varphi \wedge \langle j{:}b \rangle \varphi$, and stands for "the joint action $\langle i{:}a, j{:}b \rangle$ is going to be performed, and φ will be true afterwards". As usual $[\delta_C]\, \varphi \stackrel{\text{def}}{=} \neg \langle \delta_C \rangle \neg \varphi$.

Construction $\mathsf{K}_i \varphi$ is read as usual "agent i knows that φ", whereas the construction $[\mathsf{good}]_i\, \varphi$ is read "φ is true in all worlds which are for agent i at least as good as the current one concerning the strategy profile that is chosen". We define $\langle \mathsf{good} \rangle_i \varphi$ as an abbreviation of $\neg [\mathsf{good}]_i \neg \varphi$. Operators $[\mathsf{good}]_i$ are used in $\mathcal{EDLA}$ to define agents' preference orderings over the strategy profiles of the current game. Similar operators are studied in [6]. We use $\mathsf{EK}_C \varphi$ as an abbreviation of $\bigwedge_{i \in C} \mathsf{K}_i \varphi$, i.e. every agent in C knows φ (if $C = \emptyset$ then $\mathsf{EK}_C \varphi$ is equivalent to $\top$). Then we define by induction $\mathsf{EK}_C^k \varphi$ for every natural number $k \in \mathbf{N}$: $\mathsf{EK}_C^0 \varphi \stackrel{\text{def}}{=} \varphi$ and for all $k \geq 1$, $\mathsf{EK}_C^k \varphi \stackrel{\text{def}}{=} \mathsf{EK}_C(\mathsf{EK}_C^{k-1} \varphi)$. We define for all natural numbers $n \in \mathbf{N}$, $\mathsf{MK}_C^n \varphi$ as an abbreviation of $\bigwedge_{1 \leq k \leq n} \mathsf{EK}_C^k \varphi$. $\mathsf{MK}_C^n \varphi$ expresses C's mutual knowledge that φ up to n iterations, i.e. everyone in C knows φ, everyone in C knows that everyone in C knows φ, and so on until level n.

2.2   Semantics

Frames are tuples $F = \langle W, R, \sim, E, \preceq \rangle$ where:

- W is a nonempty set of possible worlds or states;
- $R : Agt \times Act \longrightarrow W \times W$ maps every agent-action pair $i{:}a$ to a transition relation $R_{i:a} \subseteq W \times W$ between possible worlds;
- $\sim$ is an equivalence relation on W;
- $E : Agt \longrightarrow W \times W$ maps every agent i to an equivalence relation E_i on W;
- $\preceq : Agt \longrightarrow W \times W$ maps every agent i to a reflexive, transitive relation $\preceq_i$ on W.

It is convenient to use $R_{\delta_C} = \bigcap_{i \in C} R_{\delta_i}$, and $R_{\delta_C}(w) = \{w' \in W \mid w R_{\delta_C} w'\}$. If $R_{i:a}(w) \neq \emptyset$ then i performs a at w. More generally, if $R_{\delta_C}(w) \neq \emptyset$ then coalition C performs joint action δ_C at w. If $w' \in R_{\delta_C}(w)$ then world w' results from the performance of joint action δ_C by coalition C at w.

If $w' \sim w$ then w and w' correspond to alternative strategy profiles of the same game. For short, we say that w' is alternative to w. Given a world w, we use the notation $\sim(w) = \{w' \mid w' \sim w\}$ to denote the equivalence class made up of those worlds corresponding to alternative strategy profiles of the game of which w is one of the strategy profile. Consider e.g. $Agt = \{1, 2\}$ and $Act = \{c, d, skip\}$. In the frame in Fig. 1 we have $w_1 \sim w_2$. This means that the joint action performed at w_1 (viz. $\langle 1{:}c, 1{:}c \rangle$) and the one performed at w_2 (viz. $\langle 1{:}c, 1{:}d \rangle$) are alternative strategy profiles of the same game defined by the equivalence class $\sim(w_1) = \{w_1, w_2, w_3, w_4\}$. For every $C \subseteq Agt$, if there exists $w' \in \sim(w)$ such that C performs δ_C at w' then we say that δ_C is *possible* at w (or δ_C *can be performed* at w).

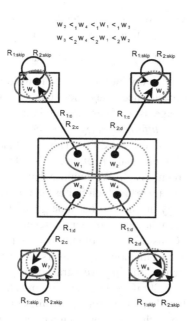

Fig. 1. The equivalence class $\{w_1, w_2, w_3, w_4\}$ represents the Prisoner's Dilemma game [18] between two players 1 and 2 (action c stands for 'cooperate' and action d stands for 'defect'). Full ellipses are epistemic relations for 1, dotted ellipses are epistemic relations for 2 (both 1 and 2 are uncertain about the other's action).

wE_iw' means that, for agent i, world w' is (epistemically) possible at w, whilst $w \preceq_i w'$ means that for agent i, world w' is at least as good as world w. We write $w =_i w'$ iff $w \preceq_i w'$ and $w' \preceq_i w$, and $w <_i w'$ iff $w \preceq_i w'$ and not $w' \preceq_i w$.

Frames have to satisfy the following semantic constraints S1-S9 in order to be $\mathcal{EDLA}$-frames. For every $w, v, v' \in W$ and $\delta, \delta' \in \Delta$ and $a \in Act$ and $i \in Agt$:

(S1) If $v \in R_\delta(w)$ and $v' \in R_{\delta'}(w)$ then $v = v'$;
(S2) $\bigcup_{\delta \in \Delta} R_\delta(w) \neq \emptyset$;
(S3) If $\delta \neq \delta'$ then $R_\delta(w) = \emptyset$ or $R_{\delta'}(w) = \emptyset$;
(S4) If for every $i \in Agt$ there is v_i such that $w \sim v_i$ and $R_{\delta_i}(v_i) \neq \emptyset$ then there is a v such that $w \sim v$ and $R_\delta(v) \neq \emptyset$;
(S5) If $w \sim v$ and $R_\delta(w) \neq \emptyset$ and $R_\delta(v) \neq \emptyset$, then $w = v$;
(S6) If wE_iv then $R_{i:a}(w) \neq \emptyset$ iff $R_{i:a}(v) \neq \emptyset$;
(S7) If $w \preceq_i v$ then $w \sim v$;
(S8) If $w \sim v$ and $w \sim v'$ then $v \preceq_i v'$ or $v' \preceq_i v$;
(S9) If wE_iw' then $w \sim w'$.

According to Constraint S1, for every world w there exists exactly one successor of w, viz. the world resulting from the execution of the strategy profile associated to w. According to the Constraints S2 and S3, every world is associated to exactly one joint action of all agents (alias strategy profile). Note that constraints S1 and S2 together ensure that for every world w there is exactly one next (future) world. We can therefore define a function Nxt in order to identify this unique next (future) world. Formally, given an arbitrary agent i we suppose that for every $w \in W$:

$$\mathsf{Nxt}(w) = v \text{ iff } \bigcup_{i \in Agt, \, a \in Act} R_{i:a}(w) = \{v\}.$$

According to the Constraint S4, if every individual action in a joint action δ is possible at w, then their simultaneous occurrence is also possible at w. We suppose determinism for the joint actions of all agents: different worlds in an equivalence class $\sim(w)$ correspond to the occurrences of *different* strategy profiles (Constraint S5). Constraint S6 just says that agents know what they are doing. This is a standard assumption in interactive epistemology and epistemic analysis of games (see [8] for instance). We also have two constraints over the relations $\preceq_i$. We suppose that a world w' is for agent i at least as good as w only if w' is a world which is possible at w, i.e. only if w' and w correspond to alternative strategy profiles of the same game (Constraint S7). Furthermore, we suppose that every agent has a complete preference ordering over the strategy profiles of the current game (Constraint S8). Finally, we suppose perfect information about the specification of the game, including the players' strategy sets (or action repertoires) and the players' preference ordering over strategy profiles. This assumption is formally expressed by Constraint S9: if world w' is epistemically possible for agent i at w, then w and w' correspond to alternative strategy profiles of the same game. Perfect information about the structure of the game is a standard assumption in game theory. In Section 5, this assumption will be relaxed in order to deal with realistic situations in which an agent might be uncertain about his own utility and other agents' utilities associated to a certain strategy profile, as well as about his own action repertoire and other agents' action repertoires.

A frame F is a $\mathcal{EDLA}$-*frame* if F satisfies constraints S1-S9. A $\mathcal{EDLA}$-*model* is a couple $M = \langle F, \pi \rangle$ where F is a $\mathcal{EDLA}$-frame (satisfying constraints S1-S9) and $\pi : Atm \longrightarrow 2^W$ is a valuation function.

Truth conditions for atomic formulas and the Boolean operators are standard. The truth conditions for the modal operators are:

- $M, w \models \langle i{:}a \rangle \varphi$ iff $M, w' \models \varphi$ for some $w' \in R_{i:a}(w)$;
- $M, w \models \Diamond \varphi$ iff $M, w' \models \varphi$ for some $w' \in \sim(w)$;
- $M, w \models \mathsf{K}_i \varphi$ iff $M, w' \models \varphi$ for all w' such that $w E_i w'$;
- $M, w \models [\mathsf{good}]_i \varphi$ iff $M, w' \models \varphi$ for all w' such that $w \preceq_i w'$.

A formula φ is *true in an $\mathcal{EDLA}$-model* M iff $M, w \models \varphi$ for every world w in M. φ is $\mathcal{EDLA}$-*valid* (noted $\models \varphi$) iff φ is true in all $\mathcal{EDLA}$-models. φ is $\mathcal{EDLA}$-*satisfiable* iff $\neg\varphi$ is not $\mathcal{EDLA}$-valid.

2.3   Axiomatization

We call $\mathcal{EDLA}$ the logic that is axiomatized by the principles given in Table 1. Note that Axiom **Indep** is the $\mathcal{EDLA}$ counterpart of the so-called *axiom of independence of agents* of STIT logic [4]. This axiom allows to express the basic game theoretic assumption that the set of strategy profiles of a game in strategic form is the cartesian product of the sets of individual actions for the agents in Agt.

Table 1. Axiomatization of $\mathcal{EDLA}$

(CPL)	All principles of classical propositional logic
(S5$_\Box$)	All principles of modal logic S5 for $\Box$
(K$_{[i:a]}$)	All principles of modal logic K for every $[i{:}a]$
(S5$_{K_i}$)	All principles of modal logic S5 for every K_i
(S4$_{[good]_i}$)	All principles of modal logic S4 for every $[good]_i$
(Alt$_{[\delta]}$)	$\langle\delta\rangle\varphi \to [\delta']\,\varphi$
(Active)	$\displaystyle\bigvee_{\delta\in\Delta} \langle\delta\rangle\top$
(Single)	$\langle\delta\rangle\top \to [\delta']\bot$ if $\delta \neq \delta'$
(Indep)	$\left(\displaystyle\bigwedge_{i\in Agt} \Diamond\langle\delta_i\rangle\top\right) \to \Diamond\langle\delta\rangle\top$
(JointDet)	$(\langle\delta\rangle\top \wedge \varphi) \to \Box(\langle\delta\rangle\top \to \varphi)$
(Aware)	$\langle i{:}a\rangle\top \to K_i\langle i{:}a\rangle\top$
(Incl$_{[good]_i,\Box}$)	$\Box\varphi \to [good]_i\,\varphi$
(PrefConnect)	$\Diamond\varphi \wedge \Diamond\psi \to \Diamond(\varphi \wedge \langle good\rangle_i\psi) \vee \Diamond(\psi \wedge \langle good\rangle_i\varphi)$
(PerfectInfo)	$\Box\varphi \to K_i\varphi$

We write $\vdash_{\mathcal{EDLA}} \varphi$ if φ is a theorem of $\mathcal{EDLA}$. We can define in $\mathcal{EDLA}$ an operator *next* as follows: $X\varphi \overset{\text{def}}{=} \bigvee_{\delta\in\Delta}\langle\delta\rangle\varphi$. Due to Axioms **Active** and **Alt$_{[\delta]}$**, X obeys the standard validity $X\varphi \leftrightarrow \neg X\neg\varphi$ of the linear-time temporal logic.

Theorem 1. *$\mathcal{EDLA}$ is determined by the class of $\mathcal{EDLA}$-frames.*

Proof. All axioms of $\mathcal{EDLA}$ are in the Sahlqvist class. Using the Sahlqvist algorithm it is routine to prove that the Axioms **Alt$_{[\delta]}$**, **Active**, **Single**, **Indep**, **JointDet**, **Aware**, **Incl$_{[good]_i,\Box}$**, **PrefConnect** and **PerfectInfo** of $\mathcal{EDLA}$ respectively correspond to the constraints S1, S2, S3, S4, S5, S6, S7, S8 and S9. Completeness of $\mathcal{EDLA}$ then follows from Sahlqvist's completeness theorem, cf. [7, Th. 2.42]. $\qquad\Box$

Theorem 2. *The satisfiability problem of $\mathcal{EDLA}$ is PSPACE-complete.*

Proof. We give a sketch of the proof. Theorem 2 is implied by the following two facts.

1. The satisfiability problem of $\mathcal{EDLA}$ is PSPACE-hard.
2. The satisfiability problem of $\mathcal{EDLA}$ is PSPACE.

Consider B the logic of the class of infinite binary trees. The logic B is PSPACE-complete (we leave to the reader the adaptation of the proof given in [7]). Then the idea is to simulate binary trees with one agent and two actions. Let φ a B-formula. We define a translation tr from the language of B to the language of $\mathcal{EDLA}$ by: $tr(\varphi) = \Box \bigwedge_{i\in md(\varphi)}(X\Box)^i(atmosta_1a_2foragent1 \wedge justactiona_1forotheragents) \wedge tr_1(\varphi)$
where:

- $atmosta_1a_2foragent1 = \bigwedge_{j\in\{3,...,n\}} [1{:}a_j]\bot$,
- $justactiona_1forotheragents = \bigwedge_{i\in Agt, i\neq 1} \bigwedge_{j\in\{2,...,n\}} [i{:}a_j]\bot$,
- $tr_1(p) = p$,
- $tr_1(\Diamond\psi) = \Diamond\mathsf{X}tr_1(\psi)$,

and tr_1 is isomorphic for other connectives. We can prove that φ is B-sat iff $tr(\varphi)$ is $\mathcal{EDLA}$-sat.

In order to prove that the satisfiability problem of $\mathcal{EDLA}$ is PSPACE, note that the satisfiability problem of the fragment of $\mathcal{EDLA}$ without time is NP-complete. The latter fragment is the set of formulas φ such that for every subformula $[i{:}a]\psi$ of φ implies $\psi = \bot$. Indeed, we can prove that every satisfiable formula is satisfiable in a model whose size is bounded by $card(Act)^{card(Agt)}$. Finally we can prove that there exists an depth first search algorithm running in PSPACE. The idea is the same as in [2].   □

It has been proved in [17] that the fragment of $\mathcal{EDLA}$ without preference and knowledge modalities embeds Coalition Logic (CL) [19]. In particular, CL cooperation modalities of the form $[C]$ can be reconstructed in our logic as follows.

$$tr([C]\,\varphi) = \bigvee_{\delta\in\Delta} (\Diamond\langle\delta_C\rangle\top \wedge \Box(\langle\delta_C\rangle\top \to \mathsf{X}\varphi))$$

That is, the CL expression "coalition C can enforce an outcome state satisfying φ" (noted $[C]\,\varphi$) is translated in our logic as "there exists a joint action δ_C of the agents in C such that the agents in C can perform δ_C, and necessarily if the agents in C perform δ_C then φ will be true in the next state, no matter what the agents outside C do".

It has also been shown in [14] that a slightly different variant of the logic presented in this paper embeds Chellas' STIT logic with agents and groups [16], under the hypothesis that the number of agents' choices is bounded. STIT logic has formulas of the form $[C\ cstit{:}\varphi]$ that are read "group C sees to it that φ". To obtain this embedding, it is sufficient to remove from $\mathcal{EDLA}$ the Axiom of joint determinism **JointDet** and to add an Axiom of the form $\langle\delta\rangle\Diamond\varphi \to \Diamond\langle\delta\rangle\varphi$ which allows to capture the so-called semantic property *no choice between undivided histories* on STIT frames (see [4]). The translation of STIT modalities of the form $[C\ cstit{:}]$ into our logic would be the following:

$$tr([C\ cstit{:}\varphi]) = \bigvee_{\delta\in\Delta}(\langle\delta_C\rangle\top \wedge \Box(\langle\delta_C\rangle\top \to \varphi))$$

That is, the STIT expression "group C sees to it that φ" is translated into $\mathcal{DLA}$ as "there exists a joint action δ_C of the agents in C such that the agents in C perform δ_C, and necessarily if the agents in C perform δ_C then φ will be true, no matter what the agents outside C do".

3   A Logical Account of Epistemic Games

3.1   Best Response and Nash Equilibrium

The modal operators $[good]_i$ and $\Box$ allow to capture in $\mathcal{EDLA}$ a notion of comparative goodness over formulas of the kind "φ is for agent i at least as good as ψ", noted $\psi \leq_i \varphi$:

$$\psi \leq_i \varphi \overset{\text{def}}{=} \Box(\psi \to \langle good\rangle_i\varphi).$$

$\psi \leq_i \varphi$ is a total preorder: the formulas $\psi \leq_i \psi$, $(\varphi_1 \leq_i \varphi_2) \wedge (\varphi_2 \leq_i \varphi_3) \rightarrow (\varphi_1 \leq_i \varphi_3)$ and $(\varphi_1 \leq_i \varphi_2) \vee (\varphi_2 \leq_i \varphi_1)$ are valid in $\mathcal{EDLA}$. We note $\psi <_i \varphi \overset{\text{def}}{=} (\psi \leq_i \varphi) \wedge \neg(\varphi \leq_i \psi)$ and $\delta \leq_i \delta' \overset{\text{def}}{=} \langle \delta \rangle \top \leq_i \langle \delta' \rangle \top$. Finally we note $\delta <_i \delta' \overset{\text{def}}{=} (\delta \leq_i \delta') \wedge \neg(\delta' \leq_i \delta)$.

Some basic concepts of game theory can be expressed in $\mathcal{EDLA}$ in terms of comparative goodness. We first consider *best response*. Agent i's action a is said to be a best response to the other agents' joint action δ_{-i}, noted $\mathsf{BR}(i{:}a, \delta_{-i})$, if and only if i cannot improve his utility by deciding to do something different from a while the others choose the joint action δ_{-i}, that is:

$$\mathsf{BR}(i{:}a, \delta_{-i}) \overset{\text{def}}{=} \bigwedge_{b \in Act}(((\langle i{:}b \rangle \top \wedge \langle \delta_{-i} \rangle \top) \leq_i (\langle i{:}a \rangle \top \wedge \langle \delta_{-i} \rangle \top)).$$

Given a certain strategic game, the strategy profile (or joint action) δ is said to be a *Nash equilibrium* if and only if for every agent $i \in Agt$, i's action δ_i is a best response to the other agents' joint action δ_{-i}:

$$\mathsf{Nash}(\delta) \overset{\text{def}}{=} \bigwedge_{i \in Agt} \mathsf{BR}(\delta_i, \delta_{-i}).$$

From Axiom **PerfectInfo** and S5 for $\square$, the following theorems are provable expressing perfect information about the players' preferences ordering over strategy profiles, perfect information about the existence of a Nash equilibrium, and perfect information about the players' repertoires: $(\psi \leq_i \varphi) \leftrightarrow \mathsf{MK}_{Agt}^n(\psi \leq_i \varphi)$, $\mathsf{Nash}(\delta) \leftrightarrow \mathsf{MK}_{Agt}^n \mathsf{Nash}(\delta)$ and $\Diamond \langle \delta_i \rangle \top \leftrightarrow \mathsf{MK}_{Agt}^n \Diamond \langle \delta_i \rangle \top$, for every $n \in \mathbf{N}$.

3.2 Epistemic Rationality

The following $\mathcal{EDLA}$ formula characterizes a notion of rationality which is commonly supposed in the epistemic analysis of games (see, e.g., [3,5]):

$$\bigwedge_{a,b \in Act}\left(\langle i{:}a \rangle \top \rightarrow \bigvee_{\delta \in \Delta}\left(\widehat{\mathsf{K}}_i \langle \delta_{-i} \rangle \top \wedge ((\langle \delta_{-i}, i{:}b \rangle \leq_i \langle \delta_{-i}, i{:}a \rangle))\right)\right).$$

This means that an agent i is rational if and only if, if he chooses a particular action a then for every alternative action b, there exists a joint action δ_{-i} of the other agents that he considers possible such that, playing a while the others play δ_{-i} is for i at least as good as playing b while the others play δ_{-i}. As in $\mathcal{EDLA}$ $\delta \leq_i \delta'$ and $\mathsf{K}_i(\delta \leq_i \delta')$ are equivalent, the previous definition of rationality can be rewritten in the following equivalent form:

$$\mathsf{Rat}_i \overset{\text{def}}{=} \bigwedge_{a,b \in Act}\left(\langle i{:}a \rangle \top \rightarrow \bigvee_{\delta \in \Delta}\left(\widehat{\mathsf{K}}_i \langle \delta_{-i} \rangle \top \wedge \mathsf{K}_i(\langle \delta_{-i}, i{:}b \rangle \leq_i \langle \delta_{-i}, i{:}a \rangle)\right)\right).$$

Theorem 3. *For all $i \in Agt$:*

(3a) $\vdash_{\mathcal{EDLA}} \mathsf{Rat}_i \leftrightarrow \mathsf{K}_i \mathsf{Rat}_i$

(3b) $\vdash_{\mathcal{EDLA}} \neg\mathsf{Rat}_i \leftrightarrow \mathsf{K}_i \neg\mathsf{Rat}_i$

Theorem 3 highlights that the concepts of rationality and irrationality are introspective. The following theorem specifies some sufficient epistemic conditions for guaranteeing that the chosen strategy profile is a Nash equilibrium: if all agents are rational and every agent knows the choices of the other agents, then the selected strategy profile is a Nash equilibrium. A similar theorem has been stated for the first time in [1,9].

Theorem 4. *For all $n \in N$, for all $\delta \in \Delta$:*

$$\vdash_{\mathcal{EDLA}} \left(\left(\bigwedge_{i \in Agt} \mathsf{Rat}_i \right) \wedge \bigwedge_{i \in Agt} \mathsf{K}_i \langle \delta_{-i} \rangle \top \right) \rightarrow \mathsf{Nash}(\delta)$$

3.3 Iterated Deletion of Strictly Dominated Strategies

A strategy a for agent i is a *strictly dominated strategy*, noted $\mathsf{SD}^{\leq 0}(i{:}a)$, if and only if, if a can be performed then there is another strategy b such that, no matter what joint action δ_{-i} the other agents choose, playing b is for i strictly better than playing a:

$$\mathsf{SD}^{\leq 0}(i{:}a) \stackrel{\text{def}}{=} \Diamond \langle i{:}a \rangle \top \rightarrow$$

$$\bigvee_{b \in Act} \left(\Diamond \langle i{:}b \rangle \top \wedge \bigwedge_{\delta \in \Delta} (\Diamond \langle \delta_{-i} \rangle \rightarrow (\langle \delta_{-i}, i{:}a \rangle <_i \langle \delta_{-i}, i{:}b \rangle)) \right).$$

An example of strictly dominated strategy is cooperation in the Prisoners Dilemma (PD) game: whether ones opponent chooses to cooperate or defect, defection yields a higher payoff than cooperation. Therefore, a rational player will never play a dominated strategy. So when trying to predict the behavior of rational players, we can rule out all strictly dominated strategies. The so-called Iterated Deletion of Strictly Dominated Strategies (IDSDS) (or iterated strict dominance) [18] is a procedure that starts with the original game and, at each step, for every player i removes from the game all i's strictly dominated strategies, thereby generating a subgame of the original game, and that repeats this process again and again. IDSDS can be inductively characterized in our logic $\mathcal{EDLA}$ by defining a concept of strict dominance in the subgame of depth at most n, noted $\mathsf{SD}^{\leq n}(i{:}a)$. For every $n \geq 1$:

$$\mathsf{SD}^{\leq n}(i{:}a) \stackrel{\text{def}}{=} \neg\mathsf{SD}^{\leq n-1}(i{:}a) \rightarrow$$

$$\bigvee_{b \in Act} \left(\neg\mathsf{SD}^{\leq n-1}(i{:}b) \wedge \bigwedge_{\delta \in \Delta} \left(\neg\mathsf{SD}^{\leq n-1}(\delta_{-i}) \rightarrow (\langle \delta_{-i}, i{:}a \rangle <_i \langle \delta_{-i}, i{:}b \rangle) \right) \right).$$

where $\mathsf{SD}^{\leq k}(\delta_C)$ is an abbreviation of $\bigvee_{i \in C} \mathsf{SD}^{\leq k}(\delta_i)$ for every $k \geq 0$ and for every δ_C. According to this definition, a is a strictly dominated strategy for agent i in a subgame of depth at most n, noted $\mathsf{SD}^{\leq n}(i{:}a)$, if and only if, if a is not strictly dominated for i in all subgames of depth $k < n$ then there is another strategy b such that b is not strictly dominated for i in all subgames of depth $k < n$ and, no matter what joint action δ_{-i} the other agents choose, if the elements in δ_{-i} are not dominated in all subgames of depth $k < n$ then playing b is for i strictly better than playing a. In other terms $\mathsf{SD}^{\leq n}(i{:}a)$ means that strategy $i{:}a$ does not survive after n rounds of IDSDS.

It has been shown that common knowledge of rationality implies that players choose strategies which survive IDSDS ([8,3,9]). This latter principle can be derived in our logic $\mathcal{EDLA}$. According to the following Theorem 5, if there is mutual knowledge of rationality among the players to n levels and the agents play the strategy profile δ then, for every agent i, δ_i survives IDSDS until the subgame of depth $n+1$.

Theorem 5. *For all $\delta \in \Delta$,* $\vdash_{\mathcal{EDLA}} \left(\left(\mathsf{MK}^n_{Agt} \bigwedge_{i \in Agt} \mathsf{Rat}_i \right) \wedge \langle \delta \rangle \top \right) \rightarrow \neg\mathsf{SD}^{\leq n}(\delta)$.

4   Game Transformation

We provide in this section an alternative and more compact characterization of the procedure IDSDS in our logic $\mathcal{EDLA}$. To this aim, we introduce special events whose effect is to delete a strictly dominated strategy from the current game. These events are similar to announcements in Dynamic Epistemic Logic (DEL) [11].

$\mathcal{L}_{\mathcal{AN}}$ is the set of *announcable formulas* and is defined by the following BNF:

$$\chi ::= \Box\psi \to [i{:}a]\bot \mid \chi \wedge \chi$$

where $\psi \in \mathcal{L}_{\mathcal{EDLA}}$, $i \in Agt$ and $a \in Act$. Thus, announcable formulas are of the form 'if property ψ necessarily holds in the current game, then action a should not be performed by agent i'.

We extend the $\mathcal{EDLA}$ language with announcements operators of the form $[\chi!]$ with $\chi \in \mathcal{L}_{\mathcal{AN}}$. We call $\mathcal{EDLA}^{AN}$ the extended logic. The truth condition for $[\chi!]\varphi$ is:

$$M, w \models [\chi!]\varphi \text{ iff } M, w \models \chi \text{ implies } M^\chi, w \models \varphi$$

with $M^\chi = \langle W^\chi, R^\chi, \sim^\chi, E^\chi, \preceq^\chi, \pi^\chi \rangle$ and:

$W^\chi = \|\chi\|_M$

$R_{i:a}^\chi = [R_{i:a} \cap (W^\chi \times W^\chi)] \cup \{(w,w) \mid R_{i:a}(w) \neq \emptyset, \ w \in \|\chi\|_M, \ \mathsf{Nxt}(w) \notin \|\chi\|_M\}$

$\sim^\chi = \sim \cap (W^\chi \times W^\chi)$

$E_i^\chi = E_i \cap (W^\chi \times W^\chi)$

$\preceq_i^\chi = \preceq_i \cap (W^\chi \times W^\chi)$

$\pi^\chi(p) = \pi(p) \cap W^\chi$

Thus, an event $\chi!$ removes from the model M all worlds in which χ is false. The epistemic relations E_i and preference orderings $\preceq_i$ are restricted to the worlds in which χ is true. Moreover, if v is the world which results from the execution of a by i at w, and w and v are not removed from the model by the event $\chi!$ then, after the occurrence of $\chi!$, v is still the world which results from the execution of a by i at w; if v is the world which results from the execution of a by i at w, w is not removed from the model by the event $\chi!$ and v is removed then, after the occurrence of $\chi!$, we impose that w itself is the world which results from the execution of a by i at w. This ensures that the constraint S2 will be preserved after the model transformation.

Theorem 6. *If M is a $\mathcal{EDLA}$ model then M^χ is a $\mathcal{EDLA}$ model.*

We have reduction axioms for $\chi!$ which guarantee the completeness of $\mathcal{EDLA}^{AN}$.

Theorem 7. *The following schemata are valid in $\mathcal{EDLA}^{AN}$.*

> **R1.** $[\chi!]p \leftrightarrow (\chi \to p)$
>
> **R2.** $[\chi!]\neg\varphi \leftrightarrow (\chi \to \neg[\chi!]\varphi)$
>
> **R3.** $[\chi!](\varphi_1 \wedge \varphi_2) \leftrightarrow ([\chi!]\varphi_1 \wedge [\chi!]\varphi_2)$
>
> **R4.** $[\chi!]\Box\varphi \leftrightarrow (\chi \to \Box[\chi!]\varphi)$
>
> **R5.** $[\chi!]\mathsf{K}_i\varphi \leftrightarrow (\chi \to \mathsf{K}_i[\chi!]\varphi)$
>
> **R6.** $[\chi!][\mathsf{good}]_i\varphi \leftrightarrow (\chi \to [\mathsf{good}]_i[\chi!]\varphi)$
>
> **R7.** $[\chi!][i{:}a]\varphi \leftrightarrow (\neg\chi \vee ([i{:}a]\chi \wedge [i{:}a][\chi!]\varphi) \vee (\neg[i{:}a]\chi \wedge [\chi!]\varphi))$

Proof. The proofs of **R1-R6** go as in DEL (see [11]). We here prove **R7**.

CASE 1. $\neg\chi$ and $[\chi!]\bot$ are equivalent in $\mathcal{EDLA}^{AN}$. Therefore,

(A) $\neg\chi \rightarrow ([\chi!]\,[i{:}a]\,\varphi \leftrightarrow \neg\chi)$

is valid in $\mathcal{EDLA}^{AN}$.

CASE 2. Suppose $M, w \models [i{:}a]\,\chi \wedge \chi$.
$M, w \models [\chi!]\,[i{:}a]\,\varphi$
IFF if $M, w \models \chi$ then $M^\chi, w \models [i{:}a]\,\varphi$
IFF if $M, w \models \chi$ then, if $v \in R^\chi_{i{:}a}(w)$ then $M^\chi, v \models \varphi$
IFF if $M, w \models \chi$ then, if $v \in R_{i{:}a}(w)$ then $M^\chi, v \models \varphi$ (because $M, w \models [i{:}a]\,\chi \wedge \chi$
implies $R^\chi_{i{:}a}(w) = R_{i{:}a}(w)$),
IFF if $M, w \models \chi$ then, if $v \in R_{i{:}a}(w)$ then $M^\chi, v \models \varphi$ and $M, v \models \chi$ (by $M, w \models$
$[i{:}a]\,\chi$),
IFF if $M, w \models \chi$ then, if $v \in R_{i{:}a}(w)$ then $M, v \models [\chi!]\varphi$ and $M, v \models \chi$
IFF if $M, w \models \chi$ then $M, w \models [i{:}a]\,([\chi!]\varphi \wedge \chi)$
IFF if $M, w \models \chi \rightarrow ([i{:}a]\,([\chi!]\varphi \wedge \chi)$
IFF if $M, w \models \chi \rightarrow [i{:}a]\,[\chi!]\varphi$ (by the hypothesis $M, w \models [i{:}a]\,\chi$).
This proves that $([i{:}a]\,\chi \wedge \chi) \rightarrow ([\chi!]\,[i{:}a]\,\varphi \leftrightarrow (\chi \rightarrow [i{:}a]\,[\chi!]\varphi))$ is valid in $\mathcal{EDLA}^{AN}$.
It follows that
(B) $([i{:}a]\,\chi \wedge \chi) \rightarrow ([\chi!]\,[i{:}a]\,\varphi \leftrightarrow [i{:}a]\,[\chi!]\varphi)$

is valid in $\mathcal{EDLA}^{AN}$ too.

CASE 3. Suppose $M, w \models \neg\,[i{:}a]\,\chi \wedge \chi$ which is equivalent to $M, w \models \langle i{:}a\rangle\neg\chi \wedge \chi$.
$M, w \models [\chi!]\,[i{:}a]\,\varphi$
IFF if $M, w \models \chi$ then $M^\chi, w \models [i{:}a]\,\varphi$
IFF if $M, w \models \chi$ then, if $v \in R^\chi_{i{:}a}(w)$ then $M^\chi, v \models \varphi$
IFF if $M, w \models \chi$ then $M^\chi, w \models \varphi$ (because $M, w \models \langle i{:}a\rangle\neg\chi \wedge \chi$ implies $R^\chi_{i{:}a}(w) = \{w\}$),
IFF if $M, w \models \chi$ then $M^\chi, w \models \varphi$ and $M, w \models \chi$ (by the hypothesis $M, w \models \chi$),
IFF if $M, w \models \chi$ then $M^\chi, w \models [\chi!]\varphi$
IFF if $M, w \models \chi \rightarrow [\chi!]\varphi$
IFF if $M, w \models [\chi!]\varphi$ (by the hypothesis $M, w \models \chi$). Therefore
(C) $(\neg\,[i{:}a]\,\chi \wedge \chi) \rightarrow ([\chi!]\,[i{:}a]\,\varphi \leftrightarrow [\chi!]\varphi)$

is valid in $\mathcal{EDLA}^{AN}$. From (A), (B) and (C) it follows that:
$$[\chi!]\,[i{:}a]\,\varphi \leftrightarrow (\neg\chi \vee ([i{:}a]\,\chi \wedge [i{:}a]\,[\chi!]\varphi) \vee (\neg\,[i{:}a]\,\chi \wedge [\chi!]\varphi))$$
is valid in $\mathcal{EDLA}^{AN}$. $\qquad\square$

Theorem 8. *The logic $\mathcal{EDLA}^{AN}$ is completely axiomatized by the axioms and inference rules of $\mathcal{EDLA}$ together with the schemata of Theorem 7 together with the rule of replacement of proved equivalence.*

Now, consider the following formula:

$$\chi_{\mathsf{SD}} \overset{\text{def}}{=} \bigwedge_{i \in Agt, a \in Act}(\square\mathsf{SD}^{\leq 0}(i{:}a) \rightarrow [i{:}a]\,\bot).$$

The effect of $\chi_{\mathsf{SD}}!$ is to delete from every game $\sim(w)$ in the model M all worlds in which a strictly dominated strategy is played by some agent.

As the following Theorem 9 highlights, the procedure IDSDS that we have charac-terized in Section 3.3 in the static $\mathcal{EDLA}$ can be characterized in a more compact way in $\mathcal{EDLA}^{AN}$. Suppose δ is the selected strategy profile. Then, for every agent i, δ_i survives IDSDS until the subgame of depth $n+1$ if and only if, the event $\chi_{SD}!$ can occur $n+1$ times in sequence.

Theorem 9. *For all* $\delta \in \Delta$, *for all* $n \geq 0$,

$$\vdash_{\mathcal{EDLA}^{AN}} \langle\delta\rangle\top \rightarrow \left(\neg SD^{\leq n}(\delta) \leftrightarrow \langle\chi_{SD}!\rangle^{n+1}\top\right).$$

Finally, here is a reformulation of Theorem 5 in $\mathcal{EDLA}^{AN}$.

Theorem 10. *For all* $n \geq 0$, $\vdash_{\mathcal{EDLA}^{AN}} \left(MK^n_{Agt} \bigwedge_{i \in Agt} Rat_i\right) \rightarrow \langle\chi_{SD}!\rangle^{n+1}\top$.

5 Concluding Remarks: Imperfect Information

We here consider a more general class of games which includes strategic games with imperfect information about the game structure. Apart from few exceptions (see, e.g., [13,12]), these games have been rarely explored. Indeed, most work in game theory assumed that players have common knowledge of all relevant aspects of the game. We are interested in verifying whether the results obtained in Sections 3.2 and 3.3 can be generalized to this kind of games, that is:

1. Are rationality of every player and every agent's knowledge about other agents' choices still sufficient to ensure that the selected strategy profile is a Nash equilibrium in a strategic game with imperfect information about the game structure?
2. Is mutual knowledge of rationality among the players still sufficient to ensure that the selected strategy profile survives iterated deletion of dominated strategies in a strategic game with imperfect information about the game structure?

To answer these questions, we have to remove Axiom **PerfectInfo** from $\mathcal{EDLA}$ and the corresponding semantic constraint S9 from the definition of $\mathcal{EDLA}$ frames expressing the hypothesis of perfect information about the game structure. We call $\mathcal{EDLA}^*$ the resulting logic and $\mathcal{EDLA}^*$-frames the resulting class of frames. Then we have to check whether Theorems 4 and 5 given in Sections 3.2 and 3.3 are still derivable in $\mathcal{EDLA}^*$.

We have a positive answer to the previous first question. Indeed, the formula

$$\left(\left(\bigwedge_{i \in Agt} Rat_i\right) \wedge \bigwedge_{i \in Agt} K_i\langle\delta_{-i}\rangle\top\right) \rightarrow Nash(\delta)$$

is derivable in $\mathcal{EDLA}^*$. But we have a negative answer to the second question. Indeed, the following formula is invalid in $\mathcal{EDLA}^*$ for every $\delta \in \Delta$ and for every $n \in \mathbf{N}$ such that $n > 0$:

$$\left(\left(MK^n_{Agt} \bigwedge_{i \in Agt} Rat_i\right) \wedge \langle\delta\rangle\top\right) \rightarrow \neg SD^{\leq n}(\delta).$$

6 Related Works

Although several modal logics of games in strategic forms have been proposed (see, e.g., [15,21]), few modal logics exist which support reasoning about epistemic (strategic) games. Among them we should mention [10,20,8].

De Bruin [10] has developed a very rich logical framework which enables to reason about the epistemic aspects of strategic games and of extensive games. His system deals with several game-theoretic concepts like the concepts of knowledge, rationality, Nash equilibrium, iterated strict dominance, backward induction. Nevertheless, de Bruin's approach differs from ours in several respects. First of all, our logical approach to epistemic games is *minimalistic* since it relies on few primitive concepts: knowledge, action, historical necessity and preference. All other notions such Nash equilibrium, rationality, iterated strict dominance are defined by means of these four primitive concepts. On the contrary, in de Bruin's logic all those notions are atomic propositions managed by a *ad hoc* axiomatization (see, e.g., [10, pp. 61,65] where special propositions for rationality and iterated strict dominance are introduced). Secondly, we provide a semantics and a complete axiomatics for our logic of epistemic games. De Bruin's approach is purely syntactic: no model-theoretic analysis of games is proposed nor completeness result for the proposed logic is given. Finally, de Bruin does not provide any complexity results about his logic while we prove that the satisfiability problem of a formula in our logic is PSPACE-complete.

Roy [20] has recently proposed a modal logic integrating preference, knowledge and intention. In his approach every world in a model is associated to a nominal which directly refers to a strategy profile in a strategic game. This approach is however limited in expressing formally the structure of a strategic game. In particular, in Roy's logic there is no principle like the $\mathcal{EDLA}$ Axiom **Indep** explaining how possible actions δ_i of individual agents are combined to form a strategy profile δ of the current game. Another limitation of Roy's approach is that it does not allow to express the concept of (weak) rationality that we have been able to define in Section 3.2 (see [20, pp. 101]). As discussed in the previous sections this is a crucial concept in interactive epistemology since it is used for giving epistemic justifications of several solution concepts like Nash equilibrium and IDSDS (see Theorems 4 and 5).

Bonanno [8] integrates modal operators for belief, common belief with constructions expressing agents' preferences over individual actions and strategy profiles, and applies them to the semantic characterization of solution concepts like Iterated Deletion of Strictly Dominated Strategies (IDSDS) and Iterated Deletion of Inferior Profiles (IDIP). As in [20], in Bonanno's logic every world in a model corresponds to a strategy profile of the current game. Although this logic allows to express the concept of weak rationality, it is not sufficiently general to enable to express in the object language solution concepts like Nash equilibrium and IDSDS (note that the latter is defined by Bonanno only in the metalanguage).

It is to be noted that, differently from $\mathcal{EDLA}$, most modal logics of epistemic games in strategic form (including Roy's logic and Bonanno's logic) postulate a one-to-one correspondence between models and games (i.e. every model of the logic corresponds to a unique strategic game, and worlds in the model are all strategy profiles of this game). Such an assumption is quite restrictive since it prevents from analyzing in the logic games with imperfect information about the game structure in which an agent can imagine alternative games. As shown in Section 5, this is something we can do in our logical framework by removing Axiom **PerfectInfo** from $\mathcal{EDLA}$.

Before concluding this section about related works it is to be noted that the approach to game dynamics based on Dynamic Epistemic Logic (DEL) we proposed in Section 4 is inspired by [5] in which strategic equilibrium is defined by fixed-points of operations of repeated announcement of suitable epistemic statements and rationality assertions. However, the analysis of epistemic games proposed in [5] is mainly semantical and the author does not provide a full-fledged modal language for epistemic games which allows to express in the object language solution concepts like Nash Equilibrium or IDSDS, and the concept of rationality. Moreover, van Benthem's analysis does not include any completeness result for the proposed framework and there is no proposal of reduction axioms for a combination of DEL with a static logic of epistemic games. On the contrary, these two aspects are central in our analysis.

References

1. Aumann, R.J., Brandenburger, A.: Epistemic conditions for Nash equilibrium. Econometrica 63, 1161–1180 (1995)
2. Balbiani, P., Gasquet, O., Herzig, A., Schwarzentruber, F., Troquard, N.: Coalition games over Kripke semantics. In: Festschrift in Honour of Shahid Rahman, pp. 11–32. College Publications (2008)
3. Battigalli, P., Bonanno, G.: Recent results on belief, knowledge and the epistemic foundations of game theory. Research in Economics 53, 149–225 (1999)
4. Belnap, N., Perloff, M., Xu, M.: Facing the future: agents and choices in our indeterminist world. Oxford University Press, Oxford (2001)
5. van Benthem, J.: Rational dynamics and epistemic logic in games. International Game Theory Review 9(1), 13–45 (2007)
6. van Benthem, J., Liu, F.: Dynamic logic of preference upgrade. Journal of Applied Non-Classical Logics 17(2), 157–182 (2007)
7. Blackburn, P., de Rijke, M., Venema, Y.: Modal Logic. Cambridge University Press, Cambridge (2001)
8. Bonanno, G.: A syntactic approach to rationality in games with ordinal payoffs. In: Proc. of LOFT 2008, pp. 59–86 (2008)
9. Brandenburger, A.: Knowledge and equilibrium in games. Journal of Economic Perspectives 6, 83–101 (1992)
10. de Bruin, B.: Explaining games: on the logic of game theoretic explanations. PhD thesis, University of Amsterdam (2004)
11. van Ditmarsch, H.P., van der Hoek, W., Kooi, B.: Dynamic Epistemic Logic. Kluwer Academic Publishers, Dordrecht (2007)
12. Halpern, J.Y., Rego, L.: Generalized solution concepts in games with possibly unaware players. In: Proc. of TARK 2007, pp. 253–262 (2007)
13. Harsanyi, J.: Games with incomplete informations played by 'bayesian' players: Part I, the basic model. Management Science 14(3), 159–182 (1967)
14. Herzig, A., Lorini, E.: A dynamic logic of agency I: STIT, abilities and powers. Technical Report RR2009-4FR, Institut de Recherche en Informatique de Toulouse (2009)
15. van der Hoek, W., Jamroga, W., Wooldridge, M.: A logic for strategic reasoning. In: Proc. of AAMAS 2005, pp. 157–164. ACM Press, New York (2005)
16. Horty, J.F.: Agency and Deontic Logic. Oxford University Press, Oxford (2001)
17. Lorini, E.: A dynamic logic of agency II: deterministic DLA, coalition logic, and game theory. Technical Report RR2009-5FR, Institut de Recherche en Informatique de Toulouse (2009)

18. Osborne, M.J., Rubinstein, A.: A course in game theory. MIT Press, Cambridge (1994)
19. Pauly, M.: A modal logic for coalitional power in games. Journal of Logic and Computation 12(1), 149–166 (2002)
20. Roy, O.: Thinking before acting: intentions, logic, rational choice. PhD thesis, University of Amsterdam (2008)
21. Troquard, N., van der Hoek, W., Wooldridge, M.: A logic of games and propositional control. In: Proc. of AAMAS 2009, pp. 961–968. ACM Press, New York (2009)

Dynamic Epistemic Logic of Finite Identification

Minghui Ma

Department of Philosophy, Tsinghua University,
Beijing, China, 100084
http://rwxy.tsinghua.edu.cn/xi-suo/zhe/en/index.htm

Abstract. Following the approach given by Nina Gierasimczuk [2009] to bridge dynamic epistemic logic and learning theory, the logic of finite identification in the limit in the process of learning has been explored in this paper. The main results are two complete axiomatic systems. One is defined in terms of public announcements, and the other in terms of update by event models.

Keywords: learning theory, finite identification, dynamic epistemic logic.

1 Introduction

The motivation of formal learning theory (FLT) is to construct precise models of language acquisition (E. M. Gold [1967]). Human beings have the ability to learn a language. If we can construct a precise model on how children learn a language, then it is possible to investigate theoretically how a natural language can be achieved. One important important notion in language acquisition is the *identification* of some natural language. Another related notion is empirical inquiry by a scientist. For example, there are only two possible worlds (possibilities): the p-world where p is true, and the not-p-world where p is false. The scientist knows all the possibilities. Suppose that the p-world is the actual one. But he cannot distinguish them. Then he makes plausible conjectures or hypotheses from clues given by the actual world, and he gets information or data about worlds by observation and experiments. Finally, if the proposition p became stable since redundant pieces of evidence show that p is the case, he can identify the actual world in the limit. In section 2, I will describe a paradigm in formal learning theory in terms of empirical inquiry by scientist. Then I would like to show the bridge built in Nina Gierasimczuk [2009] between formal learning theory (FLT) and dynamic epistemic logic (DEL) in section 3. In section 4, I will recast the syntax and semantics for the logic of finite identification. Then I will define a complete axiomatization in terms of public announcements. In section 5, the result in section 4 will be extended to the dynamic epistemic logic of update in the process of finite identification in the limit. In the final section, I will give conclusions and some further remarks.

X. He, J. Horty, and E. Pacuit (Eds.): LORI 2009, LNAI 5834, pp. 227–237, 2009.

2   A Fundamental Paradigm of FLT

Components of the paradigm described in these examples in section 1 which I will focus on in this paper is the paradigm of set learning other than paradigms like function learning. It can be extracted as following three ingredients. Details of formal learning theory can be found in Osherson, de Jongh, Martin and Weinstein [1997].

Possible Realities. In the language acquisition, a theoretically possible reality is a potential natural language. In the empirical inquiry, a possible reality is a possible world. Realities are represented by non-empty r.e. (recursively enumerable) subsets of natural numbers. Think of such sets as potential natural languages or possible worlds. Numbers can also be conceived as codes for objects and events founded in scientific contexts.

The Data Available about any Given Reality. The actual world provides a series of clues about the reality. The clues constitute the data according to which scientist makes hypotheses. Streams of data are often called environments. An environment ε of a reality S is an infinite sequence of elements from S such that it enumerates all and only elements from S, allowing repetitions.

Plausible Hypotheses about Any Given Reality. A Turing machine will function as a conjecture or hypothesis. For each reality or language S in C, we consider Turing machines h_n which generate S, and in that case we say that n is a index of S. Let $H(C)$ be the set of all Turing machines generating all realities in C. Each S may have infinitely many indices. Let $I_S := \{n : h_n \text{ generates } S\}$, and $I_{H(C)} = \bigcup_{S \in C} I_S$. A scientist working in a given reality will identify the reality in the limit if he successfully makes answers stable on a correct hypothesis.

The following notation is often used in literatures. Let ε_n be the n-th element of ε, and $\varepsilon|n$ the initial segment $(\varepsilon_0, \ldots, \varepsilon_{n-1})$ of ε. Let SEQ be the set of all finite initial segment of all environments, and $set(\varepsilon)$ the set of elements that occur in ε. Finally, let h_n be a hypothesis, i.e., a finite description of a set, a Turing machine generating S, and $L : SEQ \to I_{H(C)}$ a learning function from finite data sequences to indices of hypotheses. The following two fundamental definitions about identification often occurred in literatures are also contained in Gierasimczuk [2009].

Definition 1. *(Identification in the Limit). A learning function L identifies S in C in the limit on ε iff there is a number k such that for co-finitely many m, $L(\varepsilon|m) = k$ and $k \in I_S$. L identifies S in C in the limit iff it identifies S in the limit on every environment ε for S. L identifies C in the limit iff it identifies every S in C in the limit.*

Definition 2. *(Finite Identification in the Limit) A learning function L finitely identifies S in C on ε iff when inductively given ε, at some point, L outputs a single k such that $k \in I_S$, and stops. L finitely identifies S in C iff it finitely identifies S on every environment ε for S. L finitely identifies C iff it finitely identifies every S in C.*

Example 1. The following examples can also be found in Gierasimczuk [2009]. They show ways for a range of sets being identified or not.

(1) Let $C_0 = \{\{0,1\},\{0,2\},\{0,3\},\ldots\}$. $H(C_0) = \{h_1,h_2,h_3,\ldots\}$. C_0 is finitely identified by the following learning function $L : L(\varepsilon|n) =$ undefined, if $set(\varepsilon|n) = \{0\}$; Otherwise, $L(\varepsilon|n) = max(set(\varepsilon|n))$.

(2) Let $C_1 = \{\{1\},\{1,2\},\{1,2,3\}\}$, with $H(C_1) = \{h_1,h_2,h_3\}$. Then C_1 is not finitely identifiable. Suppose not. Let a learning function L identifies it. Assume that $\{1,2\}$ is chosen as the actual world. But then h_2 can never be conclusively decided to be true. For 3 might occur in the future, and he cannot decide that which of h_2 and h_3 is true. But C_1 is identifiable in the limit by the following learning function $L : SEQ \to H(C_1) : L(\varepsilon|n) = h_m$ where $m = max(set(\varepsilon|n))$.

(3) Let $C_2 = \{\{1\},\{1,2\},\{1,2,3\},\ldots\}$. $H(C_2) = \{h_1,h_2,h_3,\ldots\}$. This class can also be identified in the limit by the above function. Let $C_3 = \{\mathbb{N}\} \cap \{\{1\},\{1,2\},\{1,2,3\},\ldots\}$. $H(C_3) = \{h_0,h_1,h_2,h_3,\ldots\}$, where $h_0 = \omega$. Then C_3 is not identifiable in the limit. Suppose not. Let L identifies C_3. Then there are numbers k and n such that for all $m \geq n$, $L(\varepsilon|m) = k$. If $k > 0$, then L cannot identify $\mathbb{N}$. If $k = 0$, then it cannot identify $h_{max(set(\varepsilon|n))}$.

Another notion of learning presented in Gierasimczuk [2009] is the notion of learning by erasing. From an epistemological point of view, it often happens that some hypotheses falsified during the inductive inquiry are eliminated in the process of jumping to the correct conclusion. Thus, a formal model has been constructed in which the function each time eliminates a hypothesis, instead of making positive a conjecture. That is learning by erasing. The following two definitions are given in Gierasimczuk [2009].

Definition 3. *(Function Stablization) In learning by erasing, we say that a function stabilizes to number k on environment ε iff for co-finitely many number n, $k = min\{\mathbb{N} \setminus \{L(\varepsilon|0),\ldots,L(\varepsilon|n)\}\}$.*

Definition 4. *(Learning by Erasing) A learning function L learns S in C by erasing on ε iff L stabilizes to k on ε and $k \in I_S$. A learning function L learns S in C by erasing iff it learns S by erasing from every ε for S. A learning function L learns S by erasing iff it learns every S in C by erasing.*

3   A Bridge between FLT and DEL

Gierasimczuk [2009] bridges FLT and dynamic epistemic logic (DEL) in terms of notions from DEL. In the process of learning in the limit, we have hypotheses and pieces of incoming information. Hypotheses can be treated as the set of histories of events that it predicts. For example, let h be $\mathbb{N} \setminus \{3\}$. Then it predicts that the environment will enumerate all natural numbers except 3.

The possible worlds in our epistemic model are identified with hypotheses. Fix a class C of sets, and for each S_n in C, let h_n be a hypothesis describing S_n. In learning by erasing, we take the initial epistemic model as the background for a scientist and his uncertainty about which world is the actual one. Let $M = (H(C), \sim)$ be an epistemic frame, where $H(C)$ is a possibly infinite set of worlds, and $\sim$ is the uncertainty relation. This is the initial model. Next, some world is decided be to the actual one. Then some particular environment ε, consistent with the actual world, occurs. The occurrences of a piece of data ε_1 can be seen as the public announcement of ε_1 at the initial stage. More generally, it can be represented by the event frame $\mathfrak{E}_1 = (\{e\}, \sim, Pre)$, where $e \sim e$ and $Pre(e) = \varepsilon_1$. Then we have the product update $M \otimes \mathfrak{E}_1$. The scientist tests each hypothesis with ε_1. If a hypothesis is consistent with it, then it remains as a possibility, and if not, the hypothesis is removed. In general, this process has been described in Gierasimczuk [2009] as follows.

Definition 5. *Let M be an epistemic model, ε an environment. The model M^ε generated from M by ε is defined as $M \otimes \mathfrak{E}_1 \otimes \mathfrak{E}_2 \otimes \mathfrak{E}_3 \ldots$, where the event model corresponds to the public announcement of ε_n for each n.*

The following example is very good for understanding the process of successful identification of a single hypothesis in the limit in a certain scenario.

Example 2. Let $H(C) = \{h_1, h_2, h_3\}$ where h_i corresponds to reality $\{1, \ldots, i\}$. Assume that h_3 is chosen as the actual one. Let ε be the environment $1, 2, 1, 3$, $2, \ldots$. After 1 occurs, no hypothesis can be removed since the piece of data 1 is consistent with each hypothesis. After 2 occurs, the hypothesis h_1 is removed since it cannot produce such a piece of data. Finally, after 3 occurs, h_2 is removed and the actual world h_3 is finitely identified in the limit.

Proposition 1. *(Gierasimczuk [2009]) Finite identifiability can be modeled in dynamic epistemic logic.*

We have epistemic states for hypotheses, infinite sequences of announcements for environments, and epistemic update for the progress in eliminating uncertainty over hypotheses. Scientist succeeds in finite identification of S from ε iff there is a finite initial segment $\varepsilon | n$ such that the domain of the $\varepsilon | n$-generated model contains only one hypothesis h_k where k is a index for S. In other words, there is a finite step of the iterated epistemic update along ε that eliminates the uncertainty.

The update of the initial model along the given environment depends also on the given data. In the above example, let h_2 be chosen as the actual one. Given the environment $1, 2, 1, 2, 2, 2, \ldots$, the uncertainty will not be eliminated conclusively. Thus we can enrich the epistemic model with a preference relation $\leq$ over hypothesis. The relation does make sense. For example, scientist may choose the simplest hypothesis. Thus the initial epistemic frame is of the form $M = (H(C), \sim, \leq)$. For those inconsistent hypotheses, the procedure of erasing is the same as before. After h_1 is eliminated, h_2 is the scientists current belief (the smallest one now). Although scientist cannot distinguish h_2 and h_3, his preference makes him believe the current hypothesis. The approach by preference order is often used in dynamic doxastic logic. (See van Benthem [2009].)

Proposition 2. *(Gierasimczuk [2009]) Learning by erasing can be modeled in dynamic doxastic logic.*

We have epistemic states for hypotheses, infinite sequences of announcements for environments, epistemic update for the progress in eliminating uncertainty over the hypothesis space, preference relation for the underlying hypothesis space, and in each step of the procedure, the most preferred hypothesis for the actual positive guess of the learning function. Scientist learns S by erasing from ε iff there is n such that for every $m > n$, the most preferred state of the domain of the $\varepsilon|m$-generated epistemic model is h_k and $k \in I_S$.

4   Syntax and Semantics

In this section, we first define a language to talk about the finite identification in the limit exhibited in example 7. Before defining the formal language, we should first clarify some philosophical notions.

A World as a Set of Facts. Recall that, in example 7, we need possible worlds consisting of facts. It is convenient to think of a world as a set of propositional facts. Thus we may associate each possible world with one set of propositional letters which represent propositional facts. In the analytic philosophical tradition, the philosophy of logical atomism shows such a world-view. The open sentence of Ludwig Wittgensteins Tractatus says: "Die Welt is alles, was der Fall ist." ("The world is everything that is the case."(Wittgenstein [1963])) Atomic propositions stands for those facts. We even may include Bertrand Russell's universal facts (Rusell [1986]). If we keep staying outside, not going into the inner structure of a proposition, we may consider an universal or quantified sentence as standing for an universal fact. The nature chooses the actual world from which facts consisting of an environments comes out by experiments and observation of scientist. Thus we need two components in our model. One is the actual world designated by the nature. The other is a mapping that assigns a set of propositional letters to a world.

Talking about Possible Worlds. The end point of the process of successful finite identification in the limit is to determine which world is the actual one. For each possibility or possible world, scientist has a hypothesis saying that it may be the actual one. Thus we also need to talk about hypotheses. This is equivalent to say that the set of propositional letters assigned to the current world is actually the set of all facts occurring in that world. Hence we need a language containing symbols standing for the world. One alternative is to count the set of propositional letters assigned to a world as the proposition saying the fact that the current world is the world which having exactly certain facts.

Identification. We also need an operator to talk about the identification of the actual world. Actually, we only treat the notion of successful finite identification in the limit. In the identification model, there should be a fixed non-empty initial segment $\varepsilon|n$ of a environment ε. Finite identification in the limit is achieved at some stage in such a initial segment. The key feature of such an operator is that it only operates on the set of facts assigned to the actual world. Such an operator can also be reduced by other notions.

Dynamic Operations. The approach given by Nina Gierasimczuk [2009] shows us the road to explore logical dynamics in the process of finite identification in the limit. Public announcements, or more generally, events can be used to represent pieces of data that come from the actual world. This is the key clue for us to find out complete logics. (For general discussions about dynamic epistemic logic, see van Benthem [2009] and van Ditmarsch et al. [2007].) With those notions, we can define the language for talking about finite identification in the limit, and then give the semantics.

Definition 6. *(Finite Identification Model) Given a set of propositional letters Φ, a finite identification model $M = (W, d, \sim, \boldsymbol{w}, \varepsilon|n)$ is defined as follows: $W \neq \varnothing$ and W is countable; $d : W \rightarrow \wp(\Phi)$ is a function such that $d(u) \neq \varnothing$ for all $u \in W$; $\sim$ is a non-empty relation on W; $\boldsymbol{w} \in W$; and finally, ε is an enumeration of $d(\boldsymbol{w})$ and $\varepsilon|n$ is a nonempty initial segment of ε.*

The set W is called the set of possible worlds or possibilities. Each world $u \in W$ is called a possible world or a possibility. d is a mapping used to completely describe each given world by assigning a set of facts represented by propositional letters. The relation $\sim$ over possible worlds represents the uncertainty of scientist between different possible worlds. And finally, the designated world $\boldsymbol{w}$ is chosen by nature as the actual world. An environment in such a identification model consists of elements in $d(\boldsymbol{w})$. The reason for restricting W to be countable is technical. The definition of epistemic language needs to add each $d(u)$ as a basic proposition which represents the fact that u is the world determined by $d(u)$. The fact $d(u)$ may be seen as a meta-fact about facts in the world u.

In this section, the public announcement logic is used to talk about the process of eliminating hypotheses in finite identification. Let's define the language first, and then give the semantics and complete axiomatization.

Definition 7. *Given an identification model $M = (W, d, \sim, w, \varepsilon|n)$, the (single-agent) epistemic public announcement language $\mathcal{L}(M)$ for talking about finite identification in M has a countable set Φ of propositional letters, Boolean connectives, knowledge operator K, difference operator $\blacklozenge$, public announcements of facts in the actual world $\boldsymbol{w}$, the identification operator $I_{\varepsilon|n}$ which operates only on $d(\boldsymbol{w})$. The well-formed formulas $Form(\mathcal{L}(M))$ is given by the following rule:*

$$\varphi ::= p \mid \neg\varphi \mid \varphi \vee \psi \mid K\varphi \mid [!p]\varphi \mid [!p]\varphi \mid d(v) \mid I_{\varepsilon_n}d(\boldsymbol{w})$$

where $p \in d(\boldsymbol{w})$ and $v \in W$.

Definition 8. *Given an identification model $M = (W, d, \sim, \boldsymbol{w}, \varepsilon|n)$ and $u \in W$, define the satisfaction relation $M, u \vDash \varphi$ recursively as follows:*
   $M, u \vDash p$ *iff* $p \in d(u)$;
   $M, u \vDash \neg\varphi$ *iff* $M, u \nvDash \varphi$;

$M, u \vDash \varphi \vee \psi$ iff $M, u \vDash \varphi$ or $M, u \vDash \psi$;

$M, u \vDash K\varphi$ iff $M, v \vDash \varphi$ for all v with $u \sim v$;

$M, u \vDash \blacklozenge \varphi$ iff there exists a world $v \neq u$ in M with $M, v \vDash \varphi$;

$M, u \vDash [!p]\varphi$ iff if $M, u \vDash p$ then $M|p, u \vDash \varphi$;

$M, u \vDash d(v)$ iff $u = v$;

$M, u \vDash I_{\varepsilon|n}d(\boldsymbol{w})$ iff $u = w$ and there is $p \in \varepsilon|n$ with $dom(M|p) = \{\boldsymbol{w}\}$.

In the above definition, $M|p$ is the submodel of M induced by the domain $\{v \in W : M, v \vDash p\}$. Let $\langle K \rangle$ denote the dual of knowledge operator K. Let $\blacksquare := \neg \blacklozenge \neg$ be the dual of difference operator $\blacksquare$. Then $M, u \vDash \blacksquare \varphi$ iff $M, v \nvDash \varphi$ for all worlds $v \neq u$. Why do we add the difference operator in our language? The reason is the following. In the final step of finite identification in the limit, the actual world is identified by using an fact or proposition that can occur only in the actual world. Otherwise, if no such a fact appears, the actual world cannot be finitely identified. The property of the unique fact is that it does not appear in any other possible world. Such a property can be exactly expressed by the dual of difference operator. Next, we give the public announcement logic of finite identification. To define such a logic, we need first to identify the static part.

Proposition 3. *The following system ELD is a complete axiomatization of (single-agent) epistemic logic with difference operator:*

(1) All instances of propositional tautologies.

(2) $K(p \rightarrow q) \rightarrow (Kp \rightarrow Kq)$

(3) $Kp \rightarrow p$

(4) $Kp \rightarrow KKp$

(5) $\langle K \rangle p \rightarrow K \langle K \rangle p$

(6) $\blacksquare(p \rightarrow q) \rightarrow (\blacksquare p \rightarrow \blacksquare q)$

(7) $p \rightarrow \blacksquare \blacklozenge p$

(8) $\blacklozenge \blacklozenge p \rightarrow p \vee \blacklozenge p$

(9) MP: from φ and $\varphi \rightarrow \psi$ infer ψ

(10) K-Gen: from φ infer $K\varphi$

(11) D-Gen: from φ infer $\blacksquare \varphi$

(12) Sub: from φ infer any substitution of φ.

Proof. ELD is the fusion of basic epistemic $S5$ logic and the logic of inequality. By the following three lemmas, the completeness result of ELD follows:

- S5 is complete with respect to the class of frames with equivalence accessibility relation.

- (de Rijke [1992]) The logic axiomatized by the above axioms of difference operator, D-Gen and MP is strongly complete w.r.t the class of frames satisfying symmetry and pseudo-transitivity.

- (Kurucz [2007]) If modal logics ML_1 and ML_2 are characterized by frame classes C_1 and C_2 respectively, and if C_1 and C_2 are closed under the formation of disjoint union and isomorphic copies, then the fusion $ML_1 \oplus ML_2$ is characterized by the class $C1 \oplus C2 = \{(W, R_1, \ldots, R_n, S_1, \ldots, S_m) : (W, R_1, \ldots, R_n) \in C_1 \text{ and } (W, S_1, \ldots, S_m) \in C_2\}$.

Although some important facts, like $d(u) \rightarrow \blacksquare \neg d(u)$ and $d(u) \wedge d(v) \rightarrow u = v$, are not contained in ELD, we choose it as the static part of dynamic logics since these facts depends on the possible world it talks about and a more expressive language will be needed if they are to be embraced. Next we combine public announcement logic and the static part ELD. The key step for finding out complete axiomatization is to find out reduction axioms for static logical operators under public announcements.

Theorem 1. *The public announcement logic of finite identification PELD is completely axiomatized by the static logic ELD plus the following reduction axioms:*

(1) $[!p]q \leftrightarrow (p \rightarrow q)$
(2) $[!p]\neg\varphi \leftrightarrow (p \rightarrow \neg[!p]\varphi)$
(3) $[!p](\varphi \vee \psi) \leftrightarrow ([!p]\varphi \vee [!p]\psi)$
(4) $[!p]K\varphi \leftrightarrow (p \rightarrow K(p \rightarrow [!p]\varphi))$
(5) $[!p]d(v) \leftrightarrow (p \rightarrow d(v))$
(6) $[!p]\blacklozenge\varphi \leftrightarrow (p \rightarrow \blacklozenge(p \wedge [!p]\varphi))$
(7) $I_{\varepsilon|n}d(\boldsymbol{w}) \leftrightarrow d(\boldsymbol{w}) \wedge \bigvee_{p\in\varepsilon|n} \blacksquare\neg p.$

Proof. The soundness proof is easy. Only the validity proof of the final reduction axiom needs to be explained. Suppose that $M, u \vDash I_{\varepsilon|n}d(\mathbf{w})$. Then $u = \mathbf{w}$, and there exists $p \in \varepsilon|n$ with $dom(M|p) = \{\mathbf{w}\}$. By $u = w$, $M, u \vDash d(\mathbf{w})$. By $dom(M|p) = \{\mathbf{w}\} = \{u\}$, $M, v \vDash \neg p$ for all $v \neq u$. Thus $M, u \vDash \blacksquare\neg p$. Conversely, assume that $d(\mathbf{w}) \wedge \bigvee_{p\in\varepsilon|n} \blacksquare\neg p$. Then $u = \mathbf{w}$ since $M, u \vDash d(\mathbf{w})$. By $M, u \vDash \bigvee_{p\in\varepsilon|n} \blacksquare\neg p$, there exists $p \in \varepsilon|n$ with $M, u \vDash \blacksquare\neg p$. Thus $M, v \vDash \neg p$ for all $v \neq u$. Since $p \in \varepsilon|n$, $p \in d(\mathbf{w}) = d(u)$. Hence $dom(M|p) = \{\mathbf{w}\}$. The completeness is proved as follows. Suppose that $\vDash \varphi$, i.e., φ is valid. Then find the syntactically equivalent formula $\varphi^{\#}$ without public announcements by using reduction axioms. Then $\varphi^{\#}$ is valid in ELD by the soundness. By the completeness of ELD, $\vdash_{ELD} \varphi^{\#}$. Hence $\vdash_{PELD} \varphi$.

The above logic PELD is the public announcement logic of finite identification in the limit on certain environmental segment. More general notion is the notion of event model that have been studied in Baltag, Moss and Solecki [1998]. Also see van Benthem [2009] for more details. The approach is to consider events and models updated by them. In the next section, I find that it is easy to extend the PELD to the dynamic epistemic logic of finite identification in the limit on certain environmental segment.

5   DEL of Finite Identification

Some notions from dynamic epistemic logic should be explained first. And then we define the language and semantics. Finally, we present the dynamic epistemic logic DELD of finite identification in the limit on some given environmental segment.

Definition 9. *(Event Model) A event model* $\mathfrak{E} = (E, \sim, Pre, V)$ *consists of a set E of events, a binary uncertainty relation $\sim$ between events, a function Pre from the event set E to the set of preconditions, and a valuation V.*

In the process of finite identification in the limit, one occurrence of a piece of data can be taken as an event. Note that the precondition of such an event is a fact occurred in the actual world. Thus the preconditions of events are exactly atomic propositions representing facts in the actual world. Moreover, the notion of environment should be changed into the following. An environment ϑ is an enumeration of all events that can occur in the actual world. Accordingly, given a set of events, the precondition function determines which events can happen in the actual world. In the finite identification model, the environmental segment $\varepsilon|n$ should also be changed into the event segment $\vartheta|n$.

Definition 10. *(Finite Identification Event Model) A finite identification event model $M = (W, d, \sim, \boldsymbol{w}, \vartheta|n)$ is defined as an identification model except the following condition: ϑ is a non-empty r.e. sequence (allowing repetition), the sequence of events, each component of which is associated with exactly one fact in $d(\boldsymbol{w})$, i.e., the set all events in ϑ is equipotent to $d(\boldsymbol{w})$. $\vartheta|n$ is an initial segment of ϑ.*

Thus each event sequence can be taken as representing an enumeration of all facts in $d(\mathbf{w})$. These facts determine which events can happen in the actual world. Hence each event sequence is an environment for the actual world.

Definition 11. *Given a finite identification event model $M = (W, d, \sim, \boldsymbol{w}, \vartheta|n)$, an M-dependent event model $\mathfrak{E}_M = (E_M, \sim, Pre, V)$ is defined as follows: E_M is the set of all events in the sequence ϑ. The relation $\sim$ is binary on E_M. Pre is a function from E_M to $d(\boldsymbol{w})$.*

Definition 12. *Given a finite identification model $M = (W, d, \sim, \boldsymbol{w}, \vartheta|n)$ and M-dependent event model $\mathfrak{E}$, the production update model $M \otimes \mathfrak{E}$ under the environmental segment $\vartheta|n$ is defined as follows: $dom(M \otimes \mathfrak{E}) = \{(v, f) : f \in \vartheta|n, v \in W$ and $M, v \vDash Pre(f)\}$; $(u, e) \sim (v, f)$ iff $u \sim v$ and $e \sim f$.*

Definition 13. *Given a finite identification model $M = (W, d, \sim, \boldsymbol{w}, \vartheta|n)$ and a M-dependent event model $\mathfrak{E}$, the language $\mathcal{L}(M, \mathfrak{E}, \vartheta|n)$ of logic DELD consists primitive symbols in the language of logic ELD plus event model operators $\{[\mathfrak{E}, e]\}_{e \in \vartheta|n}$ and the finite identification operator $I_{\vartheta|n}$ which only operates on $d(\boldsymbol{w})$. The set of formulas $Form(\mathcal{L}(M, \mathfrak{E}, \vartheta|n))$ is given by the following rule:*

$$\varphi ::= p \mid \neg\varphi \mid \varphi \vee \psi \mid K\varphi \mid \blacklozenge\varphi \mid [\mathfrak{E}, e]\varphi \mid d(v) \mid I_{\vartheta|n}d(\boldsymbol{w})$$

where $p \in d(\boldsymbol{w})$, $e \in \vartheta|n$, and $v \in W$.

Definition 14. *Given a finite identification model $M = (W, d, \sim, \boldsymbol{w}, \vartheta|n)$, a M-dependent event model $\mathfrak{E}$ and $u \in W$, define the satisfaction relation $M, u \vDash \varphi$ recursively as follows. The clauses for ELD-formulas are the same as before. It* *suffices to define the following two clauses: (i) $M, u \vDash [\mathfrak{E}, e]\varphi$ iff $M, u \vDash Pre(e)$ implies $M \otimes \mathfrak{E}, (u, e) \vDash \varphi$; (ii) $M, u \vDash I_{\vartheta|n}d(\boldsymbol{w})$ iff $u = w$ and there is $e \in \vartheta|n$ with $dom(M|Pre(e)) = \{\boldsymbol{w}\}$.*

Theorem 2. *The dynamic epistemic logic of finite identification in the limit DELD is completely axiomatized by the static logic ELD plus the following reduction axioms:*

(1) $[\mathfrak{E}, e]q \leftrightarrow (Pre(e) \rightarrow q)$

(2) $[\mathfrak{E}, e]\neg\varphi \leftrightarrow (Pre(e) \rightarrow \neg[\mathfrak{E}, e]\varphi)$

(3) $[\mathfrak{E}, e](\varphi \vee \psi) \leftrightarrow [\mathfrak{E}, e]\varphi \vee [\mathfrak{E}, e]\psi$

(4) $[\mathfrak{E}, e]K\varphi \leftrightarrow (Pre(e) \rightarrow \bigwedge_{e \sim f} K[\mathfrak{E}, f]\varphi)$

(5) $[\mathfrak{E}, e]d(v) \leftrightarrow (Pre(e) \rightarrow d(v))$

(6) $[\mathfrak{E}, e]\blacklozenge\varphi \leftrightarrow (Pre(e) \rightarrow \blacklozenge(Pre(e) \wedge [\mathfrak{E}, e]\varphi))$

(7) $I_{\vartheta|n}d(\boldsymbol{w}) \leftrightarrow d(\boldsymbol{w}) \wedge \bigvee_{e \in \vartheta|n} \blacksquare\neg Pre(e)$.

Thus we gain the complete axiomatization of the logic of identification in the limit in terms of event models and updates. This approach is a sort of generalization of the public announcement logic presented in section 4. There may some philosophical implication of thinking events and their preconditions in philosophy of scientific inquires. The main role of the two logics presented here is to see what happens in the process of identifying some scientific results.

6   Conclusions and Further Remarks

In this paper, I recast Nina Gierasimczuks ideas of bridging formal learning theory and dynamic epistemic logic. The main work is to define proper languages for talking about finite identification and then explore logics in terms of public announcements, or in terms of events and product updates. I find that the situation can be described in the epistemic language by adding the difference operator. Finally two logics PELD and DELD are presented.

Gierasimczuk [2009] claims that learning by erasing can be modeled in dynamic doxastix logic. But the problem of finding dynamic doxastic logics of learning by erasing. It involves the notion of "Current Belief". We need to merge public announcement logic and belief revision, i.e., the semantics of the formula $[!p]B\varphi$ (i.e., after the public announcement of p, the agent believes that φ) should be made explicit. The semantics of belief involves many difficulties.

Another direction in merging learning and dynamic epistemic logic is to make the meaning of the verb "learn" in natural language explicit. One option is to interpret as follows: the agent learns φ means that he is informed of φ. The may be different approaches to find the logic of learning in the informational sense. One is the AGM-approach which take the information set of the agent as changing. Thus some operations may be found to reflect changes in agents information set. There seems to be many differences between the information set and agents belief set. For instance, information set can a contradict one. It cannot be contract. But one key feature of a belief set is that it should be consistent.

Another approach is the dynamic epistemic one. We may introduce the formula $L\varphi$ saying that the agent learns φ. The key problem here is to define the semantics of $L\varphi$. It is plausible to insist some principles. For instance, the following two formulas seems to be plausible: $L(\varphi \to \psi) \to (L\varphi \to L\psi)$ and $L\varphi \to LL\varphi$. The former formula says that if the agent has the ability to infer new information by MP. If he learns that (is informed of) $\varphi \to \psi$ and φ, he must learn that ψ. The latter formula says that if the agent is introspective with respect to information. If he learns that φ, he is aware of that he learns that φ.

But some principles seems to be false. For instance, $L\varphi \to \varphi$, $L\varphi \to K\varphi$, and $L\varphi \to B\varphi$. The first one says that the truth of φ follows from the agent's learning or information of φ. This contradicts the intuition that the truth of a formula is independent of agent's information of it. The second one says that the agent's knowledge of φ follows from his information of φ. This contradicts to the intuition that knowledge is true. The final one says that the agent's belief of φ follows from his information of φ. But it may often happens that the agent has heart some proposition but he does not believe it. To clarify these problems depends on making a good semantics for learning that can make the difference between learning, knowledge, and belief explicit. Such a semantics will be very helpful for us to understand learning as a way of obtaining information.

References

1. Gierasimczuk, N.: Bridging Learning Theory and Dynamic Epistemic Logic (to appear)
2. de Rijke, M.: The Modal Logic of Inequality. The Journal of Symbolic Logic 57(2), 566–584 (1992)
3. Kurucz, A.: Combining Modal Logics. In: Blackburn, P., van Benthem, J., Wolter, F. (eds.) Handbook of Modal Logic. Elsevier, Amsterdam (2007)
4. van Benthem, J.: Logical Dynamics of Information FLow and Interaction. Cambridge University Press, Cambridge (manuscript, 2009)
5. van Ditmarsch, H., de Hoek, W., Kooi, B.: Dynamic Epistemic Logic. Springer, Heidelberg (2007)
6. Baltag, A., Moss, L., Solecki, S.: The logic of public announcements: Common knowledge and private suspicions. In: Proceedings of the 7th conference on Theoretical aspects of rationality and knowledge, p. 43C56. Morgan Kaufmann Publishers Inc., Evanston Illinois (1998)
7. Osherson, D., de Jongh, D., Martin, E., Weinstein, S.: Formal Learning Theory. In: van Benthem, J., Ter Meulen, A. (eds.) Handbook of Logic and Language. MIT Press, Cambridge (1997)
8. Gold, E.: Language Identification in the Limit. Information and Control 57, 447–474 (1967)
9. Wittgenstein, L.: Tractatus Logico-philosophicus. Trans. by D. Pears and B. McGuinness. Routledge and Paul, London (1961)
10. Russell, B.: The Philosophy of Logical Atomism and Other Essays. George Allen and Unwin, London (1986)

An Epistemic Logic for Planning with Trials

Rajdeep Niyogi[1] and R. Ramanujam[2]

[1] Electronics and Computer Engineering Department
Indian Institute of Technology Roorkee 247667, India
[2] The Institute of Mathematical Sciences
CIT Campus, Taramani, Chennai 600113, India
rajdpfec@iitr.ernet.in, jam@imsc.res.in

Abstract. We suggest that in the context of planning in uncertain environments, an agent's performance of an action may be tentative and not definitive. In this view, an agent plans to merely *try* performing an action, and further planning is dependent on the success or failure of such a trial. Epistemic logics seem well suited to formalize the reasoning in such contexts. We study a simple such logic for one planning agent making bounded plans, for which we give a complete axiomatization and prove decidability. We discuss preliminary results for extensions to multi-agent plans as well as unbounded plans.

1 Introduction

A primary goal of research on agent-based systems is to design agents that are capable of planning [GNT04]. Typically a goal is given, and the planning problem is to synthesize a plan (a sequence of actions) leading to the goal. Such a plan synthesis is an algorithmic process, but is limited by the large size of the explored state space. Deductive planning [MW87, Lev05] typically reduces plan synthesis to theorem proving.

An important constraint on planning is uncertainty. This can come about due to incompleteness in the state information available to the planning agent, or due to non-deterministic outcomes of actions. In such contexts, actions are partial, in the sense that the outcome achieved may be only a partial realization of the desired outcome. It has been shown in [MW87, BS01, Lev05] that even for simple domains involving uncertainty, there exist no classical plans. In such scenarios it is meaningful to specify plans as programs with branching, and perhaps also iterative, structure. Such an approach has led to the use of formalisms similar to or inspired by propositional dynamic logic (PDL) ([HKT00]) for automated planning ([Ros81, GB96, ST00]), and agent programming ([vRFM05]).

One way to overcome uncertainty in unpredictable environments is by using sensing actions. A sensing or test action is an act of observation that does not change the state of the world. Some examples of sensing actions include litmus tests, read actions on variables or files, or Unix commands such as 'ls, pwd'. Sensing actions are considered in [EHW+92, GW96, Lev96, BS01, Lev05]. Spalazzi and Traverso ([ST00]) go further and consider planning in a reactive system

X. He, J. Horty, and E. Pacuit (Eds.): LORI 2009, LNAI 5834, pp. 238–250, 2009.

where sensing actions may change the state of the world and such actions may fail as well. Typically, when sensing actions fail, the plans (programs) involving sensing actions may abort or get stuck.

It is easy to see that sensing actions are epistemically founded. *Why* does a planning agent perform a sensing action ? Typically, the plan dictates an action a at that state, and the agent is uncertain, does not *know* whether a can be carried out or not. In such a situation, the sensing action is an attempt to *learn* and helps to increase the agent's knowledge (perhaps partially).

However, one foundational difficulty with the use of sensing actions is the possible uncertainty in whether the sensing action can be carried out or not, and we need to consider sensing of sensors, a clearly unsatisfactory state of affairs. Instead, a logical theory can work with the abstract effects of sensing by looking only at change in information states. Planning then becomes a process by which we seek not only goal states in "the world" but also goals in the agent's information states.

In what follows, we suggest that epistemic change, alongwith the notion of partial actions, or trials. An agent tries to perform an action a, based on its knowledge, and the observed success or failure of the action results in learning. Admittedly, trials are again actions, so such a refinement may seem unnecessary. But it is in the interaction between trials and epistemic attitudes that we see the scope of reasoning relevant to planning.

We can conceive of trials as an abstract *intentional* form of sensing actions, perhaps tentatively so. An agent, trying to pick up an object without precise information about its weight, may actually succeed in picking it up, or learn that it is too heavy, or fail to pick up and yet learn little. Thus it may well be possible for a trial to determine whether an action is executable without incurring the cost (and effects) of executing the action[1]. What learning actually results is an epistemic change, and we model this change explicitly.

However, considering trials also as actions, though of the *informational* kind, is important. Tests require preconditions to hold too, and their effect may be to change information states. Further, sensing actions may not, in general, be available but a robot may yet need to plan and carry out actions.

We use the standard S5 model of knowledge based on equivalence relations to model epistemic attitudes. It may be more realistic to consider belief sets rather than knowledge in the planning context ([SL03]) but this complicates the technical development in our setting.

A familiar context in which such 'trials' are seen is in relation to questions. Asking a question does not cause factual change but informational change. Further, questioning is epistemically founded in much the same manner as trials: a questioner (usually!) does not already know the answer, considers many answers to be possible, and the form of the question is determined by how much the questioner knows. These issues have been studied by Hintikka [Hin86], Harrah [Har02], Belnap and Steel [BS76], Hendricks and Symons [HS06], and other

---

[1] This was pointed out by a perceptive reviewer.

logicians. The crucial difference between questions and trials is that the latter carries an action component implicitly.

Another thread, relevant to this line of work, is that of *strategic reasoning* in game theory. Strategies are complete plans that advise a player on how to play in all possible game situations, and make sense when reasoning externally about games. However, when viewed within a game, a player's choice of strategies is local, tentative and partial; moves are trials whose outcomes determine further strategizing. Such ideas are pursued in [RS08] but we merely note here that these are connected.

In this paper, we propose a simple logic for one planning agent making bounded plans, for which we give a complete axiomatization and prove decidability. We then discuss preliminary results for extensions to multi-agent plans as well as unbounded plans. But before we proceed with the technical development, an example of planning based on trials may be relevant.[2]

2   Blocks World Example

As customary, we consider the Blocks World (BW) domain. The fluents in the BW domain are $ontable(x), clear(x), handempty, on(x,y)$. The operators are $pickup(x), putdown(x), stack(x,y), unstack(x,y)$. The operator $stack(x,y)$ is used to denote putting block x on top of block y; $unstack(x,y)$ is the reverse of $stack(x,y)$. An action is a ground instance of an operator. For example, if we have 3 blocks - A, B, C, then $pickup(A), putdown(C), stack(A,B)$, etc. are the actions. Let the initial configuration be that all the blocks are placed on the table. The final configuration is a tower of blocks with A at the bottom, C at the top, and B on top of A.

The planning problem is to find a plan that transforms the initial configuration to the final configuration.

We assume that the blocks are of different weights. In the original BW domain, the preconditions of $pickup(x)$ are $clear(x)$, $handempty$, and $ontable(x)$. We redefine $pickup(x)$ by attaching one more pre-condition: block is not heavy. The other natural condition is that a heavy block should not rest on a light block. To consider weights we can also assume another operator $apply\_lever(x,y)$ that places a block x on top of block y, when both the blocks x, y are heavy, and there is nothing on top of x, y.

We consider the situation that the planning agent does not have precise information regarding the weight of a block. So when the agent attempts to pick up a block, it either manages to do so or discovers that it is heavy. This is a very simple instance of partial actions.

Consider an example scenario given in figure 1. Let the initial configuration be that all the blocks are placed on the table. The final configuration is a tower of blocks such that the lighter blocks rest on top of heavy blocks.

---

[2] This paper is a companion to [Niy09] which defines the logic and works out detailed examples of plans, but does not address technical questions.

We introduce the operator $try\_pickup(x)$ that is quite similar to $pickup(x)$, except that unlike $pickup(x)$, $try\_pickup(x)$ may be enabled at the initial state. If it succeeds, its outcome is exactly that of $pickup(x)$, i.e., $holding(x)$ becomes true. Now suppose that $try\_pickup(x)$ fails. (In this paper we assume that the notion of failure is with respect to goal attainment and not that of actions aborting as in [ST00].) Then the agent comes to know that the block is heavy, implying that $pickup(x)$ cannot be performed at that state. So the try action helps the agent in updating its knowledge about the world.

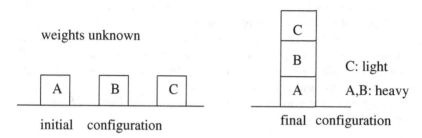

weights unknown

initial   configuration

C: light

A,B: heavy

final   configuration

Fig. 1. Modified Blocks World Domain

The table below describes the operators in the modified blocks world domain. Note that the post-condition for $try\_pickup(x)$ includes the possibility of $handempty$ even when the precondition satisfies not $heavy(x)$. This is a state where $pickup(x)$ is enabled, but a trial may yet cause no change of state. This corresponds to the notion of trial as a partial action, in this case, a partial attempt at picking up the object: we can interpret it as a tentative "tug" that does not change the object state, nor does it change the informational equivalence class for the agent.

Operators	Preconditions	Effects
$pickup(x)$	$ontable(x), clear(x),$ $handempty,$ $\neg heavy(x)$	$holding(x)$
$putdown(x)$	$holding(x)$	$ontable(x)$
$stack(x, y)$	$holding(x), clear(y)$	$on(x, y)$
$unstack(x, y)$	$on(x, y), clear(x),$ $handempty$	$holding(x)$
$apply\_lever(x, y)$	$heavy(x), heavy(y),$ $clear(y)$	$on(x, y)$
$rev\_lever(x, y)$	$on(x, y), clear(x)$	$ontable(x)$
$try\_pickup(x)$	$ontable(x), clear(x),$ $handempty$	$holding(x)$ or $handempty$

3   Transition Systems with Trials

We use the formal model of labelled transition systems to describe plans. Since we wish to study how the knowledge of the agent determines, and is in turn determined by, the agent's trials leading to plans, we enrich transition systems with information partitions.

Let act be a finite set of primitive actions, and let $\Sigma = act \cup \{try\_a | a \in act\}$. We use a, b etc to denote elements of A and x, y etc for elements of Σ.

Formally, a Σ-labelled *epistemic transition system*, often referred to as a frame, is a tuple $F = (S, \sim, \rightarrow)$, where S is a set of states, $\sim \subseteq (S \times S)$ is an equivalence relation, and $\rightarrow \subseteq (S \times \Sigma \times S)$ is the transition relation. $\sim$ is called the *indistinguishability relation* of the agent.

For example, if the weight of a block in the BW domain is not known, then the agent considers two states possible—one in which the block is heavy, and another in which it is light. $try\_pickup(x)$ is a way by which the agent can learn whether it is heavy or not and refine the partition.

A frame $F = (S, \sim, \rightarrow)$ is said to be a **trial based planning frame** if F satisfies the following conditions:

- **Condition1**.
  $\forall s_1, s_2 \in S, a \in act$, if $s_1 \overset{a}{\rightarrow} s_2$ then either $s_1 \overset{try_a}{\rightarrow} s_1$ or $s_1 \overset{try_a}{\rightarrow} s_2$.
- **Condition2**. $\forall s_1, s_2 \in S, a \in act$, if $s_1 \sim s_2$ then try_a is enabled at s_1 iff try_a is enabled at s_2.
- **Condition3**. $\forall s \in S, a \in act$, if try_a is enabled at s, then $\exists s'$ s.t. $s \sim s'$ and a is not enabled at s'.
- **Condition4**. $\forall s_1, s_2, s_1', s_2' \in S, a \in act$, if $s_1 \sim s_2$, $s_1 \overset{a}{\rightarrow} s_1'$ and $s_2 \overset{a}{\rightarrow} s_2'$ then $s_1' \sim s_2'$ as well.

The intuition underlying these conditions is as follows. The information state of the planning agent at s is captured by the equivalence class $[s]_\sim$. The agent does not know the true state of the world s. The agent, based only on the information available to it, cannot distinguish s from s', where $s \sim s'$.

Now, at any state s there are four possibilities regarding an action a.

- Neither a nor try_a is enabled at s: this is the situation when the agent knows that action a is not available as an option in state s.
- try_a is enabled at s, but a is not: this is when the agent considers the action a as possible but is unsure. Condition 3 above reflects this uncertainty: the notion of 'trying' presumes the possibility of failure. Note that when an agent tries to do a, it knows that it can try to do a: no information it has can preclude the trial itself, as asserted by Condition 1.
- a is enabled at s, but try_a is not: this clearly does not make sense and ruled out by Condition 1.
- Both a and try_a are enabled at s: Condition 1 ensures that either state does not change due to trial or both result in the same change of state. The former situation arises when the trial is partial. For example, giving a large rock a tentative tug to see whether it can be lifted.

In addition, Condition 4 asserts that when an action succeeds at indistinguishable states, the resulting states are again indistinguishable to the agent. Note that such a condition **does not** hold for try_a, which reflects the essence of learning due to trial actions. This enables hypothetical reasoning of the form: "I do not know whether the actual state satisfies p or $\neg p$ but if this trial succeeds then I will know".

Note that trial based systems can be quite *nondeterministic* with respect to actions. For instance, consider the system in Figure 2. The agent does not know whether the state is s or s'. Trying the same action in s leads to failure whereas in s' it succeeds. However, even in the latter case, the agent is uncertain whether the resulting state is s_1 or s_2.

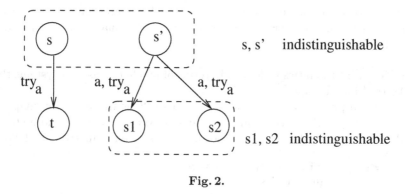

s, s' indistinguishable

s1, s2 indistinguishable

Fig. 2.

Condition 4 carries some subtlety: it only asserts that whenever an agent can perform an action after which it knows ϕ, it already knows that ϕ will hold after doing it. The converse does not hold, as we see in the system in Figure 3, which clearly satisfies Condition 4.

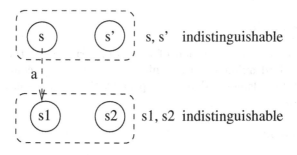

s, s' indistinguishable

s1, s2 indistinguishable

Fig. 3.

4   An Epistemic Logic with Try Actions

The logic proposed here is an action-indexed logic with a knowledge operator.

Fix $P = \{p_0, p_1, \ldots\}$, an at most countable set of atomic propositions, and Σ as above.

Syntax: From P and Σ, the set Φ of formulas defined inductively as:

$$\Phi := p \in P \mid \neg\phi \mid \phi_1 \vee \phi_2 \mid [x]\phi, \ x \in \Sigma \mid K\phi$$

Note that the syntax is an epistemic extension of standard action indexed modal logic, the only difference being the use of try_a actions. As the semantics will clarify, it is the interaction of these actions with knowledge operators that makes for new logical interest.

As usual, conjunction, implication and equivalence are defined in terms of negation and disjunction. $\langle x \rangle \phi = \neg[x]\neg\phi$ denotes the dual of $[x]\phi$, and $L\phi = \neg K\neg\phi$, the dual of $K\phi$.

Semantics: The semantics is defined in terms of labeled transition systems that describe trial based plans.

A *model* is a pair $M = (F, V)$, where $V : S \to 2^P$ is the valuation map and $F = (S, \sim, \to)$ is a trial based planning frame. Given a model M and a state s in S, the satisfaction relation $\models$ for a formula α, is given inductively as follows:

- $M, s \models p$ iff $p \in V(s)$ for $p \in P$.
- $M, s \models \neg\alpha$ iff $M, s \not\models \alpha$.
- $M, s \models \alpha_1 \vee \alpha_2$ iff $M, s \models \alpha_1$ or $M, s \models \alpha_2$.
- $M, s \models [x]\alpha$ iff   for all t such that $s \xrightarrow{x} t$, we have: $M, t \models \alpha$.
- $M, s \models K\alpha$ iff   for all s' such that $s' \sim s$, we have: $M, s' \models \alpha$.

It is easily seen that $M, s \models \langle x \rangle \alpha$ iff   there exists t, $s \xrightarrow{x} t$ and $M, t \models \alpha$.

As is usual, we say that a formula α is *satisfiable*, if there exists a model $M = (F, V)$, $F = (S, \sim, \to)$ and $s \in S$ such that $M, s \models \alpha$. We say that α is *valid* if $\neg\alpha$ is not satisfiable.

4.1   Axiomatization

We now propose an axiomatization of the valid formulas of the logic. What follows is a standard axiomatization enhanced by formulas connecting action modalities with knowledge and those capturing the interaction of knowledge with action trials.

The Axiom Schemes

1. All the substitutional instances of tautologies of boolean logic.
2. $[x](\alpha \supset \beta) \supset ([x]\alpha \supset [x]\beta)$
3. $K(\alpha \supset \beta) \supset (K\alpha \supset K\beta)$
4. $K\alpha \supset (\alpha \wedge KK\alpha)$

5. $L\alpha \supset KL\alpha$
6. $([try_a]\alpha \wedge [try_a]\beta) \supset ((\alpha \wedge \beta) \vee [a](\alpha \wedge \beta))$
7. $\langle try_a \rangle True \supset (K\langle try_a \rangle True \wedge L\neg\langle a \rangle True)$
8. $\langle a \rangle K\alpha \supset K[a]K\alpha$

Inference Rules

$$(MP) \frac{\alpha, \ \alpha \supset \beta}{\beta} \quad (NG) \frac{\alpha}{[x]\alpha} \quad (KG) \frac{\alpha}{K\alpha}$$

We say α is a **theorem** of the system, denoted $\vdash \alpha$ if it is derivable from substitution instances of the axioms by finitely many applications of inference rules. It is easily seen that every theorem is valid. It can be checked that the formula $[try_a]\alpha \supset (\alpha \vee [a]\alpha)$ is sound. However, we need the stronger form of the axiom 6 for completeness.

Theorem 4.1. *The system Ax above is a* **complete** *axiomatization of the valid formulas of the logic.*

Proof. The theorem is easily proved by the observation that the *canonical model* for the logic is a trial based planning frame. Consider $F = (\mathcal{M}, \rightarrow, \sim)$ where $\mathcal{M}$ is the set of all maximal consistent sets, and $\rightarrow$ and $\sim$ are as given below.

For a definition of maximal consistent sets (MCS), firstly, a formula α is said to be **consistent** if $\not\vdash \neg\alpha$. A finite set of formulas is consistent if the conjunction of its formulas is is consistent. A set of formulas is consistent if every finite subset is consistent. A set A is maximal consistent if whenever $A \cup \{\alpha\}$ is consistent, $\alpha \in A$. Below we use A, B, etc to denote MCS's.

Define $A \xrightarrow{x} B$ iff for every $[x]\alpha$, if $[x]\alpha \in A$ then $\alpha \in B$. Define $A \sim B$ iff for every $K\alpha$, $K\alpha \in A$ iff $K\alpha \in B$. Clearly, $\sim$ is an equivalence relation on $\mathcal{M}$.

For instance, suppose $A \xrightarrow{a} B$. To prove that $A \xrightarrow{try_a} B$ or $A \xrightarrow{try_a} A$, suppose that neither is the case. Then there exist α, β such that $[try_a]\alpha, [try_a]\beta \in A$, but $\alpha \notin A$, $\beta \notin B$. Hence $\alpha \wedge \beta \notin A$, and by axiom (A6), $[a](\alpha \wedge \beta) \in A$. Since $A \xrightarrow{a} B$, we have that $(\alpha \wedge \beta) \in B$, contradicting the fact that $\beta \notin B$.

Now suppose $A \sim B$, $A \xrightarrow{a} A'$ and $B \xrightarrow{a} B'$. To prove that $A' \sim B'$, suppose that $K\alpha \in A'$. By definition of $\xrightarrow{a}$, $\langle a \rangle K\alpha \in A$ and by axiom (A8), $K[a]K\alpha \in A$, and by definition of $\sim$, $K[a]K\alpha \in B$ and hence $[a]K\alpha \in B$ (by axiom (A4)). By definition of $\xrightarrow{a}$, $K\alpha \in B'$. Symmetrically we can prove that whenever $K\alpha \in B'$, we also have $K\alpha \in a'$. Thus $A' \sim B'$.

Define the model $M = (F, V)$, where $V(A) = A \cap P$. We can now prove by induction on the structure of formulas that, for every formula α and every MCS A, $M, A \models \alpha$ iff $\alpha \in A$.

Let α_0 be a consistent formula. Then there exists an MCS A_0 such that $\alpha_0 \in A_0$. By what we saw above, $M, A_0 \models \alpha_0$, and hence α_0 is satisfiable. Thus every consistent formula is satisfiable, establishing completeness of the axiom system. $\square$

Note that we made essential use of MCS's in proving the frame conditions. Any attempt at doing the proof with maximal consistent finite sets (say subformulas of α_0) runs into the barrier that the use of (A8) makes the set A above infinite. Thus, we need to be more careful when we attempt a decision procedure for satisfiability.

5   Decidability

We now discuss the decidability problem for the logic.

Theorem 5.1. *The satisfiability problem for the logic is* **decidable**, *and PSpace-complete.*

The action-indexed logic without epistemic operator is already PSpace-hard (by a standard Ladner-type argument), so the lower bound is clear. We now sketch the argument for the upper bound.

We first define the subformula closure of a formula ϕ, denoted $CL(\phi)$. This is the least set containing ϕ, closed under negation and subformulas, but additionally satisfies the conditions that whenever $[x]\phi' \in CL(\phi)$, $K[x]\phi' \in CL(\phi)$ as well. Additionally we assume that $K\langle x \rangle True \in CL(\phi)$, for every $x \in \Sigma$. Note that for every formula ϕ, $CL(\phi)$ is finite and its size is linear in the length of ϕ.

We also assume that the formula is rewritten so that negation is pushed inside and appears only on atomic propositions.

The notion of *successor depth* of a formula α, denoted $d(\alpha)$, is clear and defined inductively: $d(p) = 0$ for $p \in P$; $d(\neg\alpha) = d(K\alpha) = d(\alpha)$; $d(\alpha \vee \beta) = max\{d(\alpha), d(\beta)\}$; $d(\langle x \rangle \alpha) = d(\alpha) + 1$.

Now fix a given formula α_0 and let CL denote the set $CL(\alpha_0)$. Let $d_0 = d(\alpha_0)$. We work with down-closed subsets of CL. Let $A \subseteq CL$. A is down-closed if it satisfies the natural conditions that whenever $\neg\beta \in A$, $\beta \notin A$; if $\alpha \vee \beta \in A$, either $\alpha \in A$ or $\beta \in A$; if $\alpha \wedge \beta \in A$, then $\alpha \in A$ and $\beta \in A$; if $K\alpha \in A$ then $\alpha \in A$; if $\alpha \in A$ and $L\alpha \in CL$ then $L\alpha \in A$. Additionally, we ensure that whenever $\langle try_a \rangle True \in A$, we also have $K\langle try_a \rangle True \in A$ and $L\neg\langle a \rangle True \in A$.

A down-closed set A is minimal for α if it contains α and no proper subset is down-closed. For any formula α, let $DC(\alpha)$ denote the set of minimal down-closed subsets of CL for α. Similarly define $DC(B)$ for any $B \subseteq CL$.

The algorithm proceeds as follows. It attempts to build a forest whose maximum depth is at most d_0. At each level, sufficiently many nodes are added to fulfil all epistemic requirements (clearly bounded by the length of α_0).

Given $A \subseteq CL$, let $K(A) = \{K\alpha \mid K\alpha \in A\} \cup \{L\beta \mid L\beta \in A\}$, and for $x \in \Sigma$, let $N_x(A) = \{\alpha \mid [x]\alpha \in A\}$.

At the first step, guess $A \in DC(\alpha_0)$, and for each $L\beta \in A$ such that $\beta \notin A$, find $B \in DC(K(A) \cup \{\beta\})$. The algorithm fails, if it cannot find such sets. Otherwise suppose m such sets (say $B_1, \ldots, B_m$) are chosen. Then define the structure $F_0 = (S_0, \rightarrow_0, \sim_0, \lambda_0)$, where $S_0 = \{w_0, w_1, \ldots, w_m\}$, $\rightarrow_0 = \emptyset$, $\sim_0 = (S_0 \times S_0)$ and $\lambda_0 : S_0 \rightarrow 2^{CL}$ is defined by $\lambda_0(w_0) = A$ and $\lambda_0(w_j) = B_j$, where $j \in \{1, \ldots, m\}$.

Inductively, given a structure $F_k = (S_k, \rightarrow_k, \sim_k, \lambda_k)$, the algorithm tries to fulfil an x-requirement, $x \in \Sigma$: it is a $w \in S_k$ such that $\langle x \rangle True \in \lambda_k(w) = A$ but there is no $w' \in S_k$ such that $w \xrightarrow{x} w'$. If there is no such requirement, set $F_{k+1} = F_k$. Suppose there is such a requirement at w. If $x = try_a$ and $\langle a \rangle True \notin A$, the algorithm tries to pick a $B \in DC(N_x(A))$ and as before, a set B' for every $L\beta \in B$ such that $\beta \notin B$; and these many nodes are added to the structure, with one transition labelled try_a from w to a node (say w') labelled by B.

If $\langle a \rangle True \in A$, the construction is similar, but with some crucial differences: we need to check that $N_{try_a}(A) \subseteq B$, and for every $w_1, w_2 \in S_k$ such that $w_1 \sim_k w$ and $w_1 \overset{a}{\to} w_2$, $K(w_2) \subseteq B$. We set $(w_2, w') \in \sim_{k+1}$.

When such a construction is complete, we have a structure $F = (S, \to, \sim, \lambda)$ defined by pointwise union. We can check that $(S, \to, \sim)$ satisfies the frame conditions and for all $w, w' \in S$:

- If $w \overset{x}{\to} w'$ then for all $[x]\alpha$ in CL, if $[x]\alpha \in \lambda(w)$ then $\alpha \in \lambda(w')$.
- If $w \sim w'$ then for all $K\alpha$ in CL, $K\alpha \in \lambda(w)$ iff $K\alpha \in \lambda(w')$.
- If $\langle x \rangle \alpha \in \lambda(w)$ then there exists $w'' \in S$ such that $w \overset{x}{\to} w''$ and $\alpha \in \lambda(w'')$.
- If $L\alpha \in \lambda(w)$ then there exists $w'' \in S$ such that $w \sim w''$ and $\alpha \in \lambda(w'')$.

We then define a model $M = (S, \to, \sim, V)$ by: $V(w) = \lambda(w) \cap P$. We can then show that for every $w \in S$ and $\alpha \in CL$, if $\alpha \in \lambda(w)$ then $M, w \models \alpha$. Since we started with $\alpha_0 \in \lambda(w_0)$, we then have that $M, w_0 \models \alpha_0$ and hence α_0 is satisfiable.

Thus successful termination of the algorithm implies that the given formula is satisfiable. That the algorithm terminates can be seen by arguing that every time a transition is added, successor depth decreases. On the other hand, if the given formula is satisfiable, it is easily shown that a correct guess of down-closed sets is available at every stage, leading to successful termination.

That the algorithm uses only space polynomial in the length of α_0 needs careful analysis since reuse of space is needed, but follows along standard lines. This gives a nondeterministic polynomial space algorithm, and hence by Savitch's theorem, we get a PSpace decision procedure.

6 Extensions

The logic presented here is very simple, but forms the basis for very natural extensions.

6.1 Multi-agent Plans

It is clear that the framework presented here can be easily extended to multi-agent plans. Consider a system with n agents. The simplest extension consists of generalizing trial based planning frames to have n equivalence relations, set Σ to the tuple $(\Sigma_1, \ldots, \Sigma_n)$, with $\Sigma_i \cap \Sigma_j = \emptyset$ for $i \neq j$. Correspondingly the syntax of the logic is generalized to have $K_i\alpha$, for $i \in \{1, \ldots, n\}$, with the obvious semantics. It is then easy to check that the technical results extend for this logic as well (though the decision procedure does get considerably complicated).

However, such a generalization does not address the rich possibilities that make multi-agent planning interesting in the first place. For instance, in the Blocks World domain, when one agent tries to lift a block and finds it too heavy, every other agent *who has observed* the trial learns this fact as well. Then there is the differential ability among agents to lift heavy blocks. It is therefore clear that the same trial could reveal different information to distinct agents.

The next issue is whether the performance of an action by one agent can cause an enabled action to become disabled after this. For instance we can consider situations where two trials are enabled, by distinct agents, but if one of the agents performs its trial, in the resulting information state, the other agent's trial is no longer enabled.

Finally, it is interesting to consider the planning goal to be separately specified for each agent. With this we enter the realm of *game theory*. The connection between plan synthesis and strategy synthesis in games is intuitively clear, but we expect that epistemic logic [FHMV95] would play a significant role in studying their formal relationship.

This discussion suggests that we need a more refined structure in both the frames and the logic in the context of multi-agent plans. Importantly, we need to limit the *visibility neighbourhoods* of trials, and causality relationships between agents' actions.

6.2   Unbounded Plans

It is easily seen that the logic can specify branching plans of bounded duration. Since many interesting situations in planning are modelled as finite trees, this is already expressive. However, there are many situations where plans need to be unbounded. For instance, when an agent keeps trying to open a tightly sealed jam bottle, after a sequence of trials she suddenly finds that the bottle opens. While it is true that the lid has eased gradually, this is a plan of unbounded duration, typically specified by *keep trying until* a condition is satisfied.

A simple extension of the logic is to add a modality $\Diamond\alpha$, and the semantics is given by: $M, s \models \Diamond\alpha$ iff there exists s' such that $s \Rightarrow s'$ and $M, s' \models \alpha$, where $\Rightarrow = \bigcup_{x \in \Sigma} \xrightarrow{x}$. Thus $\Diamond\alpha$ holds at s if there is a state reachable from s in finitely many steps at which α holds. Let $\Box\alpha = \neg\Diamond\neg\alpha$ and $\bigcirc\alpha = \bigwedge_{x \in \Sigma} [x]\alpha$.

It is easy to see that the following formula is valid: $\Box\alpha \supset (\alpha \wedge [x]\Box\alpha)$. What is more interesting that we also have an induction axiom: $\Box(\alpha \supset \bigcirc\alpha) \supset (\alpha \supset \Box\alpha)$. Such an axiom is clearly needed, but proving that such an extension suffices for completeness seems difficult. We conjecture that these axioms, along with the Kripke axiom for $\Box$ modality, suffice for completeness.

The decidability of the logic for unbounded duration plans is also non-trivial. For one thing, we cannot work with a simple forest construction as we did in the previous section. In the decision procedure we presented, we fulfill all epistemic requirements on the addition of each node, and check for the learning condition for each successor. It turns out that we need a dual construction for this logic:

every time we add a node, we complete the construction of its reachability subgraph, checking for $\Diamond\alpha$ requirements. Then we consider epistemic requirements, and each such addition induces a subgraph to be added and checked. The termination of such a procedure and its complexity involve tricky detail. Moreover, this algorithm takes double exponential time, and perhaps can be improved. Details will be provided in the full paper; we mention this here only to point out that reasoning becomes harder for unbounded plans.

The $\Diamond$ modality is too coarse to constrain paths. For instance, we might want to consider only sequences of *try* actions until some condition becomes true. This can be achieved by extending the logic to a Propositional Dynamic Logic ([HKT00]) with tests and regular expressions over Σ. Technically this has the same complexity as the logic of unbounded plans discussed above.

Much better would be the embedding of such a logic in a Dynamic Epistemic Logic (DEL) ([vDvdHK07]) by modifying DEL to work with partial and uncertain actions. It is hoped that such an attempt will lead to new directions for plan synthesis.

Acknowledgements

The authors would like to thank the anonymous reviewers for their valuable comments and suggestions, especially for weakening Condition 1.

References

[BS76]   Belnap, N.D., Steel, T.B.: The Logic of Questions and Answers. Yale University Press (1976)

[BS01]   Baral, C., Son, T.: Formalizing sensing actions–a transition function based approach. Artificial Intelligence 125(1-2), 19–91 (2001)

[EHW+92]   Etzioni, O., Hanks, S., Weld, D., Draper, D., Lesh, N., Williamson, M.: An approach to planning with incomplete information. In: Proceedings of Knowledge Representation, pp. 115–125 (1992)

[FHMV95]   Fagin, R., Halpern, J., Moses, Y., Vardi, M.: Reasoning about Knowledge. MIT Press, Cambridge (1995)

[GB96]   Golden, R.P., Boddy, M.S.: Expressive planning and explicit knowledge. In: Proceedings of the Third Artificial Intelligence Planning Systems, pp. 110–117 (1996)

[GNT04]   Ghallab, M., Nau, D., Traverso, P.: Automated Planning: Theory and Practice. Morgan Kaufmann Publishers, San Francisco (2004)

[GW96]   Golden, K., Weld, D.: Representing sensing actions: The middle ground revisited. In: Proceedings of Knowledge Representation, pp. 213–224 (1996)

[Har02]   Harrah, D.: The logic of questions. In: Gabbay, D., Guenthner, F. (eds.) Handbook of Philosophical Logic, vol. 8, pp. 1–60 (2002)

[Hin86]   Hintikka, J.: Reasoning about knowledge in philosophy. In: Proceedings of Theoretical Aspects of Rationality and Knowledge (1986)

[HKT00]   Harel, D., Kozen, D., Tiuryn, J.: Dynamic Logic. MIT Press, Cambridge (2000)

[HS06]   Hendricks, V.F., Symons, J.: Where is the bridge? epistemology and epistemic logic. Philosophical Studies 128(1), 137–167 (2006)

[Lev96]   Levesque, H.: What is planning in the presence of sensing? In: Proceedings of AAAI, pp. 1139–1146 (1996)

[Lev05]   Levesque, H.: Planning with loops. In: Proceedings of IJCAI, pp. 311–316 (2005)

[MW87]   Manna, Z., Waldinger, R.: How to clear a block: A theory of plans. Journal of Automated Reasoning 3, 343–377 (1987)

[Niy09]   Niyogi, R.: Planning with trial and errors. In: Proceedings of International Conference on Intelligent Agents and Multi Agent Systems (IAMA 2009), Chennai, July 22-24 (2009)

[Ros81]   Rosenschien, S.: Plan synthesis: a logical perspective. In: Proceedings of IJCAI, pp. 331–337 (1981)

[RS08]   Ramanujam, R., Simon, S.: A logical structure for strategies. In: Logic and the Foundations of Game and Decision Theory. Texts in Logic and Games, vol. 3, pp. 151–176 (2008)

[SL03]   Scherl, R.B., Levesque, H.: Knowledge, action, and the frame problem. Artificial Intelligence 144(1-2), 1–39 (2003)

[ST00]   Spalazzi, L., Traverso, P.: A dynamic logic for acting, sensing, and planning. Logic Computation 10(6), 727–821 (2000)

[vDvdHK07] van Ditmarsch, H., van der Hoek, W., Kooi, B.: Dynamic Epistemic Logic. Springer, Heidelberg (2007)

[vRFM05]   van Riemsdijk, M.B., de Boer, F.S., Meyer, J.C.: Dynamic logic for plan revision in intelligent agents. In: Proceedings of the 5th International Workshop on Computational Logic in Multi agent systems, pp. 16–32 (2005)

Obligations in a Responsible World

Loes Olde Loohuis

The Graduate Center, The City University of New York, 365 Fifth Avenue, USA
L.OldeLoohuis@gmail.com

Abstract. In this paper, we suggest defining obligations under the assumption that agents are responsible. We consider how the framework developed by Horty in [2] can be modified to incorporate this idea, and how this solves some objections raised to Horty's system. Also, we discuss how the assumption of a responsible world can (not easily) be incorporated into the framework of knowledge based obligations introduced by Pacuit, Parikh and Cogan in [5].

1 Introduction

This paper is concerned with formalizing obligations to act. The normative concept of obligation is modelled in the extensively studied area of logic called deontic logic. In this paper we will focus on two existing systems of deontic logic. The first is a deontic theory of actions as developed by Horty in his book 'Deontic Logic and Agency' [2]. Horty's system models personal obligations in an environment with multiple agents acting simultaneously. The second system is a formal account of so-called 'knowledge based obligations' developed by Pacuit, Parikh and Cogan (PPC) in [5]. In this model epistemic and deontic logic are combined to capture the idea that an agent's epistemic state influences the obligations he has. For the purpose of this paper, there is no need to introduce the formal languages as proposed in both systems as we are mainly concerned with the structural concepts they define.

Our main contribution is the idea that obligations should not be defined irrespective of (all the) possible moves of other agents. Rather, we suggest defining obligations in a world in which other agents behave *responsibly, morally* or *rationally*. We suggest a modification of Horty's framework that incorporates this assumption and we will see how this solves some of the critiques raised to Hory's original system. We will also discuss how the assumption of a responsible environment cannot immediately be incorporated into the framework of PPC with multiple agents acting *simultaneously*. The reason for this being the way the notion of *goodness* is defined in PPC's model. When we consider other agents' *future* moves, however, our idea can be applied.

The paper will be organized as follows. First, we will discuss Horty's deontic theory of actions and we will mention several problems arising in this framework. Next, we will introduce the system developed by PPC and discuss some of its problems. We will compare the two systems and their respective notions of *goodness* and *obligations*. Finally, we will come to the heart of the paper where

X. He, J. Horty, and E. Pacuit (Eds.): LORI 2009, LNAI 5834, pp. 251–262, 2009.
© Springer-Verlag Berlin Heidelberg 2009

we will discuss defining obligations under the assumption of a responsible world and the way in which it can(not) be incorporated into the systems of Horty and PPC respectively.

2 Deontic Logic and Agency

In his book, "Deontic Logic and Agency" [2], Horty develops a deontic theory of what an agent ought to do using a "Seeing To It That" or STIT framework. This framework, a modal logic of action reports, was originally proposed in a series of papers by Belnap, Bartha and Horty (see [1] for an overview). The theory is developed using a model of branching time, originally due to Prior [6].

The idea of a model of branching time is that moments in time form a tree-like structure: the past is deterministic, the future branches. Each (infinite) branch of the tree is called a history (H). Propositions are evaluated with respect to a moment (m) together with a history through that moment (H). The set of histories passing through m is denoted with $\mathbf{H}_m$. The set $|A|_m^M$ denotes the set of histories from $\mathbf{H}_m$ in which the proposition A is true. Thus, propositions are regarded as sets of histories with respect to a moment.

Just like propositions, actions are also defined to be sets of histories. Thus, in a way, actions *are* propositions. At a moment m an agent α can choose his action from a set $Choice_\alpha^m$ which is a partition of $\mathbf{H}_m$. An action does not uniquely determine a single future, but generally narrows down the set of possible futures.

In order to model the notion of obligation, Horty introduces an utilitarian value-function. This function assigns to each history H a single rank-reflecting numerical value $Value(H)$ that is uniform along the history. Thus, each history has a fixed relative utility assigned to it; a utility to the world.

Within this STIT framework with values, Horty provides a formal and primitive account of what an individual ought to *do*. Or, to be more precise, of what an agent ought to *see to*.

First an order on the set of propositions is defined as follows. Given two propositions A and A', A is better than A' $(A \leq A')$ just in case $Value(H) \leq Value(H')$ for each $H \in A$ and $H' \in A'$. Also, the concept of a $State_\alpha^m$ is introduced. It is the set of agent-independent states that α faces at moment m and is defined as the set action tuples available at m to the group of agents other than α. Formally,

$$State_\alpha^m := Choice_{Agent-\alpha}^m.$$

Given this notion of a state, an ordering on the set of actions is defined as follows.

Definition 1 (Horty 2001). *Let α be an agent and m a moment from a utilitarian STIT frame, and let K, K' be members of $Choice_\alpha^m$. Then $K \preceq K'$ (K' weakly dominates K) if and only if $K \cap S \leq K' \cap S$ for each state $S \in State_\alpha^m$; and $K \prec K'$ (K' strongly dominates K) if and only if $K \preceq K'$ and it is not the case that $K' \preceq K$.*

The above definition satisfies Savage's sure-thing principle and captures the idea that K' weakly dominates K whenever the results of performing K' are at least as good as those of performing K, no matter what actions the other agents choose.

Given this definition of dominance, a set of *optimal* actions of an agent α at a moment m ($Optimal^m_\alpha$) is defined to be the set of available actions that are not strongly dominated by any other action. Thus,

$$Optimal^m_\alpha = \{K \in Choice^m_\alpha \mid \text{ there is no } K' \in Choice^m_\alpha : K \prec K'\}$$

In case there are only finitely many different utilities, Horty's definition of what an agent ought to see to is as follows:

Definition 2. *Let α be an agent and H_m an index from a utilitarian STIT model* M. *Then agent α ought to see to it that A in* M *at H_m, if and only if, $K \subseteq |A|^M_m$ for each action $K \in Optimal^m_\alpha$.*

In short, an agent ought to see to it that A, if all the undominated actions guarantee an outcome of A. From this definition it follows that an agent in obligated to *perform* an action K iff K strictly dominates each action at each state. Or, in other words, there is no action that allows for a better outcome at a particular state. In particular, an action is obligatory if it *guarantees* an outcome with maximal value.

Problems with Horty's Framework

In this section we will pose two 'objections' to Horty's framework by describing two scenarios resulting in strange Horty-obligations. The first is more of an observation rather than an important critique.

Example 1. Sam is very ill. Andrea, Sam's wife, has two choices: bring Sam to a doctor so he can get treatment, or leave him at home, in which case he might die. In the mean time, Sam's daughter Beata, can either play some music Sam likes in his room, or not.

Within Horty's framework, Beata has an obligation to play the music in Sam's room, because, no matter what Andrea does, the outcome will be better. However, in real life, we might object to saying that Beata has this obligation at the current moment. We could say; 'Beata, don't bother playing the music, Andrea will have to take Sam to the hospital, so he won't be there to enjoy it anyway'. Thus, this example depicts a situation in which an obligation to act arises in a setting where we might object to it.

The second objection is a more substantial critique posed by McNamara. It involves a scenario in which *no* obligation arises whereas there *should*. In his review of Horty's book [4], McNamara argues that dominance utilitarianism is too weak to account for actual decision making in ordinary situations of uncertainty. He gives the following illustrative example.

Example 2. [McNamara 2004.] Suppose I am obliged to be somewhere and can get there only by driving. Now consider the fact that at numerous intersections that I drive through, someone *could* choose to run the lights or stop signs and thereby kill me. These are choices other drivers could make, thus nothing I do can rule out this sort of possible history other than to not drive to my appointment.

According to Horty's theory, driving to my appointment does not dominate staying at home, and hence I will not be obliged to see to it that I keep my appointment. Given the assumption of independence of actions, McNamara argues, there will be far too few dominance oughts.

In section 5 we see how our modification of Horty's framework involving responsibility of fellow-agents will solve both examples. But first we will introduce the model of knowledge based obligations as proposed in [5].

3   Knowledge Based Obligation

In their paper '*The Logic of Knowledge Based Obligation*' [5] Pacuit, Parikh and Cogan (PPC) combine a logic of knowledge with a logic of obligation. In order to choose responsibly, they argue, one needs knowledge about the circumstances. A motivating example used in [5] is the following:

Example 3. [PPC 2006.] Uma is a physician whose neighbor [Sam] is ill. Uma does not know and has not been informed.

Because the histories in which Sam gets treated are optimal and, in this simple scenario, coincide with the action of Uma treating Sam, it follows from Horty's analysis that Uma has an obligation to treat Sam. Of course, this conclusion seems rather odd. In the system to be discussed next, oddities of this kind are eliminated.

The framework of PPC is in many ways similar to that of Horty. Just like in Horty, a model of branching time is used. And just like in Horty, each (infinite) history is assigned a value, a utility to the world. There are however, also some important differences between the two approaches. The main difference, as already mentioned above, is that in [5] notions of knowledge and uncertainty are introduced. This is done by making a distinction between *global histories* and *local histories*. A global history H includes all the (relevant) events that have taken place. An agent α's local history h, on the other hand, contains only those events from H that agent α has actually observed and non-informative clock ticks for events that α has not observed.

Furthermore, at moment m, two histories H, H' are indistinguishable for agent α, $H_m \sim_\alpha H'_m$, if and only if α's corresponding local histories h_m and h'_m are the same.

In the PPC framework, actions are events. An action a can be performed at a finite history and yields a set $a(H_m)$ of *global extensions* of H_m. Just like in [2], this is a subset of $\mathbf{H_m}$. Formally,

$$a(H_m) = \{H' | H_m a \sqsubseteq H'\},$$

where $\sqsubseteq$ denotes the initial segment relation. Note that in Horty's system agents move simultaneously, whereas in PPC's (original) model only one agent can move at a given time.

In order to formalize an agent's obligations within the framework of actions and knowledge in branching time, a notion of H-good histories $G(H)$ is introduced. $G(H)$ is defined as the set of extensions of finite history H with the highest possible value[1]. Thus, the set of H-good histories is actually the set of H-*optimal* ones. Given this notion of H-goodness, an action a is defined to be *good* if and only if $G(H) \subseteq a(H)$, i.e. every H-good history involves performing a. Note that this does *not* imply that performing a guarantees an H-good future.

Finally, the PPC notion of obligation is as follows:

Definition 3 (PPC 2006). *An agent α is* obliged *to perform an action a at global history H and moment m iff a is an action which α (only) can perform, and α knows that it is good to perform a. Formally, $(\forall H')(H_m \sim_\alpha H'_m$ and $H' \in G(H'_m)$ implies $H' \in a(H'_m))$.*

Thus, at moment m, agent α is obliged to perform a iff a is a good action and α knows this.

The above notion of knowledge based obligation eliminates Uma's obligation to treat Sam in example 3 in the following way: Assume that treating Sam when he is ill is optimal, whereas treating him when healthy is not. It follows that treating Sam is good. However, Uma does not know this and hence she has no obligation (as yet) to treat him.

Apart from the *absolute* notion of knowledge based obligation, also a weaker concept of *default obligation* is introduced in [5]. Intuitively, an agent has a default obligation to perform action a if all maximal histories that an agent considers plausible are ones in which a is performed.

Afterthoughts on Knowledge Based Obligation

In [?] several suggestions have been proposed for improving the PPC semantics of knowledge based obligations. We will only mention the one that we will come back to in section 5. This suggestion is to allow simultaneous moves, such that at each moment, every agent is allowed to choose from his set of valid actions. To capture this idea the use of action tuples is proposed. Note that this modification is in line with Horty's concurrent moves. The concept of obligation is modified as follows: an agent is obliged to perform a if all good histories are preserved, irrespective of what the other agents do.

As an aside, we would like to mention one afterthought of our own. In the PPC framework, it is assumed that all the necessary information about the possible outcomes of actions are determined by the (past) history. Now, let us consider the situation in which Uma considers performing an operation on a patient that has never been performed before (under the current circumstances). Uma is

---

[1] Some reasonable restrictions apply to the set of possible values which guarantee that a maximal possible value always exists.

unsure about the possible consequences of this action. Suppose moreover that the operation will in fact cure the patient. And thus, performing the action will preserve all good histories. According to the PPC model, this implies that Uma has the (remarkable) obligation to perform the operation.

We think that one way to deal with uncertainty about outcomes of actions is by introducing explicit uncertainty relations among *actions*. The concept of obligation then becomes: α is obliged to perform a iff α can perform a and for all worlds compatible with his local history α *knows* a is good to perform a. Unfortunately, developing this idea in further detail is not within the scope if this paper.

Problems with PPC's Framework

Though [5] is primarily concerned with how (acquiring) information influences obligations, this will not be our main focus when discussing the PPC framework. Rather, in this very brief section and the next, we will concentrate on obligation arising in situations *without* uncertainty. As we did for Horty's framework, we would like to pose two 'objections' to PPC's framework by describing two scenarios resulting in odd obligations. As before, in the first scenario an obligation arises where it should not, and in the second, no obligation arises whereas there should.

The first example is the following:

Example 4. Uma is a physician whose neighbour Sam is ill. Uma knows about this and has two options: flip a coin and treat Sam if the coin lands heads, and poison him if it lands tails. The second option is to continue watching TV (knowing that Sam might survive his illness without treatment).

In this example Uma has the, somewhat strange, obligation to flip the coin.

In the next Uma-example, Uma has *no* obligation to do what many might judge to be the only right thing.

Example 5. Uma is a physician whose neighbour Sam is ill. Uma knows about this and can decide to treat Sam or to flip a coin. If she flips the coin and it lands heads, she will treat Sam after all. If the coin lands tails, she will continue watching TV.

Unfortunately, in this case, Uma has no obligation to treat Sam. Neither does he have the obligation to flip the coin.

In section 5 we will come back to these examples and we will see how they may be interpreted under the assumption of a responsible world.

4   On Goodness and Obligations

Defining goodness or obligations in a setting in which actions are intrinsically indeterministic is not at all straightforward. And there are many possible ways of going about it. Of course, ordering actions according to their *expected value*, would be wonderful, if not ideal. But, as Horty points out in [3], the required information to do this is not always available or even meaningful, especially in situations in which free agents make independent choices. In this section we will briefly emphasize the difference between the respective notions of goodness and obligation of Horty and PPC.

First, recall that according to Horty, an action K is good if it is not strictly dominated by another action. An agent is obliged to perform an action if it strictly dominates all other actions. In particular, an action is obligatory if it is the only action that *guarantees* perfect outcomes at each state. According to PPC, on the other hand, an action is good if every optimal outcome is *preserved* by performing K. To illustrate the differences that result from these definitions, let us revisit some of our earlier examples. It turns out that in every example mentioned in the 'problems with'-sections, the Horty and PPC frameworks conflict about whether or not obligations arise. First, contrary to Horty, Beata is not PPC-obligated to play the music in example 1, since this action does not necessarily preserve all the optimal histories (in which her mother brings Sam to the hospital). Just like in Horty, however, Andrea has the PPC-obligation to bring her husband to the hospital. In example 2, I have the PPC (but not the Horty)-obligation to drive to my appointment. On the other hand, in example 4, where PPC prescribe Uma to flip the coin with the risk of poisoning her neighbor, Horty's framework tells us that both flipping the coin and watching TV is good, though neither is obligated. Finally, according to Horty's semantics, treating Sam is the only good, and in fact obligatory, action in example 5, whereas according to PPC, there are no good actions available in this situation.

Certainly, the two notions of obligation are different. But how should the differences be interpreted? One interpretation is as follows. According to Horty, an agent is obligated to perform an action only if it guarantees a good outcome *immediately*. That is, if the outcomes of this action are better than the outcomes of any other action *no matter what happens*. Here, 'no matter what happens' applies to actions of other agents in the present, but also to actions in the *future*. To see how Horty's notion of obligations fails to allow for future indeterminism or future development, consider the following example:

Example 6. Sam needs a complicated treatment performed over several weeks and by various doctors. Because Uma is the only doctor aware of Sam's illness, it is assumed that the treatment can only be initiated by Uma.

According to Horty, Uma is not obliged to start Sam's treatment if one of the future doctors to participate in the treatment is *capable* of poisoning Sam. Note that this example is a temporal variant of McNamara's driving example.

The PPC framework, to the contrary, does allow for gradually working towards a good outcome. However, as we have illustrated with example 5, it does not allow for hitting a perfect outcome right away (if there are other actions available that possibly result in another perfect outcome), or for ruling out good histories in general[2]. We think that both the definitions of Horty and that of PPC are restricted in their own way: Horty focuses too much on a guarantee in the present, whereas PPC focus too much on future development. In our view, an 'optimal' definition of goodness should allow for both present and potential future good moves. The PPC framework can quite straightforwardly be adopted to, at least partially, incorporate this idea by defining an action as good iff it preserves all good histories *or* it guarantees a good history (right away). The dominance relation of Horty is a little harder to modify. We will very briefly come back to this point when we discuss responsibilities in a moral environment in the next section.

5   Obligations in a Responsible World (RW): A New Approach

In this section we will propose a new way of defining dominance of actions and (epistemic) oughts. As we have mentioned already several times before, our main proposal is to define obligations and epistemic oughts with respect to a responsible or 'moral' world. The idea is that, in defining obligations, we assume (common knowledge of) *rationality,* or in this case, because all the agents share the same utility-functions, *responsibility* of the agents. Of course, assuming a moral responsible environment is a strong assumption, and it is certainly not always met. We think however, that in many simple daily-life situations, *it is* satisfied. Moreover, note that we are not assuming that people always *act* responsibly, we are simply trying to *define* obligations against the idealistic background of a responsible world. To see why this approach is sensible, consider the situation in which John and I both have the obligation to drive to the university to meet there. Now, if John for whatever reasons ignores this obligation and does not come, we could say that under the current circumstances, my obligation to drive to the university no longer applies. Thus, when thinking about our obligations we often assume responsibility of other agents. If this assumption is violated, that is, if someone fails to behave responsibly, we may argue that some of our obligations therefore cease to exist[3]. Given these two reasons, we feel the responsibility assumption is justified as a background for defining obligations.

We will see how the responsible world assumption (RW) can be applied to both Horty's and PPC's frameworks respectively. In Horty's framework our idea

---

[2] Note that our examples are too simplified in the sense that outcomes of a history are determined by simple actions, whereas in real life, there are many later action that will influence the value of the future. However, the point we would like to make here is that it is often possible to guarantee a good (or perfect) outcome by performing an action a that *rules out* other possible good outcomes. In a situation like this we believe it is still possible for a to be obligatory.

[3] We have to admit, however, that this reasoning does not always apply. Think for example of a mother admonishing her child about bad behavior: "I don't care that Johnny started, you are never allowed to bully other children."

can quite straightforwardly be adopted. It will turn out that, if we do so, Beata no longer has the obligation to play music in Sam's room, and I will regain the obligation to drive to my appointment. In PPC's framework, however, applying the idea of a responsible environment is a little more complicated. But we will suggest some possible ways in which the assumption can be interpreted.

RW Applied to Horty

Within Horty's framework, we can incorporate the idea of assuming responsibility of the agents by redefining the set of states as the set of all *rationalizable* action profiles, as opposed to the set of *all* possible action profiles. Given this definition, an action K dominates K' if in each *rationalizable* state, K guarantees a better outcome than K'.

We will formalize this idea by using a well-known concept from game theory: iterative deletion of dominated strategies (or actions). In doing this, we would like to suggest the following modifications to Horty's original notions of dominance and optimal actions.

First, we define a stricter notion of dominance of actions as follows. Given a set of states *State*, we say that an action K is *strictly dominated* by K' with respect to the set *State*; $K <<_{State} K'$ iff $K \cap S < K' \cap S$ for each $S \in State$. And, K is *strongly dominated* by K' with respect to *State*; $K <_{State} K'$ iff $K \cap S \leq K' \cap S$ for each $S \in State$ and $K \cap S < K' \cap S$ for some S. This last definition is an obvious generalization of Horty's dominance of actions. Next, we redifine the concepts of *states* and *choices*. Our definition will capture the idea that we iteratively delete strictly dominated (irresponsible) strategies from the set of available actions for the players. When defining the set of optimal actions, we only need to consider the remaining - non strictly dominated - actions.

At a moment m we define $rState_\alpha^i$ and $rChoice_\alpha^i$ (r standing for responsible, or rationalizable), as follows.

$$rChoice_\alpha^0 = Choice_\alpha^m,$$

where $Choice_\alpha^m$ is Horty's original choice-set for agent α at moment m. And

$$rState_\alpha^0 = rChoice_{Agent-\alpha}^0,$$

is simply the set of all action tuples of other players.

By induction, we now define:

$$rChoice_\alpha^{i+1} = \{a \mid a \in rChoice_\alpha^i \text{ and } \neg \exists a' \in rChoice_\alpha^i : a <<_{rState_\alpha^i} a'\},$$

and

$$rState_\alpha^{i+1} = rChoice_{Agent-\alpha}^{i+1}.$$

Thus, we start with the each player's full choice-set as the set of possible actions, and at each round, we delete those actions that are strictly dominated by one of the remaining ones. Finally, we define $rChoice_\alpha := rChoice_\alpha^\omega$ and $rState_\alpha := rState_\alpha^\omega$, and we define the set of optimal actions for player α as

$$rOptimal_\alpha^m = \{K \mid K \in rChoice_\alpha \text{ and } \neg \exists K' \in$$
$$rChoice_\alpha \text{ such that } K <_{rState_\alpha} K'\}.$$

Thus, an action is (r)optimal if and only if it survives iterative deletion of *strictly* dominated strategies and, finally, it is not *strongly* dominated by any other remaining action.

The last part of the definition, regarding strong dominance, is added to stay in line with Horty's notion of dominance of actions. The reason for only deleting strictly dominated actions - and not the strongly dominated ones until the very end - is that deleting strongly dominated actions may cause rationalizable actions to be deleted. Also, we don't want to have to worry about the order in which actions are deleted. Using the set $rOptimal_\alpha^m$, we define obligations analogue to Horty's original definition:

Definition 4. *Let α be an agent and H_m an index from a utilitarian STIT model* M. *Then agent α ought to see to it that A in* M *at H_m, if and only if, $K \subseteq |A|_m^M$ for each action $K \in Optimal_\alpha^m$.*

Note that this modification of Horty's definition of obligations puts an extreme burden on the degree of rationality of the players. In fact, their rationality is assumed to be unbouded. For, according to the above definition, Jack can base his actions on Carol basing her actions on Bill basing his actions on Linda etc[4]. We leave it for future work to refine our definition of obligations to allow for *bounded* rationality as well. For now, we settle the issue by pointing out that in most real-life situations such deeply nested inferences are not necessary for recognizing ones obligations.

Before considering a moral environment in the PPC framework, let us look at how the above-described modifications of Horty's framework solve examples 1 and 2. The situation as depicted in example 1 can be modeled as follows. Let H stand for Andrea taking Sam to the hospital, and M for Beata playing music in Sam's room. The situation depicted in this example can be modeled using four possible histories: $(H\&M)$, $(H\&\neg M)$, $(\neg H\&M)$, $(\neg H\&\neg M)$, with the straightforward interpretations and with values of $10, 10, 1$ and 0 respectively. In Horty's framework $\neg M \cap H \leq M \cap H$ and $\neg M \cap \neg H < M \cap \neg H$ and hence Beata has the obligation to play the music. In our framework however, we have $\neg H <<_{Choice_B} H$ and hence $rChoice_A = \{H\}$. That is, the only non-strictly dominated action for Andrea is to do H. From this, it follows that $rState_B = \{H\}$ and because both $(H\&M)$ and $(H\&\neg M)$ have value 10, neither action, M nor $\neg M$ is (strictly or strongly) dominating or obligatory for Andrea.

The driving example can be modeled similarly. Let D stands for me driving to the appointment and R for a second agent crossing the red light - for simplicity, we model the situation with only two agents. A simple model of the situation uses again four histories: $(D\&R)$, $(D\&\neg R)$, $(\neg D\&R)$, $(\neg D\&\neg R)$ with respective values $0, 10, 2$ and 5. Then clearly, $R <<_{Choice_{me}} \neg R$ and hence $rState_{me} = \{\neg R\}$. Since $\neg R \cap \neg D < \neg R \cap D$, it follows that $\neg D <<_{\{\neg R\}} D$ and hence my only non-dominated, and thus obligated, action is D.

---

[4] We thank Rohit Parikh for drawing our attention to this issue.

Ideally, we would like to extend this definition of optimal, good, or obligatory actions to also incorporate agents' *future* responsibility. This would, at least partially, solve Horty's lack of allowing for future goodness and situations like example 6. This idea can be formalized by combining backward induction with iterated deletion of dominated strategies. That is, we start at 'the end' and iteratively delete dominated strategies going back in time, until reaching the present. Of course, when histories are infinite, it is not at all clear how to decide where 'the end' starts.

RW Applied to PPC

First, let us see how our idea of a morally responsible world, or iterative deletion of dominated strategies, can *not* easily be applied to the PPC framework with multiple agents acting at the same time. The reason for this is simple. Because an agent is obliged to perform a if *all* good histories are preserved, narrowing down an other agent's scope of action, only decreases the number of potential future histories and hence the number of obligations. This holds even if these other agents act as perfectly moral beings. It seems that the only way out of this problem is by relinquishing the notion of goodness of an action as preserving *all* good histories. It is not immediately clear if or how our suggested new definition of obligation (of an action being obligated if it preserves all good histories, *or* guarantees a good history) can fully resolve this issue. Of course, we can select agents' good actions by first selecting original PPC-good actions and then applying the second Horty-like part of the definition iteratively. But the problem with the first PPC-part of the definition remains.

Other definitions of goodness are compatible with our idea as well. For example, we could define an action as good, or obligatory, if the worst histories are *avoided*. This definition can then be used (iteratively) to delete bad actions.

The assumption of a responsible environment is better applicable to or interpretable in to the original, sequential, PPC framework. To see this, let us recall example 4. Clearly, a coin cannot act responsibly. But, if in this example, the coin is replaced with an agent that can act morally responsibly, say a doctor at a hospital, he will be obliged to treat Sam (assuming there is no third doctor capable of treating Sam). And, if it is assumed that her fellow doctors are responsible, and thus comply with their obligations, it is no longer strange that Uma has the obligation to leave the decision to the other doctor (even though he is capable of poisoning Sam). A similar reasoning can be applied to example 5. If the coin is replaced with a doctor, it is no longer strange that Uma has no immediate obligation to treat Sam herself[5]. The fact that PPC-obligations 'focus on the future' may fit well with the assumption of future-rational agents. That is, an action may be obligatory despite a possible bad outcome (like in example 4), *if* this bad outcome can be ruled out by a responsible action in the future. It is not obvious however, how the idea of a morally responsible future

---

[5] We assume in this example that the extra time it costs to inform the second doctor does not decrease the value of the outcome - Sam being treated.

can be made precise. For example, as we already mentioned above, how far in the future should responsibility of agents be assumed?

Unfortunately, we have to leave formalizing this notion of assuming future responsibility for future work.

6   Conclusion

In this paper we analyzed and compared the frameworks of deontic actions of Horty as presented in [2] and of Pacuit, Parikh and Cogan [5]. We have proposed the idea that obligations of an agent should be defined under the assumption that other agents are responsible, and we have seen how this assumption can potentially solve some of the problems arising in the respective models. However, the ideas as presented in this paper are very preliminary and we would like to develop them in further detail. Hopefully, this development will involve modifying the notion of goodness and obligations of actions assuming both present and, to some extend, future responsibility.

References

1. Belnap, N., Perloff, M., Xu, M.: Facing the Future. Oxford University Press, Oxford (2001)
2. Horty, J.: Deontic Logic and Agency. Oxford University Press, Oxford (2001)
3. Horty, J.: Perspectival act utilitarianism. Working paper (Version of July 2009)
4. McNamara, P.: Review: Agency and Deontic Logic. Mind 113(449), 179–185 (2004)
5. Pacuit, E., Parikh, R., Cogan, E.: The logic of knowledge based obligation. Synthese 149, 311–341 (2006)
6. Prior, A.: Past, Present and Future. Oxford University Press, Oxford (1967)

Dynamic Epistemic Temporal Logic

Bryan Renne[1], Joshua Sack[2,*], and Audrey Yap[3]

[1] University of Groningen
http://bryan.renne.org/
[2] Reykjavík University
http://www.joshuasack.info/
[3] University of Victoria
http://web.uvic.ca/~ayap/

Abstract. We introduce a new type of arrow in the *update frames* (or "action models") of *Dynamic Epistemic Logic* in a way that enables us to reason about epistemic temporal dynamics in multi-agent systems that *need not be synchronous*. Since van Benthem and Pacuit (later joined by Hoshi and Gerbrandy) showed that standard Dynamic Epistemic Logic necessarily satisfies *synchronicity*, it follows that our arrow type is a new way of extending the domain of applicability of the Dynamic Epistemic Logic approach. Furthermore, our framework provides a new perspective on the van Benthem et al work itself. In particular, while each of our work and their work shows that epistemic temporal models generated by standard update frames necessarily satisfy certain structural properties such as synchronicity, our work clarifies the way in which these structural properties arise as a result of the inherent structure of standard update frames themselves.

1 Introduction

Dynamic Epistemic Logic [1,2,3,4,8,12] is a modal-logic approach to reasoning about belief dynamics in multi-agent systems. The characteristic feature of this approach is its use of *update modals*, which are modal operators $[U, s]$ that describe operations on Kripke models. These operations, called *updates*, represent *informational events* in which the agents receive information that may bring about changes in their beliefs. The basic idea is that an update modal $[U, s]$ describes a specific partial function $f_{[U,s]}$ that maps a pointed Kripke model (M, w) in the domain of $f_{[U,s]}$ to another pointed Kripke model that we write as $(M[U], (w, s))$. This allows us to view a sequence

$$(M_0, w_0), (M_1, w_1), (M_2, w_2), \ldots, (M_n, w_n) \tag{1}$$

of pointed Kripke models, with (M_{i+1}, w_{i+1}) generated from (M_i, w_i) by the update $f_{[U_{i+1}, s_{i+1}]}$ described by update modal $[U_{i+1}, s_{i+1}]$, as a discrete-time

---

* Joshua Sack was partly supported by the project "New Developments in Operational Semantics" (nr. 080039021) of The Icelandic Research Fund and a grant from Reykjavík University's Development Fund.

X. He, J. Horty, and E. Pacuit (Eds.): LORI 2009, LNAI 5834, pp. 263–277, 2009.

distributed multi-agent system in which the state of the system at time i is described by (M_i, w_i). Defining the *time of a world w in M_i within the sequence* (1) to be the index i, we obtain a notion of time that is *external* to the pointed Kripke model (M_i, w_i). One consequence of adopting this external notion of time is that all of the worlds that an agent considers possible relative to a world w in M_i have time i. This implies that at every world, every agent knows the current time. Systems in which the current time is known at every world are called *synchronous* [5,6]. Dynamic Epistemic Logic, which itself adopts this external notion of time, is consequently restricted to the study of *synchronous* multi-agent systems [5,6].

In this paper, we propose a simple extension to the update modals $[U, s]$ that allows us to reason about discrete-time distributed multi-agent systems that *need not be synchronous*. We achieve this by adapting the methodology of standard Dynamic Epistemic Logic so that it fits naturally within a version of *Epistemic Temporal Logic* [9,11] whose only temporal modality is a discrete one-step–past operator; this version will be called *Simple Epistemic Temporal Logic*. Simple Epistemic Temporal Logic uses *epistemic temporal models*, which are Kripke models in which one of the relational components is designated as a *time-keeping relation*. When w is related to w' according to the time-keeping relation, the intended interpretation is that w' is a possible way the system might have been one time-step before w. This provides us with an *internal* notion of time, in that the *time of a world w in an epistemic temporal model M* is determined solely based on the time-keeping relation, which is internal to the model M. Diagrammatically, we will represent this relation using arrows labeled by the symbol Y—called *Y-arrows*—where "Y" is a mnemonic for "yesterday" (so having a Y-arrow from world w to world w' is to be thought of as saying that w' is one of the possible ways w might have been "yesterday," meaning one time-step ago). In order to distinguish between Kripke models with and without a *Y-relation* (the time-keeping relation), we adopt the following terminology: *epistemic temporal models* are Kripke models with a designated Y-relation— these have an *internal* notion of time—whereas *epistemic models* are Kripke models without a designated Y-relation—these have an *external* notion of time. Since an epistemic temporal model M uses an *internal* notion of time, the ways in which the system described by M can evolve are determined in advance by the structure of the Y-relation in M; said informally, the *protocol is fixed*. In contrast, the protocol in Dynamic Epistemic Logic is *dynamic*, as it can be changed on-the-fly by using a different update modal to produce the next pointed Kripke model appearing in the sequence (1).

In extending the updates of standard Dynamic Epistemic Logic from the class of epistemic models (having *external* time) to the class of epistemic temporal models (having *internal* time), we stand to gain *dynamic protocols* for systems that *need not be synchronous*. While standard Dynamic Epistemic Logic sets each world in $M[U]$ to be one time-step ahead of any world in M, our new updates on epistemic temporal models allow us greater flexibility in modeling the passage of time. In particular, using the internal notion of time associated with the

Y-relation, our updates allow us to let worlds in $M[U]$ have any natural-number time; therefore, in certain updates that embed M into $M[U]$, each world in $M[U]$ can be seen either as a world in M or else as an arbitrarily distant possible future of a world in M. Such flexibility is essential to the study of *asynchronous* systems. To bring about this flexibility, we add a new structural component to update modals: the $\underline{Y}$-*arrow*. We use $\underline{Y}$-arrows to specify exact positions in which the update $f_{[U,s]}$ is to insert Y-arrows in the updated model $M[U]$. We then identify sufficient conditions on our new update modals $[U, s]$ that will guarantee that the update $f_{[U,s]}$ embeds M into $M[U]$ or preserves properties such as synchronicity in the resulting epistemic temporal model. We use these conditions to show that epistemic temporal models that result from sequentially applying a proper subclass of our new kinds of updates are *isomorphic* to the generated sequences of epistemic models from standard Dynamic Epistemic Logic that have been studied by a number of authors [5,6,10,14,15]. While [5,6] showed that properties such as synchronicity are necessary of sequences generated in standard Dynamic Epistemic Logic, our isomorphism result demonstrates that the necessity of these properties stems from the inherent structure of standard Dynamic Epistemic Logic update modals $[U, s]$ themselves. This provides a *new perspective* on the results of [5,6].

In the next section, we introduce the language L_{DETL} and the theory T_{DETL} of *Dynamic Epistemic Temporal Logic*. It is this theory that we use in reasoning about our new kinds of updates on epistemic temporal models. Due to space constraints, we will omit the proofs of our results; the interested reader can find full details in [13], an extended version of this paper.

2   Syntax

Notation 1 $(A, Y, \underline{Y})$. A *is a finite nonempty set of symbols not containing the symbols* Y *and* $\underline{Y}$. *The members of* A *will be called agents.*

To define L_{DETL}, we must first define the internal structure of update modals $[U, s]$. This structure is built on top of finite Kripke frames. If S is a nonempty set of symbols, then a *Kripke frame* F *(for S)* is a pair (W^F, R^F) consisting of a nonempty set W^F whose members are called *worlds* and a function $R^F : S \to (W^F \to 2^{W^F})$ mapping each symbol $m \in S$ to a function $R_m : W^F \to 2^{W^F}$; to say that F is *finite* means that W^F is finite.[1] The internal structure of update modals $[U, s]$ is given by the structure of the object U, called an *update frame*.

Definition 2. For a language L, whose formulas we call *L-formulas*, an *L-update frame* is a tuple $U = (W, R, \mathsf{p})$ satisfying the following: (W, R) is a finite Kripke frame for $A \cup \{Y, \underline{Y}\}$ that will be called the Kripke frame *underlying* U, and $\mathsf{p} : W \to L$ is a function mapping each world $s \in W$ to an L-formula

---

[1] The function R_a^F gives rise to a binary relation $\bar{R}_a^F := \{(x, y) \in W^F \times W^F \mid y \in R_a^F(x)\}$ on W^F. We will conflate R_a^F and $\bar{R}_a^F$ whenever it is convenient. We will often refer to the members of $\bar{R}_a^F$ as *a-arrows*.

p(s). A *state* in U is just a world in the Kripke frame underlying U. Notation: for an L-update frame U, we write W^U to denote the first element of the tuple U, we write R^U to denote the second element of the tuple U, and we write p^U to denote the third element of the tuple U. A *pointed L-update frame* is a pair (U, s) consisting of an L-update frame U and a state $s \in W^U$ that will be called the *point* of (U, s).

Update frames are also called "action models" (or "event models") in the Dynamic Epistemic Logic literature [1,2,3,4,8,12]. For an update frame U, a state $s \in W^U$ represents the communication of the formula $\mathsf{p}^U(s)$. For an agent $a \in A$, the relation R_a^U represents agent a's conditional uncertainty as to which formula is communicated: if $s' \in R_a^U(s)$ and the formula $\mathsf{p}^U(s)$ was in fact communicated, then agent a will think that the formula $\mathsf{p}^U(s')$ is one of the formulas that might have be communicated.

We now define our language L_{DETL} as an extension of the language L_{ETL} of Simple Epistemic Temporal Logic.

Definition 3 (L_{ETL}). L_{ETL}, the *Language of Simple Epistemic Temporal Logic*, consists of the formulas formed by the following grammar.

$$\varphi ::= \bot \mid \top \mid p_k \mid \varphi \star \varphi \mid \neg\varphi \mid [a]\varphi$$
$$k \in \mathbb{N}, \, \star \in \{\rightarrow, \vee, \wedge, \equiv\}, \, a \in A \cup \{Y\}$$

Terminology: we call $[Y]$ the *yesterday modal*. For each agent $a \in A$, we read the formula $[a]\varphi$ as "agent a believes that φ is true." We read the formula $[Y]\varphi$ as "φ is true in all possible yesterdays." Notation: for each $a \in A \cup \{Y\}$, we let $\langle a \rangle$ abbreviate $\neg[a]\neg$; we define for each $i \in \mathbb{N}$ the formula $[a]^i\varphi$ by setting $[a]^0\varphi := \varphi$ and $[a]^{i+1}\varphi := [a]([a]^i\varphi)$; for $i \in \mathbb{N}$, the formula $\langle a \rangle^i\varphi$ is defined analogously.

Definition 4 (L_{DETL}, T_{DETL}). L_{DETL} is the *Language of Dynamic Epistemic Temporal Logic*. The L_{DETL}-*formulas* are the formulas that may be formed by the grammar obtained from that in Definition 3 by adding the following formula-formation rule: if φ is an L_{DETL}-formula and (U, s) is a pointed L-update frame with $\emptyset \neq L \subseteq L_{\mathsf{DETL}}$, then $[U, s]\varphi$ is an L_{DETL}-formula. L_{DETL} consists of the L_{DETL}-formulas along with the L-update frames for which $\emptyset \neq L \subseteq L_{\mathsf{DETL}}$. Terminology: we call $[U, s]$ an *update modal*. Notation: we let $\langle U, s \rangle$ abbreviate $\neg[U, s]\neg$. We read the formula $[U, s]\varphi$ as "after update (U, s), φ is true." An *update frame* is an L_{DETL}-update frame. A *formula* is a L_{DETL}-formula. T_{DETL}, the *Theory of Dynamic Epistemic Temporal Logic*, is defined in Figure 1.

Since our interest here is in implementing update mechanisms on Kripke models with a designated Y-relation, we do not impose any of the usual properties on belief or on time that one might expect [5,6,9,10,14,15]. So T_{DETL} should be viewed as the *minimal* theory that brings update mechanisms to Simple Epistemic Temporal Logic. Future work will investigate extensions of this theory that include familiar restrictions on belief and on time, though we do address the preservation of certain time-related properties in Section 5.

BASIC SCHEMES

CL.   Schemes for Classical Propositional Logic

$K_a.$ $[a](\varphi \rightarrow \psi) \rightarrow ([a]\varphi \rightarrow [a]\psi)$ for $a \in A$

$K_Y.$ $[Y](\varphi \rightarrow \psi) \rightarrow ([Y]\varphi \rightarrow [Y]\psi)$

UA. $[U,s]q \equiv (\mathsf{p}^U(s) \rightarrow q)$ for $q \in \{p_k, \bot, \top\}$

U$\star$. $[U,s](\varphi \star \psi) \equiv ([U,s]\varphi \star [U,s]\psi)$ for $\star \in \{\rightarrow, \vee, \wedge, \equiv\}$

U$\neg$. $[U,s]\neg\varphi \equiv (\mathsf{p}^U(s) \rightarrow \neg[U,s]\varphi)$

U[a]. $[U,s][a]\varphi \equiv (\mathsf{p}^U(s) \rightarrow \bigwedge_{s' \in R_a^U(s)}[a][U,s']\varphi)$ for $a \in A$

U[Y]. $[U,s][Y]\varphi \equiv (\mathsf{p}^U(s) \rightarrow \bigwedge_{s' \in R_Y^U(s)}[Y][U,s']\varphi) \wedge$

$\qquad\qquad (\mathsf{p}^U(s) \rightarrow \bigwedge_{s' \in R_{\underline{Y}}^U(s)}[U,s']\varphi)$

RULES

$$\frac{\vdash \varphi \rightarrow \psi \quad \vdash \varphi}{\vdash \psi}\text{(MP)} \qquad \frac{a \in A \cup \{Y\} \quad \vdash \varphi}{\vdash [a]\varphi}\text{(MN)} \qquad \frac{\vdash \varphi}{\vdash [U,s]\varphi}\text{(UN)}$$

Fig. 1. The theory T_{DETL}

3   Semantics

Having defined the language L_{DETL} and theory T_{DETL} of Dynamic Epistemic Temporal Logic, we now define the semantics of L_{DETL}. A *Kripke model* M is a tuple (W^M, R^M, V^M) consisting of a Kripke frame (W^M, R^M) and a function $V^M : \{p_k \mid k \in \mathbb{N}\} \rightarrow 2^{W^M}$ called a *(propositional) valuation*. A *pointed Kripke model* is a pair (M, w) consisting of a Kripke model M and a world $w \in W^M$. The notion of L_{DETL}-truth extends the standard semantics for Dynamic Epistemic Logic [1,2,3,4,8,12] in the following way.

Definition 5 (L_{DETL}-Truth, L_{DETL}-Validity). For a pointed Kripke model (M, w) and a formula φ, we write $M, w \models_{L_{\mathsf{DETL}}} \varphi$ to mean that φ is *true at* (M, w), and we write $M, w \not\models_{L_{\mathsf{DETL}}} \varphi$ to mean that φ is not true at (or *false at*) (M, w). The notion of truth of a formula at a pointed Kripke model is defined by an induction on formula construction; we omit the Boolean cases.

- $M, w \models_{L_{\mathsf{DETL}}} [a]\varphi$ means that $M, x \models_{L_{\mathsf{DETL}}} \varphi$ for each $x \in R_a^M(w)$.
- $M, w \models_{L_{\mathsf{DETL}}} [U,s]\varphi$ means that if $M, w \models_{L_{\mathsf{DETL}}} \mathsf{p}^U(s)$, then $M[U], (w, s) \models_{L_{\mathsf{DETL}}} \varphi$, where the model $M[U]$ is defined as follows.

$$W^{M[U]} := \{(x, t) \in W^M \times W^U \mid M, x \models_{L_{\mathsf{DETL}}} \mathsf{p}^U(t)\}$$

For $a \in A:$ $R_a^{M[U]}(x, t) := \{(y, u) \in W^{M[U]} \mid y \in R_a^M(x) \text{ and } u \in R_a^U(t)\}$

$$R_Y^{M[U]}(x, t) := \{(y, u) \in W^{M[U]} \mid y \in R_Y^M(x) \text{ and } u \in R_Y^U(t)\} \cup$$

$$\{(y, u) \in W^{M[U]} \mid y = x \text{ and } u \in R_{\underline{Y}}^U(t)\}$$

$$V^{M[U]}(p_k) := \{(x, t) \in W^{M[U]} \mid M, x \models_{L_{\mathsf{DETL}}} p_k\}$$

To say that a formula φ is *valid in* a Kripke model M, written $M \models_{L_{\mathsf{DETL}}} \varphi$, means that $M, w \models_{L_{\mathsf{DETL}}} \varphi$ for each world $w \in W^M$. To say that a formula φ is *valid*, written $\models_{L_{\mathsf{DETL}}} \varphi$, means that $M \models_{L_{\mathsf{DETL}}} \varphi$ for each Kripke model M. When it ought not cause confusion, we may omit the subscript "L_{DETL}" when writing $\models_{L_{\mathsf{DETL}}}$.

Given a pointed Kripke model (M, w) representing a multi-agent situation and a pointed update frame (U, s) with $M, w \models \mathsf{p}^U(s)$, the pointed Kripke model $(M[U], (w, s))$ represents the situation after the occurrence of the update described by $[U, s]$. According to Definition 5, a world (x, t) must satisfy the property that $M, x \models \mathsf{p}^U(t)$. The set $\{x \in W^M \mid M, x \models \mathsf{p}^U(t)\}$ of worlds x in M that satisfy $\mathsf{p}^U(t)$ intuitively represents the set of worlds in M at which the formula $\mathsf{p}^U(t)$ can truthfully be communicated—these are the worlds at which t can take place.

For each $a \in A$, Definition 5 tells us that the relation $R_a^{M[U]}$ is determined by two factors: agent a's uncertainty as to which world was the case before the communication (represented by R_a^M) and agent a's uncertainty as to which communication has occurred (represented by R_a^U). In particular, suppose $(x', t') \in R_a^{M[U]}(x, t)$. Then if the communication corresponding to t actually occurred at world x, then agent a will think it possible that the communication corresponding to t' occurred at world x'.

According to Definition 5, the relation $R_Y^{M[U]}$ is determined by two factors. The first is the interaction between the relations R_Y^U and R_Y^M, which adds pairs to $R_Y^{M[U]}$ just as the interaction between R_a^U and R_a^M did to $R_a^{M[U]}$ for $a \in A$. The second factor is the relation $R_{\underline{Y}}^U$: if there is a $\underline{Y}$-arrow from state t to state t' in U, then there will be a Y-arrow from world (x, t) to world (x, t') in $M[U]$. The presence of a $\underline{Y}$-arrow from t to t' in U thus says that the communication corresponding to t' is to be thought of as occurring one time-step before the communication corresponding to t. This addition to the standard definition of updates in Dynamic Epistemic Logic [1,2,3,4,8,12] allows us to control how an update affects the time of worlds in the model $M[U]$.

Finally, we see that the valuation $V^{M[U]}$ after the update simply inherits its truth conditions from the valuation V^M before the update, making our updates *purely temporal-epistemic*.

Theorem 6 (Correctness; [13]). *For each formula φ, we have $\vdash \varphi$ if and only if $\models \varphi$.*

4   A Simple Example

Suppose Passengers a and b are traveling together by train in China. Further, suppose Passenger a understands Mandarin but that Passenger b does not, though Passenger b mistakenly believes that they are both equally ignorant of the language. Now consider two scenarios in which an announcement in Mandarin about a delay in arrival is made over the loudspeaker.

1. Passengers a and b are both awake and alert during the announcement.
2. Passenger a is awake and alert, but Passenger b, who is sleepy, dozes off and sleeps through the announcement. Waking up a few minutes later without knowing that the announcement occurred, Passenger b mistakenly thinks that instead of sleeping for a few minutes, he merely blinked.

Taking p to be a propositional letter denoting the statement about late arrival, we represent the first and the second scenarios in our framework using update frames (U_1, t_1) and (U_2, t_2), respectively pictured on the left and on the right in Figure 2.

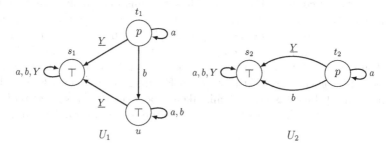

Fig. 2. Update frames for the synchronous (U_1, t_1) and asynchronous (U_2, t_2) private announcement of p to a

In the first scenario, Passenger b knows that an announcement has taken place, but it provides him with no new (non-temporal) information—nor does he believe that a gained any (non-temporal) information. In effect, this is a *synchronous private announcement to a*; after all, both a and b know that an announcement occurred—so the event is *synchronous*—but only a knows the content of the announcement—so the event is *private to a*. In Figure 2, s_1 and u are states in which no new (non-temporal) information is conveyed (since $\top$ is always true and thus conveys no new non-temporal information), while t_1 is a state in which the message p is communicated. Since t_1 and u are each connected to s_1 using a $\underline{Y}$-arrow, the communications they represent occur one time-step after the communication represented by s_1.

Since s_1 is labeled by $\top$, has a reflexive x-arrow for every $x \in \{a, b, Y\}$, and has no exiting $\underline{Y}$-arrows, we see by the definition of truth (Definition 5) that any Kripke model M is embedded into the Kripke model $M[U_1]$ by the mapping taking each world $y \in W^M$ to the world $(y, s_1) \in W^{M[U_1]}$. This embedding preserves a copy of the "past situation" M within the "current situation" $M[U_1]$, which leads us to call s_1 a "past state." So the role of the past state s_1 is to preserve a copy of a given situation M. The states t_1 and u then represent communications that occur one time-step after the situation M. At state t_1, Passenger a believes that t_1 represents the only possible communication, while Passenger b believes that u represents the only possible communication. Since both u and t_1 are one time-step after the past state s_1, the update $f_{[U_1, t_1]}$

describes the private communication of p to Passenger a in which it is common knowledge that one time-step occurs. So we see that

$$\models (\neg \langle Y \rangle \top \wedge \neg [b]p) \rightarrow [U_1, t_1]([a]\langle Y \rangle \top \wedge [a]p \wedge [b]\langle Y \rangle \top \wedge \neg [b]p).$$

That is, if no event has yet occurred and Passenger b does not believe p, then, after the occurrence of $f_{[U_1, t_1]}$, Passenger a believes that an event occurred and that p is true, whereas Passenger b believes that an event occurred but does not believe that p is true.

In contrast, the second scenario is in effect an *asynchronous private announcement to a*. After all, while Passenger a knows that an announcement occurred and she knows its content, Passenger b has two mistaken beliefs: first, that no announcement occurred, and second, that the amount of time between closing and later opening his eyes is essentially negligible. b thus *does not even think it possible* that an event has occurred. Since the announcement results in b having a mistaken belief about the number of events that have occurred, the announcement event is *asynchronous*. At state t_2 in Figure 2, Passenger a knows that p is communicated, but Passenger b mistakenly believes that *no event took place* because the only state he considers possible is the past state s_2. Accordingly, we see that

$$\models (\neg \langle Y \rangle \top \wedge \neg [b]p) \rightarrow [U_2, t_2]([a]\langle Y \rangle \top \wedge [a]p \wedge \neg [b]\langle Y \rangle \top \wedge \neg [b]p).$$

That is, if no event has yet occurred and Passenger b does not believe that p is true, then, after the occurrence of $f_{[U_2, t_2]}$, Passenger a believes that an event occurred and that p is true, whereas Passenger b believes neither that an event occurred nor that p is true.

These scenarios demonstrate the way in which our framework uses $\underline{Y}$-arrows to describe *synchronous* and *asynchronous* private communications. In particular, we see that $\underline{Y}$-arrows can be used to describe updates that *need not preserve synchronicity*, as is the case with the asynchronous private announcement.

5   Properties and Preservation

In this section, we define several properties of Kripke models and update frames and then study sufficient conditions for the preservation of these properties after the occurrence of an update.

Definition 7 (T-Runs, T-Histories, T-Depth). Fix a symbol $T \in \{Y, \underline{Y}\}$ and let $F = (W, R)$ be a Kripke frame for $A \cup \{Y, T\}$. A *T-run* (in F) is a finite nonempty sequence $\{w_i\}_{i=0}^{n}$ of worlds in F satisfying the property that $n \in \mathbb{N}$ and for each $i \in \mathbb{N}$ with $i < n$, we have that $w_{i+1} \in R_T^F(w_i)$. We say that a *T*-run $\{w_i\}_{i=0}^{n}$ *begins at* w_0 and *ends at* w_n. The *length* of a *T*-run $\{w_i\}_{i=0}^{n}$ is defined as the number n. (Observe that the length of a *T*-run is one less than the number of worlds that make up the *T*-run.) To say that a *T*-run σ' *end-extends* a *T*-run σ means that σ is a (not necessarily proper) prefix of σ'. (Note that each

T-run end-extends itself.) To say that a T-run σ is *end-maximal* (in F) means that no T-run in F end-extends σ. A T-*history* (in F) is a T-run in F that is end-maximal. (Note that a suffix of a T-history is itself a T-history.) A world appearing at the end of a T-history in F is said to be T-*terminal* (in F). We define a function $\mathrm{d}_T^F : W^F \to \mathbb{N} \cup \{\infty\}$ as follows: if there is a maximum $n \in \mathbb{N}$ such that there is a T-history in F of length n that begins at w, then $\mathrm{d}_T^F(w)$ is n; otherwise, if no such maximum $n \in \mathbb{N}$ exists, then $\mathrm{d}_T^F(w)$ is ∞. We will call $\mathrm{d}_T^F(w)$ the T-*depth of* w.

Definition 8. Fix $T \in \{Y, \underline{Y}\}$ and let $F = (W, R)$ be a Kripke frame for $A \cup \{Y, T\}$.

- T-*Depth–Defined (T-DD).* To say that F is T-*depth–defined (T-DD)* means that for each world w in F, we have that $\mathrm{d}_T^F(w) \neq \infty$.[2]
- *Non-T-Branching.* To say that F is *non-T-branching* means that for each $w \in W^F$, the set $R_T^F(w)$ has at most one member.
- T-*Synchronous.* If F is T-DD, then to say that F is T-*synchronous* means that for each $a \in A$, each $w \in W^F$, and each $w' \in R_a^F(w)$, we have that $\mathrm{d}_T^F(w') = \mathrm{d}_T^F(w)$. The negation of "$T$-synchronous" is T-*asynchronous*.

Convention: for tuples J having a Kripke frame (W^J, R^J) underlying J, any use of a property or concept from Definition 7 or Definition 8 in reference to J is meant to be a use of that property or concept in reference to the Kripke frame (W^J, R^J) underlying J. Example: for an update frame U, the expression "$\underline{Y}$-run in U" is to be identified with the expression "$\underline{Y}$-run in (W^U, R^U)."

Definition 9 (Kripke Model Properties). Let M be a Kripke model.

- *Synchronicity (under Y-DD).* If M is Y-DD, then to say that M is *synchronous* means that M is Y-synchronous. The negation of "synchronous" is *asynchronous*.
- *Non-Past-Branching.* To say that M is *non–past-branching* means that M is non-Y-branching.
- *Forest-like.* To say that M is *forest-like* means that M is Y-DD and non–past-branching.

Definition 10 (Update Frame Properties and Concepts). Let U be an update frame.

- *Path-Preserving.* A *path-preserving run* (in U) is a $\underline{Y}$-run $\{s_i\}_{i=0}^n$ in U satisfying the property that for each $i \in \mathbb{N}$ with $i < n$, we have $\models \mathsf{p}^U(s_i) \to \mathsf{p}^U(s_{i+1})$. To say that U is a *path-preserving update frame* means that each $\underline{Y}$-run in U is path-preserving.

---

[2] We observe that if F is T-depth–defined, then F is T-*converse well-founded* (that is, for every nonempty set S of worlds in F, there is a nonempty subset $S' \subseteq S$ such that for each $w \in S'$, the unique T-run in F that begins at w has length zero). However, if F is T-converse well-founded, it need not be the case that F is also T-depth–defined. So the notion of T-depth–definedness is strictly stronger than the notion of T-converse well-foundedness.

- *Depth-Respecting (under $\underline{Y}$-DD)*. If U is $\underline{Y}$-DD, then to say that U is *depth-respecting* means that for each $s \in W^U$ and each $s' \in R^U_{\underline{Y}}(s)$, we have that $d^U_Y(s') \leq d^U_{\underline{Y}}(s)$.
- *Past State, Past-Preserving*. A *past state* is a state $s \in W^U$ satisfying the property that $\mathsf{p}^U(s) = \top$, that $R^U_{\underline{Y}}(s) = \emptyset$, and that $R^U_a(s) = \{s\}$ for each $a \in A \cup \{Y\}$. To say that U is *past-preserving* means U is $\underline{Y}$-DD and path-preserving and that every $\underline{Y}$-run in U can be end-extended to a $\underline{Y}$-history in U that ends at a past state.
- *Non–Past-Splitting*. To say that U is *non–past-splitting* means that for each $s \in W^U$, we have that $R^U_{\underline{Y}}(s) \cup R^U_Y(s)$ has at most one element and that $R^U_{\underline{Y}}(s) \cap R^U_Y(s) = \emptyset$.

Having defined these properties, we investigate their preservation under the presence of updates in the following two theorems. Theorem 11 concerns the behavior of past states in update frames, and Theorem 12 concerns the preservation of properties in Kripke models.

Theorem 11 (Past State Theorem; [13]). *Let U be an update frame and M be a Kripke model.*

- *If s is a past state in U, then for each $\varphi \in \mathsf{L_{DETL}}$ and each $w \in W^M$, we have that $M[U], (w, s) \models \varphi$ if and only if $M, w \models \varphi$.*
- *If U is past-preserving and non–past-splitting, $s \in W^U$ has $\mathrm{d}^U_{\underline{Y}}(s) = n$, and $w \in W^M$ satisfies $M, w \models \mathsf{p}^U(s)$, then for each $\varphi \in \mathsf{L_{DETL}}$, we have that $M[U], (w, s) \models \langle Y \rangle^n \varphi$ if and only if $M, w \models \varphi$.*

Theorem 11 tells us that past states play the role of "maintaining a link to the past" within past-preserving, non–past-splitting update frames. In particular, if s is a past state, then the submodel of $M[U]$ consisting of the worlds of the form (w, s) for some world $w \in W^M$ is $\mathsf{L_{DETL}}$-indistinguishable from the Kripke model M itself. So the operation $(M, w) \mapsto \left(M[U], (w, s) \right)$ retains a copy of the "past" state of affairs (M, w). Furthermore, if U is past-preserving, then from any world in $W^{M[U]}$, there is a finite sequence of Y-arrows that leads back to this "past" state of affairs, thereby "maintaining a link to the past."

Let us now examine the preservation of properties of the Kripke model M in the presence of the operation $M \mapsto M[U]$.

Theorem 12 (Preservation Theorem; [13]). *Let (U, s) be a pointed update frame and (M, w) be a pointed Kripke model such that $M, w \models \mathsf{p}^U(s)$.*

- *Y-DD. If M is Y-DD and U is $\underline{Y}$-DD and depth-respecting, then $M[U]$ is Y-DD.*
- Synchronicity. *If M is synchronous (and Y-DD) and U is $\underline{Y}$-DD, depth-respecting, past-preserving, and $\underline{Y}$-synchronous, then $M[U]$ is synchronous.*
- *Non–Past-Branching. If M is non–past-branching and U is non–past-splitting, then $M[U]$ is non–past-branching.*
- *Forest-likeness. If M is forest-like and U is $\underline{Y}$-DD and non–past-splitting, then $M[U]$ is forest-like.*

6   Embedding Standard DEL

In this section, we show that standard (Temporal) Dynamic Epistemic Logic, whose update modals contain neither $\underline{Y}$- nor Y-arrows, can be embedded in our framework in a natural way. This provides clear connections between our work and the work in [5,6,10,14,15] on (Temporal) Dynamic Epistemic Logic, which will be described at the end of this section.

Definition 13 (Standard). Choose $T \in \{Y, \underline{Y}\}$. To say that a Kripke frame F for $A \cup \{Y, T\}$ is *standard* means that for each $s \in W^F$ and each $m \in \{Y, T\}$, we have $R_m^F(s) = \emptyset$. To say that a Kripke model or an L-update frame is *standard* means that the Kripke frame underlying that model or L-update frame is standard. To say that a pointed Kripke model or a pointed L-update frame is *standard* means that the Kripke model or L-update frame making up the first component of the pair is standard.

Definition 14 (L_{TDEL}; [10]). L_{TDEL} is the *Language of Temporal Dynamic Epistemic Logic*. The L_{TDEL}-*formulas* are the formulas that may be formed by the grammar obtained from that in Definition 3 (the definition of L_{ETL}) by adding the following formula-formation rule: if φ is an L_{TDEL}-formula and (U, s) is a standard pointed L-update frame with $\emptyset \neq L \subseteq L_{\mathsf{TDEL}}$, then $[U, s]\varphi$ is an L_{TDEL}-formula. L_{TDEL} consists of the L_{TDEL}-formulas along with the L-update frames for which $\emptyset \neq L \subseteq L_{\mathsf{TDEL}}$.

Notation 15 (Sequences). *Let τ be a finite possibly empty sequence. We write $\tau \cdot x$ to denote the sequence obtained from τ by adding x at the end. $|\tau|$ denotes the number of elements in τ.*

Definition 16 (Adapted from [5,6,11,14,15]). A *run* is a nonempty finite sequence $\{M_i\}_{i=0}^n$ of Kripke models satisfying the property that for each $i \in \mathbb{N}$ with $0 < i \leq n$ and each $w \in W^{M_i}$, we have that w is of the form $(\pi(w), s)$ for some world $\pi(w) \in W^{M_{i-1}}$. A *pointed run* is a pair $(r \cdot M, w)$ consisting of a run $r \cdot M$ and a world $w \in W^M$; the world w is called the *point* of $(r \cdot M, w)$. A *standard (pointed) run* is a (pointed) run whose constituent pointed Kripke models are all standard. An L *event-run* is a finite possibly empty sequence of pointed L-update frames. A *standard L event-run* is an L event-run whose constituent pointed L-update frames are all standard.

Definition 17 (L_{TDEL}-Truth; [5,6,10]). We define a notion of truth for L_{TDEL}-formulas at standard runs r by an induction on the construction of L_{TDEL}-formulas; we consider only the non-Boolean cases.

- For $a \in A$: $r \cdot M, w \models_{L_{\mathsf{TDEL}}} [a]\varphi$ means that $r \cdot M, x \models_{L_{\mathsf{TDEL}}} \varphi$ for each $x \in R_a^M(w)$.
- $r \cdot M, w \models_{L_{\mathsf{TDEL}}} [Y]\varphi$ means that if $|r| > 0$, then $r, \pi(w) \models_{L_{\mathsf{TDEL}}} \varphi$.
- $r \cdot M, w \models_{L_{\mathsf{TDEL}}} [U, s]\varphi$ means that if we have $r \cdot M, w \models_{L_{\mathsf{TDEL}}} \mathsf{p}^U(s)$, then, letting $r' := r \cdot M$, it follows that $r' \cdot r'[U], (w, s) \models_{L_{\mathsf{TDEL}}} \varphi$, where $r'[U]$ is the standard Kripke model defined as follows.

$$W^{r'[U]} := \{(x,t) \in W^M \times W^U \mid r \cdot M, x \models_{L_{\mathsf{TDEL}}} \mathsf{p}^U(t)\}$$
$$\text{For } a \in A: \ R_a^{r'[U]}(x,t) := \{(y,u) \in W^{r'[U]} \mid y \in R_a^M(x) \text{ and } u \in R_a^U(t)\}$$
$$R_Y^{r'[U]}(x,t) := \emptyset$$
$$V^{r'[U]}(p_k) := \{(x,t) \in W^{r'[U]} \mid r \cdot M, x \models_{L_{\mathsf{TDEL}}} p_k\}$$

When it ought not cause confusion, we may omit the subscript "L_{TDEL}" in writing $\models_{L_{\mathsf{TDEL}}}$.

Definition 18 (Generated Structures). Let (M,w) be a standard pointed Kripke model.

- If $\sigma = \{(U_i, s_i)\}_{i=1}^n$ is an L_{DETL} event-run, then $(M,w) *^{\mathsf{p}} \sigma$, the *pointed Kripke model that is point-generated from (M,w) by σ*, is the pointed Kripke model (M_m, w_m) appearing at the end of the sequence $\{(M_i, w_i)\}_{i=0}^m$ having the largest integer $m \le n$ subject to the following restrictions: $(M_0, w_0) = (M,w)$ and for each $j \in \mathbb{N}$ with $j < m$, we have

  - $M_j, w_j \models_{L_{\mathsf{DETL}}} \mathsf{p}^{U_{j+1}}(s_{j+1})$ and
  - $(M_{j+1}, w_{j+1}) = \big(M_j[U_{j+1}], (w_j, s_{j+1})\big).$

  Note: "$\models_{L_{\mathsf{DETL}}}$" and $M_j[U_{j+1}]$ are given by L_{DETL}-truth (Definition 5).

- If $\sigma = \{(U_i, s_i)\}_{i=1}^n$ is a standard L_{TDEL} event-run, then $(M,w) *^{\mathsf{s}} \sigma$, the *pointed run that is sequence-generated from (M,w) by σ*, is the pointed run $(\{M_i\}_{i=0}^m, w_m)$ obtained from the sequence $\{(M_i, w_i)\}_{i=0}^m$ of pointed Kripke models having the largest integer $m \le n$ subject to the following restrictions: $(M_0, w_0) = (M,w)$ and for each $j \in \mathbb{N}$ with $j < m$, we have

  - $\{M_i\}_{i=0}^j, w_j \models_{L_{\mathsf{TDEL}}} \mathsf{p}^{U_{j+1}}(s_{j+1})$ and
  - $(M_{j+1}, w_{j+1}) = \big(\{M_i\}_{i=0}^j[U_{j+1}], (w_j, s_{j+1})\big).$

  Note: "$\models_{L_{\mathsf{TDEL}}}$" and $\{M_i\}_{i=0}^j[U_{j+1}]$ are given by L_{TDEL}-truth (Definition 17).

Definition 19 ($\downarrow$). Let $(r,w) = (\{M_i\}_{i=0}^n, w)$ be the standard pointed run sequence-generated by a standard L_{TDEL} event-run from a standard pointed Kripke model. We write $(r,w){\downarrow}$ to denote the pointed Kripke model (M,w) defined in the following way.

$$W^M := \bigcup_{i=0}^n W^{M_i}$$
$$R_a^M(v) := R_a^{M_i}(v) \text{ if } i \in \mathbb{N} \text{ and } v \in W^{M_i}$$
$$R_Y^M(v) := \begin{cases} \{v'\} & \text{if } v = (v', s) \in W^{M_i} \text{ and } i > 0 \\ \emptyset & \text{otherwise} \end{cases}$$

Definition 20 ($\sharp n$, $\sharp$). For $n \in \mathbb{N}$, we define the function $\sharp n : L_{\mathsf{TDEL}} \to L_{\mathsf{DETL}}$ in Figure 3. If $\sigma = \{(U_i, s_i)\}_{i=1}^n$ is a standard L_{TDEL} event-run, then we define $\sigma^{\sharp} := \{(U_i^{\sharp(i-1)}, s_i)\}_{i=1}^n$.

$$q^{\sharp n} := q \text{ if } q \in \{p_k, \bot, \top\}$$
$$(\varphi \star \psi)^{\sharp n} := \varphi^{\sharp n} \star \psi^{\sharp n}$$
$$(\neg \varphi)^{\sharp n} := \neg(\varphi^{\sharp n})$$
$$([a]\varphi)^{\sharp n} := [a](\varphi^{\sharp n}) \text{ if } a \in A \text{ or } (a = \underline{Y} \text{ and } n = 0)$$
$$([\underline{Y}]\varphi)^{\sharp n} := [\underline{Y}]\varphi^{\sharp(n-1)} \text{ if } n > 0$$
$$([U, s]\varphi)^{\sharp n} := [U^{\sharp n}, s](\varphi^{\sharp(n+1)})$$

$$W^{U^{\sharp n}} := U^W \uplus \{\flat\} \quad \text{(disjoint union)}$$

$$\text{for } a \in A \cup \{Y, \underline{Y}\}, \qquad R_a^{U^{\sharp n}}(s) := \begin{cases} R_a^U(s) & \text{if } a \neq \underline{Y} \text{ and } s \neq \flat, \\ \{\flat\} & \text{if } a \neq \underline{Y} \text{ and } s = \flat, \\ \{\flat\} & \text{if } a = \underline{Y} \text{ and } s \neq \flat, \\ \emptyset & \text{if } a = \underline{Y} \text{ and } s = \flat. \end{cases}$$

$$\mathsf{p}^{U^{\sharp n}}(s) := \begin{cases} \left(\mathsf{p}^U(s)\right)^{\sharp n} \wedge \langle \underline{Y} \rangle^n [\underline{Y}] \bot & \text{if } s \neq \flat, \\ \top & \text{if } s = \flat. \end{cases}$$

Fig. 3. Definition of $\sharp n : L_{\mathsf{TDEL}} \to L_{\mathsf{DETL}}$ for $n \in \mathbb{N}$

Theorem 21 (Isomorphism Theorem; [13]). *Let (M, w) be a standard pointed Kripke model and let σ be a standard L_{TDEL} event-run. Defining $m := |(M, w) *^{\mathsf{p}} \sigma| - 1$, we have each of the following.*

1. *For $\varphi \in L_{\mathsf{TDEL}}$: $(M, w) *^{\mathsf{s}} \sigma \models_{L_{\mathsf{TDEL}}} \varphi$ if and only if $(M, w) *^{\mathsf{p}} \sigma^{\sharp} \models_{L_{\mathsf{DETL}}} \varphi^{\sharp m}$.*
2. *$\left((M, w) *^{\mathsf{s}} \sigma\right)\downarrow$ and $(M, w) *^{\mathsf{p}} \sigma^{\sharp}$ are isomorphic.[3]*

The Isomorphism Theorem (Theorem 21) allows us to view results about Kripke models that have been sequence-generated by standard L_{TDEL} event-runs as results about (Temporal) Dynamic Epistemic Logic—and the other way around. In particular, [5,6] studies certain structural properties of the forest structure given by a run $(M, w) *^{\mathsf{s}} \sigma$ that has been sequence-generated from a standard pointed Kripke model (M, w) by a standard L_{ETL} event-run σ. In [5,6], the authors define what it means for the run $(M, w) *^{\mathsf{s}} \sigma$ to be *synchronous* (among other properties) and then show that every run sequence-generated from a

---

[3] To say that two (pointed) Kripke models are *isomorphic* means that there exists an isomorphism between them. An *isomorphism between Kripke models* M and M' is a bijection $f : W^M \to W^{M'}$ satisfying each of the following: (i) $v \in V^M(p_k)$ if and only if $f(v) \in V^{M'}(p_k)$ for each $k \in \mathbb{N}$, and (ii) $u \in R_a^M(v)$ if and only if $f(u) \in R_a^{M'}(f(v))$ for each $a \in A \cup \{Y\}$. An *isomorphism between pointed Kripke models* (M, w) and (M', w') is an isomorphism f between M and M' for which $f(w) = w'$. See [7] for more information.

standard pointed Kripke model by a standard L_{ETL} event-run is synchronous.[4] Our Preservation Theorem (Theorem 12) works together with the Isomorphism Theorem (Theorem 21) to provide a different perspective on this synchronicity result. In particular, our work shows that the results of [5,6] can be viewed as a consequence of the structural properties that are present in an update frame $U^{\sharp n}$, produced from a standard update frame U, thereby pinpointing the source of the synchronicity result in the structure of standard update frames themselves.

7   Conclusion

In this paper, we showed how to extend the updates of Dynamic Epistemic Logic so that they operate not just on *epistemic models* but also on *epistemic temporal models* in a way that allowed us to control how an update affects the time of worlds in the model $M[U]$. This enabled us to extend the domain of applicability of the Dynamic Epistemic Logic approach to discrete-time multi-agent distributed systems that *need not be synchronous*. We then studied sufficient conditions for the preservation of various properties of Kripke models, such as synchronicity. Identifying an isomorphism that connects epistemic temporal models generated in our framework with epistemic temporal models generated by standard updates as in [5,6], we saw that the necessity of synchronicity in standardly generated epistemic temporal models stems from the structure of standard updates themselves. We then presented two scenarios contrasting synchronous and asynchronous private announcements.

In its technical essence, this paper is about adding a new type of arrow—the $\underline{Y}$-arrow—to update frames and then studying what we can do when the operation $M \mapsto M[U]$ described on epistemic models in [1,2] is extended by the $\underline{Y}$-arrow mechanism to epistemic temporal models in a way that allows us to control how the update affects the time of worlds in the model $M[U]$. Essentially, the $\underline{Y}$-arrow describes a *sufficient condition for the creation of Y-arrows* in the model $M[U]$ resulting from the occurrence of an update. Namely, when there is a $\underline{Y}$-arrow from state s to state s' in update frame U, then there should be a Y-arrow from state (x, s) to state (x, s') in $M[U]$. While this is one possible sufficient condition for the creation of a certain kind of arrow, there other conditions we may wish to consider. In particular, examining the hybrid scheme

$$[U, s][a]\varphi \equiv \mathsf{p}^U(s) \to \bigwedge_{s' \in W^U} \forall z.\big(\mathsf{a}_a^U(s, s') \to @_z(\mathsf{p}^U(s') \to [U, s']\varphi)\big) \qquad (2)$$

in which a_a^U is a function mapping pairs (s, s') of states in U to a formula (possibly containing z), we see that the function a_a^U allows us to express a *precondition for the creation of a-arrows* in the model $M[U]$ produced by a *generalized update*

---

[4] If $(M, w) *^{\mathfrak{s}} \sigma$ is a run sequence-generated from a standard pointed Kripke model (M, w) by a standard L_{ETL} event-run σ, then the definition in [5,6] would have us say that $(M, w) *^{\mathfrak{s}} \sigma$ satisfies *synchronicity* if and only if $\big((M, w) *^{\mathfrak{s}} \sigma\big){\downarrow}$ is synchronous (according to our Definition 9).

frame $U = (W, \mathsf{p}, \mathsf{a})$. The hybrid language of such generalized update frames, called the *arrow-precondition language*, allows us to describe a wide variety of arrow-creation conditions, including all of those mentioned in this paper [13]. Though there is much to be studied about this generalization, it may prove useful in extending Dynamic Epistemic Logic to a much wider class of applications.

References

1. Baltag, A., van Ditmarsch, H.P., Moss, L.S.: Epistemic logic and information update. In: Adriaans, P., van Benthem, J. (eds.) Handbook on the Philosophy of Information, pp. 369–463. Elsevier, Amsterdam (2008)
2. Baltag, A., Moss, L.S.: Logics for epistemic programs. Synthese 139(2), 165–224 (2004)
3. Baltag, A., Moss, L., Solecki, S.: The logic of public announcements, common knowledge and private suspicions. In: Gilboa, I. (ed.) TARK 1998, pp. 43–56 (1998)
4. van Benthem, J., van Eijck, J., Kooi, B.: Logics of communication and change. Information and Computation 204(11), 1620–1662 (2006)
5. van Benthem, J., Gerbrandy, J., Hoshi, T., Pacuit, E.: Merging frameworks for interaction. Journal of Philosophical Logic (in press, 2009)
6. van Benthem, J., Gerbrandy, J., Pacuit, E.: Merging frameworks for interaction: DEL and ETL. In: Proceedings of the 11th Conference on Theoretical Aspects of Rationality and Knowledge (TARK XI), pp. 72–81 (2007)
7. Blackburn, P., de Rijke, M., Venema, Y.: Modal Logic. Cambridge University Press, Cambridge (2001)
8. van Ditmarsch, H., van der Hoek, W., Kooi, B.: Dynamic Epistemic Logic. Synthese Library. Springer, Heidelberg (2007)
9. Fagin, R., Halpern, J.Y., Moses, Y., Vardi, M.Y.: Reasoning about Knowledge. MIT Press, Cambridge (1995)
10. Hoshi, T., Yap, A.: Dynamic epistemic logic with branching temporal structures. Synthese (in press, 2009)
11. Parikh, R., Ramanujam, R.: A knowledge based semantics of messages. Journal of Logic, Language, and Information 12, 453–467 (2003)
12. Renne, B.: A survey of Dynamic Epistemic Logic. Manuscript (July 2008)
13. Renne, B., Sack, J., Yap, A.: Dynamic Epistemic Temporal Logic. Extended Manuscript (June 2009)
14. Sack, J.: Temporal languages for epistemic programs. Journal of Logic, Language, and Information 17(2), 183–216 (2008)
15. Yap, A.: Dynamic epistemic logic and temporal modality. Forthcoming in Proceedings of Dynamic Logic Montréal (2007)

Measurement-Theoretic Foundation of Preference-Based Dyadic Deontic Logic

Satoru Suzuki

Faculty of Arts and Sciences, Komazawa University, Japan
bxs05253@nifty.com

Abstract. The contemporary development of deontic logic since von Wright has been based on the study of the analogies between normative and alethic modalities. The weakest deontic logic called standard deontic logic (SDL) is the modal system of type KD. Jones and Sergot argued that contrary-to-duty (CTD) reasoning was necessary to represent the legal codes in legal expert systems. This reasoning invites such CTD paradoxes as Chisholm's Paradox of SDL that is monadic. Hansson's dyadic deontic logic can avoid CTD paradoxes. But it introduces such dilemmas as the Considerate Assassin's Dilemma. Prakken and Sergot, and van der Torre and Tan proposed preference-based dyadic deontic logics that can explain away this dilemma. However, these logics face the Fundamental Problem of Intrinsic Preference. The aim of this paper is to propose a new non-modal logical version of complete and decidable preference-based dyadic deontic logic–conditional expected utility maximiser's deontic logic (CEUMDL) that can avoid Chisholm's Paradox and explain away the Considerate Assassin's Dilemma. In the model of CEUMDL we can explain an agent's preferences in terms of his degrees of belief and degrees of desire via conditional expected utility maximisation, which can avoid the Fundamental Problem of Intrinsic Preference and furnish a solution to the Gambling Problem. We provide CEUMDL with a Domotor-type model that is a kind of measurement-theoretic and decision-theoretic one.

Keywords: deontic logic, preference, measurement theory, representation theorem, conditional expected utility maximisation, projective geometry.

1 Introduction

Much of the recent work on deontic logic has been based on the view that deontic logic is a branch of modal logic, and that the concepts of obligation, permission and prohibition are related to each other in the same way as the alethic necessity, possibility and impossibility. The contemporary development of deontic logic since von Wright ([26]) has been based on the study of the analogies between normative and alethic modalities. Following the Chellas classification ([2]), the weakest deontic logic called standard deontic logic (SDL) is the weakest normal modal system of type KD. Jones and Sergot ([9]) argued that contrary-to-duty (CTD) reasoning was necessary to represent the legal codes in legal expert systems. This reasoning invites such CTD paradoxes as *Chisholm's Paradox* ([3])

X. He, J. Horty, and E. Pacuit (Eds.): LORI 2009, LNAI 5834, pp. 278–291, 2009.

of SDL that is *monadic*. Hansson's dyadic deontic logic ([6]) and Lewis' dyadic deontic logic ([11]) can avoid CTD paradoxes. But they introduce such dilemmas as the *Considerate Assassin's Dilemma* ([17], [18]). Prakken and Sergot ([18]) and van der Torre and Tan ([25]) proposed preference-based dyadic deontic logics that can explain away this dilemma. However, these logics face the *Fundamental Problem of Intrinsic Preference* ([28]) mentioned later.

The aim of this paper is to propose a new *non-modal logical* version of complete and decidable preference-based dyadic deontic logic–*conditional expected utility maximiser's deontic logic* (CEUMDL) that can avoid Chisholm's Paradox and explain away the Considerate Assassin's Dilemma. In the model of CEUMDL we can explain an agent's preferences in terms of his degrees of belief and degrees of desire via conditional expected utility maximisation, which can avoid the Fundamental Problem of Intrinsic Preference and furnish a solution to the Gambling Problem ([8]) mentioned later.

Von Wright ([27]) divided preferences into two categories: *extrinsic* and *intrinsic* preference. An agent is said to prefer φ_1 extrinsically to φ_2 if φ_1 is better than φ_2 in some explicit respect. So we can explain extrinsic preference from some explicit point of view. If we cannot explain preference from any explicit point of view, we call it intrinsic. Most preference logics that have been proposed are intrinsic but little attention has been paid to extrinsic preference. Von Wright ([28]) posed the following fundamental problem intrinsic preference logics were doomed to face. The development of a satisfactory logic of preference has turned out to be unexpectedly problematic. The evidence for this lies in the fact that almost every principle which has been proposed as fundamental to one preference logic has been rejected by another one. We call it the *Fundamental Problem of Intrinsic Preference*. For example, the status of such logical properties as (transitivity), (contraposition), (conjunctive expansion), (disjunctive distribution) and (conjunctive distribution) is as follows:

Example 1 (Variety of Preferences)

	von Wright ([27])	Martin ([14])	Chisholm and Sosa ([4])
Transitivity	+	+	+
Contraposition	−	+	−
Conjunctive Expansion	+	−	−
Disjunctive Distribution	−	−	−
Conjunctive Distribution	+	−	−

'+' denotes the property in question being provable in the logic in question. '−' denotes the property in question not being provable in the logic in question. (Conjunctive expansion) says that an agent does not prefer φ_1 to φ_2 iff he does not prefer $\varphi_1 \wedge \neg\varphi_2$ to $\varphi_2 \wedge \neg\varphi_1$. (Disjunctive distribution) says that if he does not prefer $\varphi_1 \vee \varphi_2$ to φ_3, then he does not prefer φ_1 to φ_3 or does not prefer φ_2 to φ_3. (Conjunctive distribution) says that if he does not prefer φ_1 to φ_2 and does not prefer φ_3 to φ_2, then he does not prefer $\varphi_1 \vee \varphi_3$ to φ_2.

Mullen ([15]) analysed the cause of the Fundamental Problem as follows. Different theories, such as ethics, welfare economics, consumer demand theory, game theory and decision theory make different demands upon the fundamental properties of preference. The adequacy criteria for preference principles considered by preference logicians have been whether the principles are consistent with our *intuitions* of reasonableness. But intuitions cannot be used as a major factor in the evaluation of the principles of a preference logic. This is because conformity to intuition imposes almost no restraint upon the principles of a theory. For example, a principle may conform to intuition and useless, or it may be inconsistent with intuition and essential to the theoretical solution to a particular problem. Mullen came to the conclusion that preference logic rested upon the mistaken belief that concept construction of preference could satisfactorily be carried out in isolation from *theory construction*. In order to adopt conditional expected utility maximisation as a theory that makes demands upon the fundamental properties of preference, we resort to *measurement theory*.[1] There are two fundamental problems with measurement theory: (1) the representation problem–justifying the assignment of numbers to objects or phenomena, and (2) the uniqueness problem–specifying the transformation up to which this assignment is unique. A solution to the former can be furnished by a *representation theorem*, which establishes that the chosen numerical system preserves the relations of the relational system. From a measurement-theoretic viewpoint of decision theory, there is a tradition to explain an agent's degrees of belief and degrees of desire in terms of his preferences [and vice versa]. This explanation takes the form of a representation theorem of [conditional] expected utility maximisation:

> If [and only if] an agent's preferences satisfy such-and-such conditions, there exist a probability function and a utility function such that he should act as a [conditional] expected utility maximiser.

In mono-set measurement theories where probability functions and utility functions are functions of propositions and preference relations are relations between propositions, Domotor's representation theorem is the only known one of conditional expected utility maximisation that has the "only if" part. Mono-set measurement theories are more suitable for the semantics of logic than non-mono-set ones like Savage's ([20]), for regarding propositions as the semantic values of sentences is simpler than regarding entities like acts (that is, functions from the set of possible worlds to the set of consequences) as those when we wish to provide logic with its semantics. So only by virtue of Domotor's representation theorem, we can explain, in a mono-set measurement theory, an agent's preferences in terms of his degrees of belief and degrees of desire via conditional expected utility maximisation, which can avoid the Fundamental Problem of Intrinsic Preference. We provide CEUMDL with a model based on Domotor's representation theorem. On the other hand, the weak preference relations in both Prakken and Sergot's logic ([25]) and van der Torre and Tan's logic ([18])

---

[1] [19] gives a comprehensive survey of measurement theory. The mathematical foundation of measurement had not been studied before Hölder developed his axiomatisation for the measurement of mass ([7]). [10], [22] and [13] are seen as milestones in the history of measurement theory.

are *pre-orders* (reflexive and transitive relations). It is impossible to construct a meaningful theory and prove a representation theorem of a pre-order based on it. So both Prakken and Sergot's logic and van der Torre and Tan's logic cannot avoid the Fundamental Problem of Intrinsic Preference.

Castañeda ([1]) made a distinction between two kinds of ought: the ought-to-be and the ought-to-do. According to Horty ([8]), the idea of analysing what an agent out to do as what it ought to be that he does was advanced by Meinong, Chisholm and others. Horty observed that the Meinong/Chisholm analysis of the ought-to-do was vulnerable to the *Gambling Problem* as follows:

> Imagine that an agent α is faced with two options at the moment m: to gamble the sum of five dollars, or to refrain from gambling. If α gambles, we suppose that there is a history in which he wins ten dollars, and another in which he loses his stake; but of course, α cannot determine whether he wins or loses. If α does not gamble, we suppose that he preserves his original stake of five dollars no matter how things turn out. ([[8]: p. 55])

He went on to say as follows:

> The Meinong/Chisholm analysis of what an agent ought to do thus tells us unambiguously that, in this situation, the agent ought to gamble: the most valuable history, with a utility of 10, is that in which he gambles and wins, and it is a necessary condition for achieving this outcome that he should gamble. But this is a strange conclusion; for by gambling, the agent risks achieving an outcome with a utility of 0, while he is able to guarantee a utility of 5 by refraining from the gamble. ([[8]: pp. 56–57])

Here we classify decision problems into the following three types. We shall say that we are in the realm of decision making under:

1. *Certainty* if each action is known to lead invariably to a specific outcome,
2. *Risk* if each action leads to one of a set of possible specific outcomes, each outcome occurring with a known probability,[2]
3. *Uncertainty* if either action or both has as its consequence a set of possible specific outcomes, but where the probabilities of these outcomes are completely unknown or are not even meaningful. ([[12]: p. 13])

Horty considered the situation of the Gambling Problem to a case of uncertainty and dealt with it by means of the dominance ordering among actions given by the sure-thing principle. On the other hand, we deal with it by means of conditional expected utility maximisation. Indeed when we specify a probability function that can represent a conditional expected utility maximiser's belief state, conditional expected utility maximisation may be a valid decision rule only for decision makings under certainty and risk. But because in the model of CEUMDL, by virtue of Domotor's representation theorem, it is not necessary to specify a probability, CEUMDL enables us to treat obligations defined by preferences resulting from decision makings under certainty, risk and uncertainty.

---

[2] Of course, certainty is a degenerate case of risk where the probabilities are 0 or 1.

2   Measurement-Theoretic Settings

2.1   Projective-Geometric Concepts

We need some projective-geometric concepts to state Domotor's representation theorem. We define the preliminaries to the measurement-theoretic settings as follows:

Definition 1 (Preliminaries). $\mathbf{W}$ *is a nonempty set of possible worlds. Let $\mathcal{F}$ denote a Boolean field of subsets of $\mathbf{W}$. We call $A \in \mathcal{F}$ a proposition.*

We define a characteristic function as follows:

Definition 2 (Characteristic Function I). *A characteristic function $\hat{\ } : \mathcal{F} \to \{0,1\}^{\mathbf{W}}$ is one where for any $A \in \mathcal{F}$ we have $\hat{A} : \mathbf{W} \to \{0,1\}$ such that*

$$\hat{A}(w) := \begin{cases} 1 & \text{if } w \in A, \\ 0 & \text{otherwise}, \end{cases}$$

for any $w \in \mathbf{W}$.

Because it is impossible to characterise multiplication of probabilities and utilities in terms of union, intersection and preferences, we need a Cartesian product $\times$. $\hat{\ }$ is defined also on Cartesian products of propositions:

Definition 3 (Characteristic Function II)

$$(A \times B)\hat{\ }(w_1, w_2) := \begin{cases} 1 & \text{if } w_1 \in A \text{ and } w_2 \in B, \\ 0 & \text{otherwise}, \end{cases}$$

for any $w_1, w_2 \in \mathbf{W}$.

By means of $\times$, we define an exterior product $\hat{A} \circ \hat{B}$ as follows:

Definition 4 (Exterior Product). $\hat{A} \circ \hat{B}$ *is a 3-valued random variable defined by*

$$\hat{A} \circ \hat{B} := (A \times B)\hat{\ } - (B \times A)\hat{\ }.$$

We combine exterior products by means of a symmetric product $\hat{A} \odot \hat{B}$ as follows:

$(\hat{A} \circ \hat{B}) \odot (\hat{C} \circ \hat{D})$
$:= (\hat{A} \circ \hat{B}) \circ (\hat{C} \circ \hat{D}) + (\hat{C} \circ \hat{D}) \circ (\hat{A} \circ \hat{B}) =$
$(A \times B \times C \times D)\hat{\ } + (B \times A \times D \times C)\hat{\ } + (C \times D \times A \times B)\hat{\ } + (D \times C \times B \times A)\hat{\ }$
$-(A \times B \times D \times C)\hat{\ } - (B \times A \times C \times D)\hat{\ } - (C \times D \times B \times A)\hat{\ } - (D \times C \times A \times B)\hat{\ }.$

By means of symmetric products, we define a four-fold exterior product $\triangle$ $(\hat{A}, \hat{B}, \hat{C}, \hat{D})$ as follows:

Definition 5 (Four-Fold Exterior Product). $\triangle(\hat{A}, \hat{B}, \hat{C}, \hat{D})$ *is a 25-valued random variable defined by*

$$\triangle(\hat{A}, \hat{B}, \hat{C}, \hat{D}) :=$$
$$(\hat{A} \circ \hat{B}) \odot (\hat{C} \circ \hat{D}) + (\hat{A} \circ \hat{C}) \odot (\hat{D} \circ \hat{B}) + (\hat{A} \circ \hat{D}) \odot (\hat{B} \circ \hat{C}) =$$
$$(A \times B \times C \times D)\hat{} + (B \times A \times D \times C)\hat{} + (C \times D \times A \times B)\hat{} + (D \times C \times B \times A)\hat{}$$
$$-(A \times B \times D \times C)\hat{} - (B \times A \times C \times D)\hat{} - (C \times D \times B \times A)\hat{} - (D \times C \times A \times B)\hat{}$$
$$+(A \times C \times D \times B)\hat{} + (C \times A \times B \times D)\hat{} + (D \times B \times A \times C)\hat{} + (B \times D \times C \times A)\hat{}$$
$$-(A \times C \times B \times D)\hat{} - (C \times A \times D \times B)\hat{} - (D \times B \times C \times A)\hat{} - (B \times D \times A \times C)\hat{}$$
$$+(A \times D \times B \times C)\hat{} + (D \times A \times C \times B)\hat{} + (B \times C \times A \times D)\hat{} + (C \times B \times D \times A)\hat{}$$
$$-(A \times D \times C \times B)\hat{} - (D \times A \times B \times C)\hat{} - (B \times C \times D \times A)\hat{} - (C \times B \times A \times D)\hat{}.$$

2.2 Deontic Preference Space and Deontic Preference Space Assignment

We define deontic preference space and deontic preference space assignment as follows:

Definition 6 (Deontic Preference Space and Deontic Preference Space Assignment). $\preceq_w$ *is a deontic weak preference relation on* $\mathcal{F}^2$. $A \preceq_w B$ *is interpreted to mean that* A *is not deontically preferred to* B *in* w. $\sim_w$ *and* $\prec_w$ *are defined as follows:*

- $A \sim_w B := A \preceq_w B$ *and* $B \preceq_w A$,
- $A \prec_w B := A \preceq_w B$ *and* $A \not\preceq_w B$.

For any $w \in \mathbf{W}$, $(\mathbf{W}, \mathcal{F}, \preceq_w, \hat{} , \times, +, -)$ *is called a deontic preference space. Let* **PS** *denote the set of all deontic preference spaces.* $\rho : \mathbf{W} \to \mathbf{PS}$ *is called a deontic preference space assignment.*

2.3 Conditions for Representation

We can state necessary and sufficient conditions for representation as follows:

1. $A \preceq_w B$ or $B \preceq_w A$ (**Connectedness**),
2. If $(A_i \preceq_w B_i$ and $C_i \preceq_w D_i$ for any $i < n)$,
   then (if $A_n \preceq_w B_n$, then $D_n \preceq_w C_n$),
   where $\sum_{i \leq n} (\hat{A}_i \circ \hat{B}_i) \odot (\hat{C}_i \circ \hat{D}_i) = \triangle(\hat{A}_n, \hat{B}_n, \hat{C}_n, \hat{D}_n)$ (**Projectivity**).

2.4 Domotor's Representation Theorem

We can prove Domotor's representation theorem as follows:[3]

---

[3] In Theorem 1, we do not obtain the uniqueness result. But it does not matter when we provide CEUMDL with its model.

Theorem 1 (Representation). *For any $w \in \mathbf{W}$, $(\mathbf{W}, \mathcal{F}, \preceq_w, \hat{\ }, \times, +, -)$ satisfies Connectedness and Projectivity iff there are $P_w : \mathcal{F} \to \mathbb{R}$ and $U_w : \mathcal{F} \backslash \emptyset \to \mathbb{R}$ such that the following conditions hold for any $A, B \in \mathcal{F} \backslash \emptyset$:*

- *$(\mathbf{W}, \mathcal{F}, P_w)$ is a finitely additive probability space,*
- *$A \preceq_w B$ iff $U_w(A) \leq U_w(B)$,*
- *If $A \cap B = \emptyset$, $U_w(A \cup B) = P_w(A|A \cup B)U_w(A) + P_w(B|A \cup B)U_w(B)$,*
- *When $A \in \mathcal{F}$, if $P_w(A) = 0$, then $A = \emptyset$.*

Proof. Except that the proof is relative to world, it is similar to that of [[5]: 184–194].

3   Conditional Expected Utility Maximiser's Deontic Logic **CEUMDL**

3.1   Language

The language $\mathcal{L}_{\mathsf{CEUMDL}}$ of CEUMDL is defined as follows:

Definition 7 (Language). *Let $\mathbf{S}$ denote a set of sentential variables, $\mathbf{WPR}$ a deontic weak preference relation symbol, and $\mathbf{FCP}$ a four-fold Cartesian product symbol. $\mathcal{L}_{\mathsf{CEUMDL}}$ is given by the following rule:*

$$\varphi ::= s \mid \top \mid \neg\varphi \mid \varphi_1 \wedge \varphi_2 \mid \mathbf{WPR}(\varphi_1, \varphi_2) \mid \mathbf{FCP}(\varphi_1, \varphi_2, \varphi_3, \varphi_4),$$

where $s \in \mathbf{S}$, and nestings of $\mathbf{FCP}$ do not occur. $\bot, \vee, \to$ and $\leftrightarrow$ are introduced by the standard definitions. We define a deontic indifference relation symbol $\mathbf{IND}$ and a deontic strict preference relation symbol $\mathbf{SPR}$ as follows:

$$\mathbf{IND}(\varphi_1, \varphi_2) := \mathbf{WPR}(\varphi_1, \varphi_2) \wedge \mathbf{WPR}(\varphi_2, \varphi_1),$$
$$\mathbf{SPR}(\varphi_1, \varphi_2) := \mathbf{WPR}(\varphi_1, \varphi_2) \wedge \neg\mathbf{IND}(\varphi_1, \varphi_2).$$

We define a deontic relation symbol $\mathbf{O}(\ \mid\ )$ as follows:

$$\mathbf{O}(\varphi_1 | \varphi_2) := \mathbf{SPR}(\neg\varphi_1 \wedge \varphi_2, \varphi_1 \wedge \varphi_2).$$

The set of all well-formed formulae of $\mathcal{L}_{\mathsf{CEUMDL}}$ will be denoted by $\Phi_{\mathcal{L}_{\mathsf{CEUMDL}}}$.

3.2   Semantics

DAG   In order to state $\sum_{i \leq n}(\hat{A_i} \circ \hat{B_i}) \odot (\hat{C_i} \circ \hat{D_i}) = \triangle(\hat{A_n}, \hat{B_n}, \hat{C_n}, \hat{D_n})$ of (projectivity) in logical terms, we use $\mathbf{FCP}$. To provide $\mathbf{FCP}$ with a truth definition, we use a *directed acyclic graph (DAG)*. We got a hint about this idea from [16]. We define directedness as follows:

Definition 8 (Directedness). *A graph G is directed if G consists of a nonempty set $\mathbf{W}$ of vertices (possible worlds) and an irreflexive accessibility relation R on $\mathbf{W}$. G is denoted as $(\mathbf{W}, R)$.*

We define a path as follows:

Definition 9 (Path). *A sequence $[w_1, \ldots, w_{n+1}]$ of vertices is a path of length n in G from w_1 to w_{n+1} if $(w_i, w_{i+1}) \in R$ for $i = 1, \ldots, n$.*

By means of a path, we define a cycle.

Definition 10 (Cycle). *A cycle of length n is a path $[w_1, \ldots, w_n, w_1]$ from w_1 to w_1.*

By means of a circle, we define acyclicity as follows:

Definition 11 (Acyclicity). *G is acyclic if G contains no cycles.*

By means of directedness and acyclicity, we define a directed acyclic graph (DAG) as follows:

Definition 12 (DAG). *G is a directed acyclic graph (DAG) if G is both directed and acyclic.*

Remark 1. DAGs can be considered to be a generalisation of trees in which certain subtrees can be shared by different parts of the tree.

Model. By means of a DAG, we define a Domotor-type structured model $\mathcal{M}$ for preference as follows:

Definition 13 (Model). *$\mathcal{M}$ is a quintuple $(\mathbf{W}, R, L, V, \rho)$, where $\mathbf{W}$ is a nonempty set of possible worlds, R is an accessibility relation on $\mathbf{W}^2$, $(\mathbf{W}, R)$ is a DAG, $L : R \to \{\pi_1, \pi_2, \pi_3, \pi_4\}$ is a function that assigns labels to the edges of the graph, any two edges leaving the same vertex have different labels, any vertex either has π_1-, π_2-, π_3- and π_4-labeled outgoing edges or none of them, V is a truth assignment to each $s \in \mathbf{S}$ for each $w \in \mathbf{W}$, and ρ is a deontic preference space assignment that assigns to each $w \in \mathbf{W}$ $(\mathbf{W}, \mathcal{F}, \preceq_w, \hat{\ }, \times, +, -)$ that satisfies Connectedness and Projectivity. For any $w_1 \in \mathbf{W}$, by $\pi_i(w_1)$ $(i = 1, 2, 3, 4)$ we mean the unique $w_2 \in \mathbf{W}$ such that $R(w_1, w_2)$ and $L(w_1, w_2) = \pi_i$ if such world exists.*

Truth Definition. We can provide CEUMDL with the following truth definition:

Definition 14 (Truth). *The notion of $\varphi \in \Phi_{\mathcal{L}_{\mathsf{CEUMDL}}}$ being true at $w \in W$ in $\mathcal{M}$, in symbols $(\mathcal{M}, w) \models_{\mathsf{CEUMDL}} \varphi$ is inductively defined as follows:*

- $(\mathcal{M}, w) \models_{\mathsf{CEUMDL}} s$ *iff* $V(w)(s) = \mathbf{true}$,
- $(\mathcal{M}, w) \models_{\mathsf{CEUMDL}} \top$,
- $(\mathcal{M}, w) \models_{\mathsf{CEUMDL}} \varphi_1 \wedge \varphi_2$ *iff* $(\mathcal{M}, w) \models_{\mathsf{CEUMDL}} \varphi_1$ *and* $(\mathcal{M}, w) \models_{\mathsf{CEUMDL}} \varphi_2$,
- $(\mathcal{M}, w) \models_{\mathsf{CEUMDL}} \neg\varphi$ *iff* $(\mathcal{M}, w) \not\models_{\mathsf{CEUMDL}} \varphi$,
- $(\mathcal{M}, w) \models_{\mathsf{CEUMDL}} \mathbf{FCP}(\varphi_1, \varphi_2, \varphi_3, \varphi_4)$ *iff* $(\mathcal{M}, \pi_1(w)) \models_{\mathsf{CEUMDL}} \varphi_1$ *and* $(\mathcal{M}, \pi_2(w)) \models_{\mathsf{CEUMDL}} \varphi_2$ *and* $(\mathcal{M}, \pi_3(w)) \models_{\mathsf{CEUMDL}} \varphi_3$ *and* $(\mathcal{M}, \pi_4(w)) \models_{\mathsf{CEUMDL}} \varphi_4$,
- $(\mathcal{M}, w) \models_{\mathsf{CEUMDL}} \mathbf{WPR}(\varphi_1, \varphi_2)$ *iff* $[\![\varphi_1]\!] \preceq_w [\![\varphi_2]\!]$,

where $\llbracket \varphi \rrbracket := \{w \in \mathbf{W} : (\mathcal{M}, w) \models_{\text{CEUMDL}} \varphi\}$. *If* $(\mathcal{M}, w) \models_{\text{CEUMDL}} \varphi$ *for all* $w \in \mathbf{W}$, *we write* $\mathcal{M} \models_{\text{CEUMDL}} \varphi$ *and say that* φ *is valid in* $\mathcal{M}$. *If* φ *is valid in all Domotor-type structured models for preference, we write* $\models_{\text{CEUMDL}} \varphi$ *and say that* φ *is valid.*

Semantic Properties. In CEUMDL, the conditional versions of **K** and **D** are valid.

Proposition 1 (Conditional K and D)

- $\models_{\text{CEUMDL}} \mathbf{O}(\varphi \to \psi | \chi) \to (\mathbf{O}(\varphi | \chi) \to \mathbf{O}(\psi | \chi))$ (*Conditional K*),
- $\models_{\text{CEUMDL}} \mathbf{O}(\varphi | \psi) \to \neg \mathbf{O}(\neg \varphi | \psi)$ (*Conditional D*)

Deontic Detachment is concerned with Chisholm's Paradox. Restricted Downward Inheritance is concerned with the Considerate Assassin's Dilemma.

Proposition 2 (Deontic Detachment and Restricted Downward Inheritance)

- $\models_{\text{CEUMDL}} \mathbf{O}(\psi | \varphi) \to (\mathbf{O}(\varphi | \top) \to \mathbf{O}(\psi | \top))$ (*Deontic Detachment*),
- $\models_{\text{CEUMDL}} (\mathbf{WPR}(\varphi \wedge \neg \psi, \neg \varphi \wedge \neg \psi) \vee \mathbf{SPR}(\neg \varphi \wedge \psi, \varphi \wedge \psi)) \to (\mathbf{O}(\varphi | \top) \to \mathbf{O}(\varphi | \psi))$

(*Restricted Downward Inheritance*)

Remark 2. What Restricted Downward Inheritance says is as follows.

1. $\mathbf{WPR}(\varphi \wedge \neg \psi, \neg \varphi \wedge \neg \psi)$, that is, the negation of $\mathbf{SPR}(\neg \varphi \wedge \neg \psi, \varphi \wedge \neg \psi)$ is sufficient for $\mathbf{SPR}((\neg \varphi \wedge \psi) \vee (\neg \varphi \wedge \neg \psi), (\varphi \wedge \psi) \vee (\varphi \wedge \neg \psi)) \to \mathbf{SPR}(\neg \varphi \wedge \psi, \varphi \wedge \psi)$ that is equivalent by definition to Downward Inheritance ($\mathbf{O}(\varphi | \top) \to \mathbf{O}(\varphi | \psi)$).
2. $\mathbf{SPR}(\neg \varphi \wedge \psi, \varphi \wedge \psi)$ is trivially sufficient for Downward Inheritance because the consequent of Downward Inheritance is equivalent by definition to $\mathbf{SPR}(\neg \varphi \wedge \psi, \varphi \wedge \psi)$.

Remark 3. Prakken and Sergot ([17],[18]) observed that the condition concerning alethic modalities that $\Diamond(\varphi \wedge \psi) \wedge \neg \Box(\neg \varphi \to \psi)$ was sufficient for Downward Inheritance. On the other hand, we have proposed the above condition concerning deontic preference relations sufficient for Downward Inheritance.

3.3 Syntax

Syntactic Counterpart of Projectivity. We devise a syntactic counterpart of Projectivity. By developing the idea of [21], we define $\mathbf{DC}_i$ (Disjunction of Conjunctions) as follows:

Definition 15 (Disjunction of Conjunctions). *For any* i $(1 \leq i \leq 4n + 4)$, $\mathbf{DC}_i$ *is defined as the disjunction of all the following conjunctions:*

$$\bigwedge_{j=1}^{n-1} d_j \mathbf{FCP}(\varphi_j, \psi_j, \chi_j, \tau_j)$$

$$\wedge d_n \mathbf{FCP}(\varphi_n, \chi_n, \psi_n, \tau_n)$$

$$\wedge d_{n+1} \mathbf{FCP}(\varphi_n, \tau_n, \chi_n, \psi_n)$$

$$\wedge \bigwedge_{j=n+2}^{2n} d_j \mathbf{FCP}(\psi_{j-n-1}, \varphi_{j-n-1}, \tau_{j-n-1}, \chi_{j-n-1})$$

$$\wedge d_{2n+1} \mathbf{FCP}(\chi_n, \varphi_n, \tau_n, \psi_n)$$

$$\wedge d_{2n+2} \mathbf{FCP}(\tau_n, \varphi_n, \psi_n, \chi_n)$$

$$\wedge \bigwedge_{j=2n+3}^{3n+1} d_j \mathbf{FCP}(\chi_{j-2n-2}, \tau_{j-2n-2}, \varphi_{j-2n-2}, \psi_{j-2n-2})$$

$$\wedge d_{3n+2} \mathbf{FCP}(\tau_n, \psi_n, \chi_n, \varphi_n)$$

$$\wedge d_{3n+3} \mathbf{FCP}(\psi_n, \chi_n, \tau_n, \varphi_n)$$

$$\wedge \bigwedge_{j=3n+4}^{4n+2} d_j \mathbf{FCP}(\tau_{j-3n-3}, \chi_{j-3n-3}, \psi_{j-3n-3}, \varphi_{j-3n-3})$$

$$\wedge d_{4n+3} \mathbf{FCP}(\psi_n, \tau_n, \varphi_n, \chi_n)$$

$$\wedge d_{4n+4} \mathbf{FCP}(\chi_n, \psi_n, \varphi_n, \tau_n)$$

$$\wedge \bigwedge_{j=1}^{n-1} e_j \mathbf{FCP}(\varphi_j, \psi_j, \tau_j, \chi_j)$$

$$\wedge e_n \mathbf{FCP}(\varphi_n, \chi_n, \tau_n, \psi_n)$$

$$\wedge e_{n+1} \mathbf{FCP}(\varphi_n, \tau_n, \psi_n, \chi_n)$$

$$\wedge \bigwedge_{j=n+2}^{2n} e_j \mathbf{FCP}(\psi_{j-n-1}, \varphi_{j-n-1}, \chi_{j-n-1}, \tau_{j-n-1})$$

$$\wedge e_{2n+1} \mathbf{FCP}(\chi_n, \varphi_n, \psi_n, \tau_n)$$

$$\wedge e_{2n+2} \mathbf{FCP}(\tau_n, \varphi_n, \chi_n, \psi_n)$$

$$\wedge \bigwedge_{j=2n+3}^{3n+1} e_j \mathbf{FCP}(\chi_{j-2n-2}, \tau_{j-2n-2}, \psi_{j-2n-2}, \varphi_{j-2n-2})$$

$$\wedge e_{3n+2} \mathbf{FCP}(\tau_n, \psi_n, \varphi_n, \chi_n)$$

$$\wedge e_{3n+3} \mathbf{FCP}(\psi_n, \chi_n, \varphi_n, \tau_n)$$

$$\wedge \bigwedge_{j=3n+4}^{4n+2} e_j \mathbf{FCP}(\tau_{j-3n-3}, \chi_{j-3n-3}, \varphi_{j-3n-3}, \psi_{j-3n-3})$$

$$\wedge e_{4n+3} \mathbf{FCP}(\psi_n, \tau_n, \chi_n, \varphi_n)$$

$$\wedge e_{4n+4} \mathbf{FCP}(\chi_n, \psi_n, \tau_n, \varphi_n)$$

such that exactly i of the d_j's and i of the e_j's are the negation symbols, the rest of them being the empty string of symbols.

By means of $\mathbf{DC}_i$, we define $\mathbf{DDC}$ as follows:

Definition 16 (Disjunction of Disjunctions of Conjunctions)

$$\mathbf{DDC}_{i=1}^{n}(\varphi_i, \psi_i, \chi_i, \tau_i) := \bigvee_{i=1}^{4n+4} \mathbf{DC}_i.$$

Proof System. We provide CEUMDL with the following proof system.

Definition 17 (Proof System)

- *Axioms of* CEUMDL

(A1)   *All tautologies of classical sentential logic,*

(A2) $\mathbf{WPR}(\varphi_1, \varphi_2) \vee \mathbf{WPR}(\varphi_2, \varphi_1)$
(***Syntactic Counterpart of Connectedness***),

$\mathbf{DDC}_{i=1}^n(\varphi_i, \psi_i, \chi_i, \tau_i) \rightarrow$
(A3) $(\wedge_{i=1}^n(\mathbf{WPR}(\varphi_i, \psi_i) \wedge \mathbf{WPR}(\chi_i, \tau_i)) \rightarrow (\mathbf{WPR}(\varphi_n, \psi_n) \rightarrow \mathbf{WPR}(\tau_n, \chi_n)))$
(***Syntactic Counterpart of Projectivity***),

(A4) $\mathbf{FCP}(\top, \top, \top, \top)$ (***Tautology and Four-Fold Cartesian Product***),

$\mathbf{FCP}(\varphi_1 \wedge \varphi_2, \psi_1 \wedge \psi_2, \chi_1 \wedge \chi_2, \tau_1 \wedge \tau_2)$
(A5) $\rightarrow (\mathbf{FCP}(\varphi_1, \psi_1, \chi_1, \tau_1) \wedge \mathbf{FCP}(\varphi_2, \psi_2, \chi_2, \tau_2))$
(***Conjunction and Four-Fold Cartesian Product 1***),

(A6) $(\mathbf{FCP}(\varphi_1, \mu, \nu, \xi) \wedge \mathbf{FCP}(\varphi_2, \mu, \nu, \xi)) \rightarrow \mathbf{FCP}(\varphi_1 \wedge \varphi_2, \mu, \nu, \xi)$
(***Conjunction and Four-Fold Cartesian Product 2***),

(A7) $(\mathbf{FCP}(\lambda, \psi_1, \nu, \xi) \wedge \mathbf{FCP}(\lambda, \psi_2, \nu, \xi)) \rightarrow \mathbf{FCP}(\lambda, \psi_1 \wedge \psi_2, \nu, \xi)$
(***Conjunction and Four-Fold Cartesian Product 3***),

(A8) $(\mathbf{FCP}(\lambda, \mu, \chi_1, \xi) \wedge \mathbf{FCP}(\lambda, \mu, \chi_2, \xi)) \rightarrow \mathbf{FCP}(\lambda, \mu, \chi_1 \wedge \chi_2, \xi)$
(***Conjunction and Four-Fold Cartesian Product 4***),

(A9) $(\mathbf{FCP}(\lambda, \mu, \nu, \tau_1) \wedge \mathbf{FCP}(\lambda, \mu, \nu, \tau_2)) \rightarrow \mathbf{FCP}(\lambda, \mu, \nu, \tau_1 \wedge \tau_2)$
(***Conjunction and Four-Fold Cartesian Product 5***),

$\neg\mathbf{FCP}(\varphi, \psi, \chi, \tau)$
(A10) $\leftrightarrow (\mathbf{FCP}(\neg\varphi, \psi, \chi, \tau) \vee \mathbf{FCP}(\varphi, \neg\psi, \chi, \tau)$
$\vee \mathbf{FCP}(\varphi, \psi, \neg\chi, \tau) \vee \mathbf{FCP}(\varphi, \psi, \chi, \neg\tau))$
(***Negation and Four-Fold Cartesian Product***).

- *Inference Rules of* CEUMDL

(R1) $\dfrac{\varphi_1 \quad \varphi_1 \rightarrow \varphi_2}{\varphi_2}$ (***Modus Ponens***),

(R2) $\dfrac{\varphi \wedge \psi \wedge \chi \wedge \tau}{\mathbf{FCP}(\varphi, \psi, \chi, \tau)}$ (***Four-Fold Cartesian Product Necessitation***).

A proof of $\varphi \in \Phi_{\mathsf{CEUMDL}}$ is a finite sequence of $\mathcal{L}_{\mathsf{CEUMDL}}$-formulae having φ as the last formula such that either each formula is an instance of an axiom, or it can be obtained from formulae that appear earlier in the sequence by applying an inference rule. If there is a proof of φ, we write $\vdash_{\mathsf{CEUMDL}} \varphi$.

3.4   Metalogic

We can prove the soundness of CEUMDL.

Theorem 2 (Soundness). *For every* $\varphi \in \Phi_{\mathcal{L}_{\text{CEUMDL}}}$, *if* $\vdash_{\text{CEUMDL}} \varphi$, *then* $\models_{\text{CEUMDL}} \varphi$.

Proof. The nontrivial part of the proof is to show that (A3) is true in every model.

We can prove the completeness of CEUMDL.

Theorem 3 (Completeness). *For every* $\varphi \in \Phi_{\mathcal{L}_{\text{CEUMDL}}}$, *if* $\models_{\text{CEUMDL}} \varphi$, *then* $\vdash_{\text{CEUMDL}} \varphi$.

Proof. By Lindenbaum Lemma and Truth Lemma.

We can prove the decidability of CEUMDL.

Theorem 4 (Decidability). CEUMDL *is decidable.*

Proof. by Finite Model Property Lemma.

4   Chisholm's Paradox and Considerate Assassin's Dilemma

By means of CEUMDL we analyse Chisholm's Paradox as follows:

Example 2 (Chisholm's Paradox).

1. Jones ought to go to help his neighbors ($\mathbf{O}(\text{help}|\top)$).
2. Jones ought to tell his neighbors he is coming if he is going to help them ($\mathbf{O}(\text{tell}|\text{help})$).
3. If Jones does not go to help his neighbors, he ought not to tell them he is coming ($\mathbf{O}(\neg\text{tell}|\neg\text{help})$).
4. Jones does not go to help his neighbors ($\neg\text{help}$).

Factual Detachment: $\mathbf{O}(\psi|\varphi) \rightarrow (\varphi \rightarrow \mathbf{O}(\psi|\top))$ is necessary for (1)–(4) to lead to a contradiction. Although Deontic Detachment is valid in CEUMDL, Factual Detachment is not valid. So (1)–(4) do not lead to a contradiction.

By means of CEUMDL we analyse the Considerate Assassin's Dilemma as follows:

Example 3 (Considerate Assassin's Dilemma).

1. You should not offer cigarettes ($\mathbf{O}(\neg\text{offer}|\top)$).
2. If you kill the witness, you should offer him a cigarette ($\mathbf{O}(\text{offer}|\text{kill})$).

Van der Torre and Tan ([25]) observed that (1) and (2) without additional assumptions led to a contradiction. By means of CEUMDL, on the other hand, we diagnose as follows. On the assumption that

$$\mathbf{WPR}(\neg\text{offer} \wedge \neg\text{kill}, \text{offer} \wedge \neg\text{kill}) \vee \mathbf{SPR}(\text{offer} \wedge \text{kill}, \neg\text{offer} \wedge \text{kill}),$$

(1) and (2) lead to a contradiction because of Restricted Downward Inheritance.

5   Conclusion

In this paper we have proposed a new non-modal logical version of complete and decidable preference-based dyadic deontic logic–conditional expected utility maximiser's deontic logic (CEUMDL) that can avoid Chisholm's Paradox and explain away the Considerate Assassin's Dilemma. In the model of CEUMDL we can explain an agent's preferences in terms of his degrees of belief and degrees of desire via conditional expected utility maximisation, which can avoid the Fundamental Problem of Intrinsic Preference and furnish a solution to the Gambling Problem.

Acknowledgements. We would like to thank two anonymous reviewers for their helpful comments.

References

1. Catanēda, H.-N.: On the Seamantics of the Ought-to-Do. Synthese 21, 449–468 (1970)
2. Chellas, B.J.: Modal Logic: An Introduction. Cambridge UP, Cambridge (1980)
3. Chisholm, R.M.: Contrary-to-Duty Imperatives and Deontic Logic. Analysis 24, 33–36 (1963)
4. Chisholm, R.M., Sosa, E.: On the Logic of Intrinsically Better. American Philosophical Quarterly 3, 244–249 (1966)
5. Domotor, Z.: Axiomatisation of Jeffrey Utilities. Synthese 39, 165–210 (1978)
6. Hansson, B.: An Analysis of Some Deontic Logics. Noûs 3, 373–398 (1969)
7. Hölder, O.: Die Axiome der Quantität und die Lehre von Mass. Berichte über die Verhandlungen der Königlich Sächsischen Gesellschaft der Wissenschaften zu Leipzig. Mathematisch-Physikaliche Classe 53, 1–64 (1901)
8. Horty, J.F.: Agency and Deontic Logic. Oxford UP, Oxford (2001)
9. Jones, A.J.I., Sergot, M.: Deontic Logic in the Representation of Law: Towards a Methodology. Artificial Intelligence and Law 1, 45–64 (1992)
10. Krantz, D.H., et al.: Foundations of Measurement, vol. I. Academic Press, New York (1971)
11. Lewis, D.: Semantic Analysis for Dyadic Deontic Logic. In: Stunland, S. (ed.) Logical Theory and Seamtnical Analysis, pp. 1–14. Reidel, Dordrecht (1974)
12. Luce, R.D., Raiffa, H.: Games and Decisions. John Wiley & Sons, Inc., New York (1957)
13. Luce, R.D., et al.: Foundations of Measurement, vol. III. Academic Press, San Diego (1990)
14. Martin, R.M.: Intension and Decision. Prentice-Hall, Inc., Englewood Cliffs (1963)
15. Mullen, J.D.: Does the Logic of Preference Rest on a Mistake? Metaphilosophy 10, 247–255 (1979)
16. Naumov, P.: Logic of Subtyping. Theoretical Computer Science 357, 167–185 (2006)
17. Prakken, H., Sergot, M.J.: Contrary-to-Duty Obligations. Studia Logica 57, 91–115 (1996)
18. Prakken, H., Sergot, M.J.: Dyadic Deontic Logic and Contrary-to-Duty Obligations. In: Nute, D. (ed.) Defeasible Deontic Logic, pp. 223–262. Kluwer, Dordrecht (1997)

19. Roberts, F.S.: Measurement Theory. Addison-Wesley, Reading (1979)
20. Savage, L.: The Foundations of Statistics, Second Revised Edition. Dover, New York (1972)
21. Segerberg, K.: Qualitative Probability in a Modal Setting. In: Fenstad, J.E. (ed.) Proceedings of the Second Scandinavian Logic Symposium, pp. 341–352. North-Holland, Amsterdam (1971)
22. Suppes, P., et al.: Foundations of Measurement, vol. II. Academic Press, San Diego (1989)
23. Suzuki, S.: Preference Logic and Its Measurement-Theoretic Semantics. Accepted Paper of 8th Conference on Logic and the Foundations of Game and Decision Theory, LOFT 2008 (2008)
24. Suzuki, S.: Prolegomena to Dynamic Epistemic Preference Logic. In: Hattori, H., et al. (eds.) New Frontiers in Artificial Intelligence. LNCS(LNAI), vol. 5447, pp. 177–192. Springer, Heidelberg (2009)
25. Van der Torre, L., Tan, Y.-H.: Contrary-to-Duty Reasoning with Preference-Based Dyadic Obligations. Annals of Mathematics and Artificial Intelligence 27, 49–78 (1999)
26. Von Wright, G.H.: Deontic Logic. Mind 60, 1–15 (1951)
27. Von Wright, G.H.: The Logic of Preference. Edinburgh UP, Edinburgh (1963)
28. Von Wright, G.H.: The Logic of Preference Reconsidered. Theory and Decision 3, 140–169 (1972)

An Update Operator for Strategic Ability

Paolo Turrini[1], Jan Broersen[1], Rosja Mastop[2], and John-Jules Meyer[1]

[1] Department of Information and Computing Sciences, Utrecht University
{paolo,broersen,jj}@cs.uu.nl
[2] Department of Philosophy, Utrecht University
rosja.matop@phil.uu.nl

Abstract. Coalition Logic does not explicitly talk about the effects of a coalitional move on the strategic ability of the remaining players, while in Game Theory reasoning patterns involving this concept often occur. To fill this gap, we study an update operator for strategic ability update in coalition structures. Its formal connections with the update operators known from Dynamic Epistemic Logic will be discussed.

1 Introduction

Ever since the work of Rohit Parikh on the logic of games [9] the research on the characterization of game-theoretical notions in terms of a logical language has grown rapidly. In Cooperative Game Theory for instance results on the correspondence between strategic games and neighbourhood models - such as Pauly Representation Theorem for Coalition Logic [10] or the completeness of Alternating-Time Temporal Logic (ATL) [7] - have opened the possibility of studying cooperative interactions by means of modal logic. ATL and Coalition Logic reason on what coalitions can achieve by cooperating, however they do not explicitly describe what the effects of a given coalitional action or strategy are on the moves of the remaining players. Game Theory instead deals with reasoning structures, as for instance that of Dominant Strategy Equilibrium [8], in which players consider all the possible reactions of their opponents and choose the best strategy given all such reactions.

As affirmed in [11], p.1:

> Much of game theory is about the question whether strategic equilibria exist. But there are hardly any explicit languages for defining, comparing, or combining strategies as such - the way we have them for actions and plans, maybe the closest intuitive analogue to strategies. True, there are many current logics for describing game structure - but these tend to have existential quantifiers saying that "players have a strategy" for achieving some purpose, while descriptions of these strategies themselves are not part of the logical language.

In order to capture the reasoning structure behind Dominant Strategy Equilibrium and many other solution concepts, we intuitively need a language able to

X. He, J. Horty, and E. Pacuit (Eds.): LORI 2009, LNAI 5834, pp. 292–301, 2009.

talk about strategic ability update and consequently to make the role of strategic ability explicit. Updates are not new to the realm of modal logics. Formalizations of dynamics of information flow, like Dynamic Epistemic Logic [15] (DEL), reason about how agents' knowledge is updated after an epistemic event, for instance a public announcement, takes place.

Logics for strategic ability using a model update have already been studied, ranging from the use of counterfactuals in CATL [14], to the action expressions used in Coalition Action Logic [4] and the first order strategy terms in Strategy Logic [6]. Nevertheless all these extensions use arbitrary strategy terms that do not allow to reduce strategy execution to strategic ability. The reduction of the language of Public Announcement Epistemic Logic to Epistemic Logic is instead one of the most elegant results in Dynamic Epistemic Logic.

The idea of this paper is to extend the *update paradigm* of public announcements to account for the changes that moves in a game induce on players' strategic ability and to study strategies reducing them to the choice structures under which they can be executed.

1.1 Motivating Example

To provide a clearer intuition of the notion of strategic ability update, we resort to the well known gametheoretical example of the Prisoners' Dilemma [8], that is an interactive situation in which the advantages of cooperation are overruled by the incentive for individual players to defect. In Table 1 a Prisoners' Dilemma is described, where players i and j, that we assume to be rational, can choose between a cooperative move C and a defective move D, yielding an outcome (x_i, x_j), x_k being the payoff for each $k \in \{i, j\}$. If we focus on player i we can observe that, after the choice C by j, the choice D becomes preferable to the choice C - yielding $(4, 0)$ instead of $(3, 3)$ - and the same holds in case j moved D - yielding $(1, 1)$ instead of $(0, 4)$. Our rationality assumption warrants player i to reason on the updates of his own choices brought about by player j, and to select his best response in each such scenario.

Our aim is to formally capture the reasoning structure of players in strategic interaction, in which players consider the best action to take, *given* what their opponents do. This should not be confused with the reasoning patterns in extensive games, in which players reason on the best action to take *after* their opponents have moved, neither with the notion of ability to guarantee an outcome *independently* of what the other players do, which is the typical reading of the operators in the various game logics. To make these intuitions precise we

Table 1. A Prisoners' Dilemma

i＼ j	C	D
C	(3, 3)	(0, 4)
D	(4, 0)	(1, 1)

will provide a semantics for the notion of game restriction induced by the moves of the players in a strategic interaction. We will work on cooperative structures, where players can form coalitions to achieve their goals [2]. In our treatment we will focus on coalitional ability, abstracting away from players' preferences.

The paper is structured as follows: in the first part we introduce Coalition Logic, that we use to model strategic ability; in the second part we introduce an operator to talk about the model transformations induced by the choices of coalitions: the subgame operator. Finally we give reduction axioms for the subgame operator and discuss the links with Public Announcement Logic.

2   Coalition Logic and Strategic Ability

In Game Theory players may be able to force the interaction to end up in an outcome satisfying certain properties. An abstract representation of this notion is given by the dynamic effectivity function, first described in [10], which we adopt to model strategic ability.

Definition 1 (Dynamic Effectivity Function)
Given a finite set of agents Agt and a set of states W, a dynamic effectivity function *is a function $E : W \rightarrow (2^{Agt} \rightarrow 2^{2^W})$.*

Any subset of *Agt* will henceforth be called a *coalition*. The elements of W are called *states* or *worlds*; the sets of states $X \in E(w)(C)$ are called the *choices* of coalition C in state w. The set $E(w)(C)$ is called the *choice set* of C in w. The complement of a set X is indicated as $\overline{X}$ and calculated relative to the expected domain. A dynamic effectivity function can be seen as a "formal description of the power structure in a society" [1]; it assigns, in each world, to every coalition a set of sets of states that represents the strategic ability of that coalition. Intuitively, if $X \in E(w)(C)$, C is said to be able from w to *force* the interaction to end up in some member of X. Every effectivity function has the property of **outcome monotonicity**: for all $X \subseteq W, Y \subseteq W, w \in W, C \in 2^{Agt}$, if $X \in E(w)(C)$ and $X \subseteq Y$, then $Y \in E(w)(C)$. Said in other words, if a coalition is able to force the the interaction to end up in some member of X then is also able to force the interaction to end up in some member of any supersets of X. Together with outcome monotonicity we will assume the properties of **regularity**: if $X \in E(w)(C)$, then $\overline{X} \notin E(w)(\overline{C})$; and **closed-worldness**: $E(w)(\emptyset) = \{W\}$. Regularity means that disjoint coalitions do not make choices that contradict each other, while closed-worldness requires the empty coalition not to influence the interaction. For an in depth discussion on the desirability of these properties see the results in [5].

2.1   Models and Language

The models we refer to are structures of the form

$$\langle W, E, V \rangle$$

where W is a nonempty set of states, E an outcome monotonic, regular and closed-world effectivity function, $V : W \to 2^P$ a valuation function that assigns to each state a subset of a countable set of atomic propositions P, to be interpreted as true at that state. The formulas for the basic language are of the form

$$p \mid \neg\phi \mid \phi \wedge \psi \mid [C]\phi \mid A\phi$$

where p is any atomic proposition in P, $[C]\phi$ is the coalitional operator expressing the fact that coalition C can force or bring about the formula ϕ; $A\phi$ is the global modality, which talk about a formula that holds in every world in the model. Their interpretation is standard [10] [3] [12] and it is given as follows:

$$M, w \models p \text{ iff } p \in V(w)$$
$$M, w \models \neg\phi \text{ iff } \text{ not } M, w \models \phi$$
$$M, w \models \phi \wedge \psi \text{ iff } M, w \models \phi \text{ and } M, w \models \psi$$
$$M, w \models [C]\phi \text{ iff } \phi^M \in E(w)(C)$$
$$M, w \models A\phi \text{ iff } M, v \models \phi, \text{ for all } v \in W$$

where $\phi^M = \{w \in W \mid M, w \models \phi\}$ is the *truth set* of ϕ.

What we can say in Coalition Logic. The Prisoners' Dilemma can intuitively be rewritten as a coalition model. Here coalition $\{i\}$ can force that $\{i\}$ defects and can force that $\{i\}$ cooperates, but $\{i\}$ cannot force that $\{j\}$ cooperates (and equivalently it cannot force that $\{j\}$ defects). In any world w, we have therefore that $PD, w \models [\{i\}](\text{ i defects }) \wedge \neg[\{i\}](\text{ j defects })$. On the other hand we cannot express what i can do given that j defects. This would mean i to have a strategy forcing that i defects and j defects and a strategy forcing that i cooperates and j defects. This at the model level is $PD, w \models [\{i\}](\text{ i defects and j defects }) \wedge [\{i\}](\text{ i cooperates and j defects })$. By the property of outcome monotonicity, we would then get $PD, w \models [\{i\}](\text{ j defects })$, which is at odds with our initial statement. The reason of this limitation is to be found in the interpretation of the coalition logic operator, that expresses what a coalition can achieve *independently* of what its opponents do. Reasoning about how the strategic ability (to force some outcome) of a coalition depends on the possible moves of its opponents requires that we can express in our language that a coalition can force some outcome *given* what its opponents do.

3   Strategic Ability Update

To model strategic ability update we introduce an operator $[C \downarrow \psi]\phi$ whose informal reading is: "after coalition C chooses ψ, ϕ holds". We define the dual $\langle C \downarrow \psi \rangle \phi$ as an abbreviation of $\neg[C \downarrow \psi]\neg\phi$. Intuitively what we do is to talk about the model *restrictions* that are caused by the possible move ψ of coalition C. For this reason it will be called *the subgame operator*. Its formal interpretation goes as follows:

$$M, w \models [C \downarrow \psi]\phi \Leftrightarrow \psi^M \in E(w)(C) \text{ implies } M \downarrow_{(C, \psi^M, w)}, w \models \phi$$

The interpretation of the operator has a conditional reading: if a coalition C has a certain choice ψ^M at w, then the model where this choice is actually executed makes a certain proposition ϕ true. The capacity of C to choose ψ^M is seen here as a precondition for C to actually execute ψ^M.

The restricted models $M \downarrow_{(C,\psi^M,w)}$ are so defined:

$$M \downarrow_{(C,\psi^M,w)} \doteq \langle W, E \downarrow_{(C,\psi^M,w)}, V \rangle$$

They inherit the domain and the valuation function from the original coalition model while they update the coalitional relation[1] $E \downarrow_{(C,\psi^M,w)}$ in the following way:

$$E \downarrow_{(C,\psi^M,w)} (w)(D) \doteq (\{\psi^M\})^{\mathrm{sup}} \qquad \text{for } D \cap C \neq \emptyset$$
$$E \downarrow_{(C,\psi^M,w)} (w)(D) \doteq (E(w)(D) \sqcap \psi^M)^{\mathrm{sup}} \quad \text{for } D \cap C = \emptyset \text{ and } D \neq \emptyset$$
$$E \downarrow_{(C,\psi^M,w)} (w')(D) \doteq E(w')(D) \qquad\qquad \text{for } w' \neq w \text{ or } D = \emptyset$$

where for a set of sets $\mathcal{X}$, $(\mathcal{X})^{\mathrm{sup}} = \{X \subseteq W |$ there is $Y \in \mathcal{X}$ and $Y \subseteq X \subseteq W\}$. In words, $()^{\mathrm{sup}}$ is the superset closure of a set of sets. Moreover taken two sets of sets $\mathcal{X}, \mathcal{P}$, $\mathcal{X} \sqcap \mathcal{P} = \{\xi \cap \psi | \xi \in \mathcal{X}$ and $\psi \in \mathcal{P}\}$.

The way the relation is updated deserves some comment. A distinction is made between the strategic ability update of the players who made a certain choice ϕ and all the other players. After coalition C has made a choice ϕ, all the coalitions involving agents belonging to C are given $(\phi^M)^{\mathrm{sup}}$ as a choice set. This view maintains that a coalition comprising players in a coalition that has already formed cannot further influence the outcome of the game. This fact implies that the subgame operator is not coalition monotonic, in the sense given in [10], that is bigger coalitions need not have bigger power. Said in other words, we do not allow players to make a choice within a certain coalition and then, at the same time, to make a choice within different coalitions. The models of reference are strategic games, in which strategies are decided in the beginning once and for all [8]. The other (nonempty) coalitions instead *truly update* their choice set having it restricted by the choice of C. Restriction is implemented in this case by intersecting the effectivity function with the move that has been carried out. If for instance C chooses to force ψ and $\overline{C}$ were able to decide on ξ, then given the choice by C, $\overline{C}$ is able to force $\xi \wedge \psi$. The coalitional relation at worlds different from the one where the choice is made remains instead unchanged. This means that the update is local. Again, the references are strategic games, where the sequential structure of strategies is substantially ignored. Notice that by the last condition the empty coalition never gains power. In sum the strategic ability update is governed by three principles: the **irrelevance of hybrid coalitions**, that does not allow members of the coalition that moved to further influence the interaction, the **restriction of opponents' choices**, that truly updates

---

[1] Here the word *functional relation* would be more appropriate. In fact the Effectivity Function behaves as a relation in a Neighbourhood model and our restriction uniquely associates to an Effectivity Function the restriction imposed by a coalitional choice.

Table 2. Proof System

	Axioms
	Regularity
A1	$[C]\phi \rightarrow \neg[\overline{C}]\neg\phi$
	Closed-Worldness
A2	$[\emptyset]\phi \leftrightarrow A\phi$
	Global Modality Axioms
A3	$\phi \rightarrow E\phi$
A4	$EE\phi \rightarrow E\phi$
A5	$\phi \rightarrow AE\phi$
A6	$A(\phi \rightarrow \psi) \rightarrow (A\phi \rightarrow A\psi)$
	Strategic Ability Update Axioms
A7	$[C \downarrow \xi]p \leftrightarrow ([C]\xi \rightarrow p)$
A8	$[C \downarrow \xi]\neg\phi \leftrightarrow ([C]\xi \rightarrow \neg[C \downarrow \xi]\phi)$
A9	$[C \downarrow \xi](\phi \wedge \psi) \leftrightarrow ([C \downarrow \xi]\phi \wedge [C \downarrow \xi]\psi)$
A10	$[C \downarrow \xi]A\phi \leftrightarrow ([C]\xi \rightarrow A\phi)$
A11	$[C \downarrow \xi][D]\phi \leftrightarrow ([C]\xi \rightarrow [D](\xi \rightarrow \phi))$ (for $D \cap C = \emptyset$ and $D \neq \emptyset$)
A12	$[C \downarrow \xi][D]\phi \leftrightarrow A(\xi \rightarrow \phi)$ (for $D \cap C \neq \emptyset$)
A13	$[C \downarrow \xi][D]\phi \leftrightarrow ([C]\xi \rightarrow [D]\phi)$ (for $D = \emptyset$)
	Rules
R1	$\phi \wedge (\phi \rightarrow \psi) \Rightarrow \psi$
R2	$\phi \rightarrow \psi \Rightarrow [C]\phi \rightarrow [C]\psi$
R3	$\phi \Rightarrow A\phi$
R4	$\phi \Rightarrow [C \downarrow \xi]\phi$
R5	$\phi \leftrightarrow \psi \Rightarrow [C \downarrow \xi]\chi \leftrightarrow [C \downarrow \xi]\chi[\phi/\psi]$

the effectivity function of the coalitions opposing the one that moved, and the **locality of the update**, that leaves the coalitional power at different worlds untouched.

The following relevant fact can be easily verified:

Proposition 1. *For every C,w, $\psi^M \in E(w)(C)$, we have that $E \downarrow_{(C,\psi^M,w)}$ is outcome monotonic, regular and closed-world.*

The proposition represents the basis for our reduction results. Whatever update is carried out a model is obtained that obeys the properties that have been assumed for coalition models.

Even though the interpretation of the update operator may look complex, its structural behaviour is rather simple. The validities in Table 2 allow to translate every sentence where the operator is occurring to a sentence where the operator is not occurring, provided an appropriate law for substitution of equivalent formulas (as $R5$ in the Table). Resemblance to Public Announcement Logic is no coincidence. The axioms reduce in fact the update operator to the global modality and the coalition logic operator. So the operator adds no expressivity to the language and completeness of the language with the update operator follows from the completeness of the language without it. A completeness proof

Table 3. Proof System for Public Announcement Logic

	Axioms
	Public Announcement Axioms
A1	$[\phi]p \leftrightarrow (\phi \to p)$
A2	$[\phi]\neg\psi \leftrightarrow (\phi \to \neg[\phi]\psi)$
A3	$[\phi](\xi \wedge \psi) \leftrightarrow ([\phi]\xi \wedge [\phi]\psi)$
A4	$[\phi]\Box_a\psi \leftrightarrow (\phi \to \Box_a[\phi]\psi)$
	Rules
R1	$\xi \wedge (\xi \to \psi) \Rightarrow \psi$
R2	$\xi \Rightarrow [\phi]\xi$

for Closed-World coalition logic, where the global modality interacts with the coalition logic modality by means of the axiom $[\emptyset]\phi \leftrightarrow A\phi$ is provided in [5].

3.1 Back to the Game

With the new operator it becomes possible to formalize the conditional aspect of strategic reasoning. In the structure PD we have that $PD, w \models [\{i\} \downarrow$ i defects $]([\{j\}]($ j defects and i defects $) \wedge [\{j\}]($ j cooperates and i defects $))$. Nothing changes at the level of grand coalition, since $PD \models [\emptyset \downarrow \phi][Agt]\psi \leftrightarrow [Agt]\psi$.

4 Discussion: Choices as Announcements

Public Announcement Logic formalizes the effect of the announcement of a true formula in each agent's a epistemic relation $R(a)$, defined as a partition on a domain W. The standard operator $[\phi]\psi$ says that ψ holds after ϕ is announced. Its semantics is given as follows:

$$M, w \models [\phi]\psi \Leftrightarrow M, w \models \phi \text{ implies } M|\phi, w \models \psi$$

where $M|\phi = (W', R'(a), V')$ takes these values:

- $W' = \phi^M$
- $R'(a) = R(a) \cap (W \times \phi^M)$
- $V'(p) = V(p) \cap \phi^M$

The model restriction of public announcement *throws worlds away*. In fact, as shown for instance in [13], public announcements can be defined by only updating the epistemic relation. A reduction can be shown in which every sentence from the modal language with the $S5$ knowledge relation and the public announcement operator can be translated into a sentence from the same language without the public announcement operator occurring in it. We report the reduction axioms in Table 3.

If we compare the public announcement operator to the subgame operator, we can observe the structure of the two axiom systems is very similar in the atomic and boolean case, but very different in the modal case. A subtle difference can be though observed in the atomic clause. If Public Announcement Logic reduces the atomic announcement to an implication between atoms ($[q]p \leftrightarrow (q \rightarrow p)$), the subgame operator reduces it to an implication between an atom and a choice ($[C \downarrow q]p \leftrightarrow ([C]q \rightarrow p)$). This fact witnesses that we are really reducing strategy execution to strategic ability. The appendix will make it clear that the similarity of the logics applies to the proof techniques as well, that are at least for the basic cases identical to those of Public Announcement Logic [15]. The specific differences are given, once again, by the way the coalitional relation is updated.

5 Conclusion and Future Work

We have built a logic for strategic ability update, where we can represent the effects of a coalitional choice on the players' strategic ability, extending the *update paradigm* of Dynamic Epistemic Logic to account for the dynamics of strategic ability in Coalition Structures. Our framework explicitly expresses how a coalitional move modifies the ability of all the players involved in the interaction, providing a useful framework for capturing coalitional reasoning in strategic settings. Our results are limited to Coalition Logic. Further study is needed to analyze whether the same characterizations are possible in different frameworks for strategic ability, for instance the Consequentialist-STIT framework, ATL and the full Game Logic. Further work can also be done in characterizing within this framework a number of other gametheoretical concepts like Nash Equilibrium and the Core for Cooperative Games without transferable utility.

References

1. Abdou, J., Keiding, H.: Effectivity Functions in Social Choice. Kluwer Academic Publishers, Dordrecht (1991)
2. Aumann, R.J., Peleg, B.: Von Neumann-Morgenstern solutions to cooperative games without side payments. Bulletin of the American Mathematical Society 66, 173–179 (1960)
3. Blackburn, P., de Rijke, M., Venema, Y.: Modal Logic. Cambridge Tracts in Theoretical Computer Science (2001)
4. Borgo, S.: Coalitions in action logic. In: IJCAI, pp. 1822–1827 (2007)
5. Broersen, J., Mastop, R., Meyer, J.-J.C., Turrini, P.: A logic for closed-world interaction. In: Hölldobler, S., Lutz, C., Wansing, H. (eds.) JELIA 2008. LNCS (LNAI), vol. 5293, pp. 89–99. Springer, Heidelberg (2008)
6. Chatterjee, K., Henzinger, T., Piterman, N.: Strategy logic. Technical Report UCB/EECS-2007-78, University of California, Berkeley (May 2007)
7. Goranko, V., van Drimmelen, G.: Complete axiomatization and decidability of alternating-time temporal logic. Theor. Comput. Sci. 353(1-3), 93–117 (2006)
8. Osborne, M., Rubinstein, A.: A course in Game Theory. MIT Press, Cambridge (1994)

9. Parikh, R.: The logic of games and its applications. In: Selected papers of the international conference on "foundations of computation theory" on Topics in the theory of computation, New York, NY, USA, pp. 111–139. Elsevier North-Holland, Inc., Amsterdam (1985)
10. Pauly, M.: Logic for Social Software. ILLC Dissertation Series (2001)
11. van Benthem, J.: In praise of strategies. Research Report (2007), http://www.illc.uva.nl/Publications/ResearchReports/PP-2008-03.text.pdf
12. van Benthem, J.: Where is logic going, and should it? Topoi 25, 117–122 (2006)
13. van Benthem, J., Liu, F.: Dynamic logic of preference upgrade. Journal of Applied Non-Classical Logics 14 (2004)
14. van der Hoek, W., Jamroga, W., Wooldridge, M.: A logic for strategic reasoning. In: AAMAS 2005: Proceedings of the fourth international joint conference on Autonomous agents and multiagent systems, pp. 157–164. ACM, New York (2005)
15. van Ditmarsch, H., van der Hoek, W., Kooi, B.: Dynamic Epistemic Logic. Synthese Library (2007)

A   Proofs for Reduction Axioms

Atomic and Boolean Cases

$$[C \downarrow \xi]p \leftrightarrow ([C]\xi \rightarrow p)$$

Take arbitrary M, w. $M, w \models [C \downarrow \xi]p \Leftrightarrow M, w \models [C]\xi$ implies that $M \downarrow_{(C,\xi^M,w)}, w \models p \Leftrightarrow M, w \models [C]\xi$ implies that $M, w \models p \Leftrightarrow M, w \models [C]\xi \rightarrow p$. Q.E.D.

$$[C \downarrow \xi]\neg\phi \leftrightarrow ([C]\xi \rightarrow \neg[C \downarrow \xi]\phi)$$

Take arbitrary M, w. $M, w \models [C \downarrow \xi]\neg\phi \Leftrightarrow M, w \models [C]\xi$ implies that $M \downarrow_{(C,\xi^M,w)}, w \models \neg\phi \Leftrightarrow M, w \models [C]\xi$ implies that $(M, w \models [C]\xi$ and $M \downarrow_{(C,\xi^M,w)}, w \models \neg\phi) \Leftrightarrow M, w \models [C]\xi$ implies that not$(M, w \models [C]\xi$ implies $M \downarrow_{(C,\xi^M,w)}, w \not\models \neg\phi) \Leftrightarrow \models [C]\xi$ implies that not$(M, w \models [C]\xi$ implies $M \downarrow_{(C,\xi^M,w)}, w \models \phi) \Leftrightarrow M, w \models [C]\xi$ implies that $M, w \not\models [C \downarrow \xi]\phi \Leftrightarrow M, w \models [C]\xi \rightarrow \neg[C \downarrow \xi]\phi$ Q.E.D.

$$[C \downarrow \xi](\phi \wedge \psi) \leftrightarrow ([C \downarrow \xi]\phi \wedge [C \downarrow \xi]\psi)$$

Take arbitrary M, w. $M, w \models [C \downarrow \xi](\phi \wedge \psi) \Leftrightarrow M, w \models [C]\xi$ implies that $M \downarrow_{(C,\xi^M,w)}, w \models \phi \wedge \psi \Leftrightarrow M, w \models [C]\xi$ implies that $(M \downarrow_{(C,\xi^M,w)}, w \models \phi$ and $M \downarrow_{(C,\xi^M,w)}, w \models \psi) \Leftrightarrow (M, w \models [C]\xi$ implies that $M \downarrow_{(C,\xi^M,w)}, w \models \phi)$ and $(M, w \models [C]\xi$ implies that $M \downarrow_{(C,\xi^M,w)}, w \models \psi) \Leftrightarrow (M, w \models [C \downarrow \xi]\phi)$ and $(M, w \models [C \downarrow \xi]\psi) \Leftrightarrow M, w \models ([C \downarrow \xi]\phi \wedge [C \downarrow \xi]\psi)$ Q.E.D.

Interaction with Global Modality

$$[C \downarrow \xi]A\phi \leftrightarrow ([C]\xi \rightarrow A\phi)$$

Take an arbitrary M, w. $M, w \models [C \downarrow \xi]A\phi \Leftrightarrow M, w \models [C]\xi$ implies that $M \downarrow_{(C,\xi^M,w)}, w \models A\phi \Leftrightarrow M, w \models [C]\xi$ implies that $M \downarrow_{(C,\xi^M,w)}, w \models [\emptyset]\phi \Leftrightarrow M, w \models [C]\xi$ implies that $M, w \models [\emptyset]\phi \Leftrightarrow M, w \models [C]\xi$ implies that $M, w \models A\phi \Leftrightarrow M, w \models [C]\xi \rightarrow A\phi$

Interaction with Coalition Modality

$$[C \downarrow \xi][D]\phi \leftrightarrow ([C]\xi \rightarrow [D](\xi \rightarrow \phi))(\text{ for } D \cap C = \emptyset \text{ and } D \neq \emptyset)$$

Proof by contraposition.

$\Leftarrow$: Suppose, for some $D \neq \emptyset$, that $[C]\chi \rightarrow [D](\chi \rightarrow \phi)$ and $M, w \not\models [C \downarrow \chi][D]\phi$ for some C such that $(C \cap D) = \emptyset$. The semantic clauses then tell us that (if $\chi^M \in E(w)(C)$ then $(\chi \rightarrow \phi)^M \in E(w)(D)$) and $\chi^M \in E(w)(C)$ and $\phi^M \notin E'(w)(D)$. [I write E' for $E \downarrow_{(C,\chi^M)}$.] By modus ponens $\phi^M \notin E'(w)(D)$.

By the definition of update, $E'(w)(D) = (E(w)(D) \sqcap \chi^M)^{\text{sup}}$. So, $((\chi \rightarrow \phi)^M \cap \chi^M) \in E'(w)(D)$. By elementary set theory this just says that $\phi^M \in E'(w)(D)$. Contradiction.

$\Rightarrow$: Suppose, for some $D \neq \emptyset$, that $M, w \models [C \downarrow \chi][D]\phi$ and $M, w \not\models [C]\chi \rightarrow [D](\chi \rightarrow \phi)$ for some C such that $(C \cap D) = \emptyset$. The semantic clauses then tell us that (if $\chi^M \in E(w)(C)$ then $\phi^M \in E'(w)(D)$) and $\chi^M \in E(w)(C)$ and $(\chi \rightarrow \phi)^M \notin E(w)(D)$. By modus ponens we are assuming that $\phi^M \in E'(w)(D)$ and $(\chi \rightarrow \phi)^M \notin E(w)(D)$.

By the definition of update, $E'(w)(D) = (E(w)(D) \sqcap \chi^M)^{\text{sup}}$. Because $\phi^M \in E'(w)(D)$, there must be some $X \in E(w)(D)$, such that $(X \cap \chi^M) \subseteq \phi^M$. By elementary set theory, it must be the case that $X \subseteq (\chi \rightarrow \phi)^M$.

Hence, by outcome monotonicity of E, if $X \in E(w)(D)$, then $(\chi \rightarrow \phi)^M \in E(w)(D)$. Contradiction.

$$[C \downarrow \xi]([D]\phi \leftrightarrow A(\xi \rightarrow \phi))(\text{ for } D \cap C \neq \emptyset)$$

Proof. Take arbitrary M, w, and arbitrary $\xi^M \in E(w)(C)$. Consider a coalition D with $D \cap C \neq \emptyset$. We have that $E \downarrow_{(C,\xi^M,w)} (w)(D) = (\xi^M)^{\text{sup}}$ by semantics. This means that $\xi^M \subseteq \phi^M$ iff $\phi^M \in E \downarrow_{(C,\xi^M,w)} (w)(D)$. It is easy to conclude that $M, w \models [C \downarrow \xi]([D]\phi \leftrightarrow A(\xi \rightarrow \phi))$. Notice that this also means $M, w \models [C \downarrow \xi][D]\phi \leftrightarrow A(\xi \rightarrow \phi)$. Q.E.D.

$$[C \downarrow \xi][D]\phi \leftrightarrow ([C]\xi \rightarrow [D]\phi)(\text{ for } D = \emptyset)$$

It follows directly from the semantics of the update operator for the case of $D = \emptyset$. Q.E.D.

Strategy Elimination in Games with Interaction Structures[*]

Andreas Witzel[1,2], Krzysztof R. Apt[1,2], and Jonathan A. Zvesper[1,2]

[1] University of Amsterdam, Science Park 904, 1098XH Amsterdam
[2] CWI, Kruislaan 413, 1098SJ Amsterdam, The Netherlands

Abstract. We study games in the presence of an interaction structure, which allows players to communicate their preferences, assuming that each player initially only knows his own preferences. We study the outcomes of iterated elimination of strictly dominated strategies (IESDS) that can be obtained in any given state of communication.

We also give epistemic foundations for these "intermediate" IESDS outcomes. This involves firstly describing the knowledge that the players would have in any state of communication, using the framework from Apt et al. [3]. We then prove that when there is common knowledge of rationality, each intermediate outcome is entailed by the knowledge in the relevant state of communication.

1 Introduction

1.1 Background and Motivation

There is a substantial amount of research within game theory on the implications of assumptions concerning players' *knowledge* and *beliefs* [5]. In particular, Tan and Werlang [16] have shown that if payoffs are commonly known and all players are *rational* and commonly believe in each other's rationality, they will only play strategies that survive iterated elimination of strictly dominated strategies (IESDS). In this context rationality means that one does not choose strictly dominated strategies.

Another line of research stresses the relevance of *locality* in strategic games. For example, in *graphical games* [14] the locality assumption is formalized by assuming a graph structure over the set of players and using payoff functions which depend only on the strategies of players' neighbors.

In this paper we study a game-theoretic framework which combines *locality* and *interaction*. The locality assumption refers to the *information* about payoffs (or more generally, preferences), rather than to the payoffs themselves. In turn, interaction takes place by means of *communication* within (possibly overlapping) groups of players. The framework is realized by incorporating the notion of a strategic game into the setting of interaction structures discussed in [3].

---

[*] Proofs are omitted for space reasons. The full version is available at http://arxiv.org/abs/0908.2399v1

X. He, J. Horty, and E. Pacuit (Eds.): LORI 2009, LNAI 5834, pp. 302–315, 2009.
© Springer-Verlag Berlin Heidelberg 2009

An *interaction structure* consists of (possibly overlapping) groups of players within which synchronous communication is possible. We assume that players' preferences are *not* commonly known. Instead, the initial information of each player only covers *his own* preferences, and the players can communicate this information only within the limits of the interaction structure.

More precisely, we make the following assumptions:

- the players initially know their own preferences;
- they are rational;
- they are part of an interaction structure and can communicate their own preferences within any group they belong to;
- communication is truthful and synchronous, as in [3];
- the players have no knowledge other than what follows from these assumptions, and this is common knowledge.

In this setting we then study the outcome of iterated elimination of strictly dominated strategies started in some intermediate state of communication, in particular in the state in which all communication permitted by the interaction structure has taken place. We use the results from our previous work [3] to prove that this outcome can be described by analyzing what the players know in the considered state.

It is important to note that we do *not* examine strategic or normative aspects of the *communication* here. So we do not allow players to lie and do not examine *why* they communicate or *what* they should communicate. Rather, we examine what happens *if* they do communicate, assuming that they are rational and have reasoning powers.

To justify this focus, we can think of a setting in which the strategic aspects of communication are not relevant. One possibility is when communication is not a deliberate act, but rather occurs through observing somebody's behavior. Such communication is certainly more difficult to manipulate and more laborious to fake than mere words. In a sense it is inherently credible, and research in social learning argues along similar lines [8, Ch. 3].

This also helps to explain another assumption we make, corresponding to the framework we examined in [3]: players only communicate their *own* preferences, since information about *others'* preferences is either difficult to obtain or communication about them is not credible. One may also assume that communicating about preferences of third parties is less common for *privacy* reasons. From this perspective the groups of the interaction structure can be viewed as the ones who can commonly observe each other, for example colleagues sharing lunch at work.

In other settings, for example that of artificial agents communicating by means of messages, it may be more difficult to view communication as something non-deliberate. Here, ignoring strategic aspects of communication can be interpreted as bounds on the players' rationality or reasoning capabilities—they simply lack the capabilities to deal with all the consequences of such an inherently rich phenomenon as communication.

In general, strategic communication is a research topic on its own, with controversial discussions (see, e.g., [15]) and many questions widely open. Crawford and Sobel [10] have considered the topic in a probabilistic setting, and Farrell and Rabin [12] have looked at related issues under the notion of *cheap talk*. Also within epistemic logic, formalizations of the information content of strategic communication have been suggested, e.g., by Gerbrandy [13].

Finally, it is useful to clarify the relation between strategic games with interaction structures and *pre-Bayesian games*, see, e.g., Ashlagi et al. [4]. In these games, too, each player knows his payoff but does not know the payoffs of the other players and makes no assumptions about them. In our setup this private knowledge aspect of pre-Bayesian games can be trivially modelled by the empty interaction structure, or viewed as corresponding to our initial situation. Due to the different nature of these frameworks, however, the questions of interest are also different.

1.2   Plan of the Paper

This paper is organized as follows. In the following Sect. 2, we review the basic definitions concerning strategic games, optimality notions and operators on restrictions of games. Next, in Sect. 3, we study the outcome of IESDS in the presence of an interaction structure. We first look at the outcome that is arrived at after all communication permitted in the given interaction structure has taken place, and then detail the outcomes obtained in any particular intermediate state of communication. The formulations we consider make no direct use of the notion of knowledge. The connection with knowledge is made in 4, where we prove the outcomes we have obtained to be correct with respect to the epistemic framework from [3], in the sense that the outcomes capture exactly what the players can do given their partial knowledge of the game structure in any particular state. Finally, in Sect. 5, we suggest some future research directions.

2   Preliminaries

Following [2], by a **strategic game with parametrized preferences** (in short, a **game**) for players $N = \{1, \ldots, n\}$, where $n > 1$, we mean a tuple $(S_1, \ldots, S_n, \succ_1, \ldots, \succ_n)$, where for each $i \in N$,

- S_i is the non-empty, finite set of **strategies** available to player i. We write S to abbreviate the set of **strategy profiles**: $S = S_1 \times \cdots \times S_n$.
- $\succ_i$ is the strict **preference relation** for player i, so $\succ_i \subseteq S \times S$.

This qualitative approach precludes the use of mixed strategies, but they will not be needed in our considerations.

As usual we denote player i's strategy in a strategy profile $s \in S$ by s_i, and the tuple consisting of all other strategies by s_{-i}, i.e., $s_{-i} = (s_1, \ldots, s_{i-1}, s_{i+1}, \ldots, s_n)$. Similarly, we use S_{-i} to denote $S_1 \times \cdots \times S_{i-1} \times S_{i+1} \times \cdots \times S_n$, and for $s_i' \in S_i$ and $s_{-i} \in S_{-i}$ we write (s_i', s_{-i}) to denote $(s_1, \ldots, s_{i-1}, s_i', s_{i+1}, \ldots, s_n)$. Finally, we use $s_i' \succ_{s_{-i}} s_i$ as a notational alternative for $(s_i', s_{-i}) \succ_i (s_i, s_{-i})$.

Fix now an *initial* strategic game $\mathcal{G} := (S_1, \ldots, S_n, \succ_1, \ldots, \succ_n)$. We say that $(S'_1, \ldots, S'_n)$ is a **restriction** of $\mathcal{G}$ if each S'_i is a non-empty subset of S_i. We identify the restriction $(S_1, \ldots, S_n)$ with $\mathcal{G}$.

To analyze iterated elimination of strategies from the initial game $\mathcal{G}$, we view such procedures as operators on the set of restrictions of $\mathcal{G}$. This set together with component-wise set inclusion forms a complete lattice.

For any restriction $\mathcal{G}' := (S'_1, \ldots, S'_n)$ of $\mathcal{G}$ and strategies $s_i, s'_i \in S_i$, we say that s_i is **strictly dominated by** s'_i **on** S'_{-i} if $s'_i \succ_{s'_{-i}} s_i$ for all $s'_{-i} \in S'_{-i}$. Then we introduce the following abbreviations (ℓ stands for "local" and g stands for "global"; the terminology is from Apt [1]):

- $sd^\ell(s_i, \mathcal{G}')$ which holds iff strategy s_i of player i is not strictly dominated on S'_{-i} by any strategy from S'_i (i.e., $\neg \exists s'_i \in S'_i \, \forall s'_{-i} \in S'_{-i} \, s'_i \succ_{s'_{-i}} s_i$),
- $sd^g(s_i, \mathcal{G}')$ which holds iff strategy s_i of player i is not strictly dominated on S'_{-i} by any strategy from S_i (i.e., $\neg \exists s'_i \in S_i \, \forall s'_{-i} \in S'_{-i} \, s'_i \succ_{s'_{-i}} s_i$).

So in sd^g, the global version of strict dominance introduced by [9], it is stipulated that a strategy is not strictly dominated by a strategy *from the initial game*.

We call each relation of the form sd^ℓ or sd^g an **optimality notion**. We say then that the optimality notion ϕ used by player i is **monotonic** if for all restrictions $\mathcal{G}''$ and $\mathcal{G}'$ and strategies s_i, $\mathcal{G}'' \subseteq \mathcal{G}'$ and $\phi(s_i, \mathcal{G}'')$ implies $\phi(s_i, \mathcal{G}')$.

As noted in [1, 7], sd^g is monotonic, while sd^ℓ is not (though in finite games their respective outcomes coincide, as discussed in the proof of Theorem 1).

Given an operator T on a finite lattice $(D, \subseteq)$ with the largest element $\top$ and $k \geq 0$, we denote by T^k the k-fold iteration of T, where $T^0 = \top$ (so the iterations start "at the top") and put $T^\infty := \bigcap_{k \geq 0} T^k$. We call T **monotonic** if for all D', D'', we have that $D' \subseteq D''$ implies $T(D') \subseteq T(D'')$.

Finally, as in [3], an **interaction structure** H is a *hypergraph* on N, i.e., a set of non-empty subsets of $A \subseteq N$, called *hyperarcs*.

3 Iterated Strategy Elimination

In this section we define procedures for iterated elimination of strictly dominated strategies. Let us fix a strategic game $\mathcal{G} = (S_1, \ldots, S_n, \succ_1, \ldots, \succ_n)$ for players N, an interaction structure $H \subseteq 2^N \setminus \{\emptyset\}$, and an optimality notion ϕ. In Sect. 3.1, we look at the outcome reached after all communication permitted by H has taken place, that is, when within each hyperarc of H all of its members' preferences have been communicated. In Sect. 3.2, we then look at the outcomes obtained in any particular intermediate state of communication. We stress that in general there is no relation between the preferences $\succ_i$ and H.

The formulations we give here make no direct use of a formal notion of knowledge. The connection with a formal epistemic model is made in Sect. 4.

All iterations of the considered operators start at the initial restriction $(S_1, \ldots, S_n)$.

3.1   Completed Communication

Let us assume that within each hyperarc $A \in H$, all its members have shared all information about their preferences. We leave the exact definition of communication to Sect. 3.2 and the epistemic formalization to Sect. 4, and focus here on an operational description.

For each group of players $G \in N$, let S_G denote the set of those restrictions of $\mathcal{G}$ which only restrict the strategy sets of players from G. That is,

$$S_G := \{(S'_1, \ldots, S'_n) \mid S'_i \subseteq S_i \text{ for } i \in G \text{ and } S'_i = S_i \text{ for } i \notin G\}.$$

Now we introduce an elimination operator T_G on each such set S_G, defined as follows. For each $\mathcal{G}' = (S'_1, \ldots, S'_n) \in S_G$, let $T_G(\mathcal{G}') := (S''_1, \ldots, S''_n)$, where for all $i \in N$,

$$S''_i := \begin{cases} \{s_i \in S'_i \mid \phi(s_i, \mathcal{G}')\} & \text{if } i \in G \\ S'_i & \text{otherwise.} \end{cases}$$

We call T_G^∞ the **outcome of iterated elimination (of non-ϕ-optimal strategies) on** G. We then define the restriction $\mathcal{G}(H)$ of $\mathcal{G}$ as[1] $\mathcal{G}(H) := (\mathcal{G}(H)_1, \ldots, \mathcal{G}(H)_n)$, where for all $i \in N$,

$$\mathcal{G}(H)_i := T_{\{i\}}\left(\bigcap_{A: i \in A \in H} T_A^\infty\right)_i.$$

That is, the ith component of $\mathcal{G}(H)$ is the ith component of the result of applying $T_{\{i\}}$ to the intersection of T_A^∞ for all $A \in H$ containing i. We call $\mathcal{G}(H)$ the **outcome of iterated elimination (of non-ϕ-optimal strategies) with respect to** H. Note that $\mathcal{G}(H)$ implicitly depends on ϕ.

Let us "walk through" this definition to understand it better. Given a player i and a hyperarc $A \in H$ such that $i \in A$, T_A^∞ is the outcome of iterated elimination on A, starting at $(S_1, \ldots, S_n)$. The strategies of players from outside of A are not affected by this process. This elimination process is performed simultaneously for each hyperarc that i is a member of. By intersecting the outcomes, i.e., by considering the restriction $\bigcap_{A: i \in A \in H} T_A^\infty$, one arrives at a restriction in which all such "groupwise" iterated eliminations have taken place. However, in this restriction some of the strategies of player i may be non-ϕ-optimal. They are eliminated using one application of the $T_{\{i\}}$ operator. We illustrate this process, and in particular this last step, in the following example.

Example 1. Consider local strict dominance, sd^ℓ, in the following three-player game $\mathcal{G}$ where the payoffs of players 1 and 2 and those of players 1 and 3 respectively depend on each other's actions, but the payoffs of player 2 and 3 are independent:

		Pl. 2, 3			
		L, l	L, r	R, l	R, r
Pl. 1	U	1, 1, 1	0, 1, 0	0, 0, 1	0, 0, 0
	D	0, 1, 1	1, 1, 0	1, 0, 1	1, 0, 0

---

[1] Here and elsewhere the outer subscript 'i' refers to the preceding restriction.

Fig. 1. Illustrating Example 1. Hyperarcs are shown in gray. Callouts attached to hyperarcs represent communicated, and thus commonly known, information. The thought bubble represents private information, in this case obtained from the combination of information only available to player 1.

So, for example, the payoffs for the strategy profile (U, L, r) are, respectively, 0, 1, and 0. Now assume the interaction structure $H = \{\{1, 2\}, \{1, 3\}\}$. We obtain $T^{\infty}_{\{1,2\}} = (\{U, D\}, \{L\}, \{l, r\})$ and $T^{\infty}_{\{1,3\}} = (\{U, D\}, \{L, R\}, \{l\})$. The restriction defined by these two outcomes is $(\{U, D\}, \{L\}, \{l\})$, and in the final step player 1 eliminates his strategy D by one application of $T_{\{1\}}$. The outcome of the whole process is thus $\mathcal{G}(H) = (\{U\}, \{L\}, \{l\})$. See Fig. 1 for an illustration of this situation. □

In this example, the outcome with respect to the given interaction structure coincides with the outcome of the customary IESDS on the fully specified game matrix. We should emphasize that this is not the case in general, and the purpose of this example is simply to illustrate how the operators work. Example 2 later on shows in a different setting how the interaction structure can influence the outcome.

Note that when H consists of the single hyperarc N that contains all the players, then for each player i, $\bigcap_{A:i \in A \in H} T^{\infty}_A$ reduces to T^{∞}_N, and this is closed under application of each operator $T_{\{i\}}$. So then, indeed, $\mathcal{G}(H) = T^{\infty}_N$, that is, $\mathcal{G}(H)$ in this special case coincides with the customary outcome of iterated elimination of non-ϕ-optimal strategies.

In general, this customary outcome is included in the outcome w.r.t. any hypergraph H. This result is established in Theorem 1, and Example 2 shows a case where the inclusion is proper.

Theorem 1. *For $\phi \in \{sd^{\ell}, sd^g\}$ and for all hypergraphs H, we have $T^{\infty}_N \subseteq \mathcal{G}(H)$.*

The inclusion proved in this result cannot be reversed, even when each pair of players shares a hyperarc. The following example also shows that the hypergraph structure is more informative than the corresponding graph structure.

Example 2. Consider the following strategic game with three players. The payoffs of player 1 and 2 depend here only on each other's choices, and the payoffs of player 3 depend only on the choices of player 2 and 3:

<table>
<tr><td></td><td colspan="3">Pl. 2</td></tr>
<tr><td></td><td></td><td>L</td><td>R</td></tr>
<tr><td rowspan="2">Pl. 1</td><td>U</td><td>0,1</td><td>0,0</td></tr>
<tr><td>D</td><td>1,0</td><td>1,1</td></tr>
</table>

<table>
<tr><td></td><td colspan="2">Pl. 2</td></tr>
<tr><td></td><td>L</td><td>R</td></tr>
<tr><td rowspan="2">Pl. 3</td><td></td><td></td></tr>
</table>

Payoff of players 1 and 2   Payoff of player 3

So, for example, the payoffs for the strategy profile (U, L, A) are, respectively, 0, 1, and 0. If we assume the hypergraph H that consists of the single hyper-arc $\{1, 2, 3\}$, then the outcome of iterated elimination of non-ϕ-optimal strategies w.r.t. H is the customary outcome which equals $(\{D\}, \{R\}, \{A\})$. Indeed, player 1 can eliminate his strictly dominated strategy U, then player 2 can eliminate L, and subsequently player 3 can eliminate B.

In contrast, if the hypergraph consists of all pairs of players, so $H = \{\{1, 2\}, \{2, 3\}, \{1, 3\}\}$, then the outcome of iterated elimination of non-ϕ-optimal strategies w.r.t. H equals $(\{D\}, \{R\}, \{A, B\})$.

Informally, the reason for this difference is that in the latter case, player 3 can eliminate B only using the fact that player 2 eliminated L, but this information is available only to players 1 and 2.   □

To familiarize ourselves further with our definitions, we establish the following intuitive monotonicity result. We say that H' **extends** H if for each $A \in H$ there is $A' \in H'$ such that $A \subseteq A'$.

Proposition 1. *If H' extends H and T is monotonic, then $\mathcal{G}(H') \subseteq \mathcal{G}(H)$.*

3.2   Intermediate States

The setting considered in Sect. 3.1 corresponds to a state in which in all hyperarcs all players have shared all information about their preferences. Given the game $\mathcal{G}$ and the hypergraph H, the outcome $\mathcal{G}(H)$ there defined thus reflects which strategies players can eliminate if initially they know only their own preferences and they communicate all their preferences in H. We now define formally what communication we assume possible, and then look at intermediate states, where only certain preferences have been communicated.

Each player i can communicate his preferences to each $A \in H$ with $i \in A$. We take a **message** by i to consist of a preference statement $s'_i \succ_{s_{-i}} s_i$ for $s_i, s'_i \in S_i$ and $s_{-i} \in S_{-i}$. We denote such a message by $(i, A, s'_i \succ_{s_{-i}} s_i)$, and require that $i \in A$ and that it is **truthful** with respect to the given initial game $\mathcal{G}$, that is, indeed $s'_i \succ_{s_{-i}} s_i$ in $\mathcal{G}$. Note that the fact that i is the sender is, strictly speaking, never used. Thus, in accordance with the interpretation of communication described in Sect. 1.1, we could drop the sender and simply write "the players in A commonly observe that $s'_i \succ_{s_{-i}} s_i$." An **intermediate state** is now given by the set M of messages which have been communicated.

We now adjust the definition of an optimality notion to account for intermediate states. An **intermediate optimality notion** $\phi_{G,M}$ (derived from an optimality notion ϕ) uses only information shared among the group G in the intermediate state given by M. So with singleton $G = \{i\}$ only i's preferences are used, and with larger G only preferences contained in messages to a superset of G are used. Thus in the case of sd^g we have that $sd^g_{G,M}(s_i, \mathcal{G}')$ holds iff

$$\neg \exists s'_i \in S_i \; \forall s_{-i} \in S'_{-i} \; s'_i \succ_{s_{-i}} s_i \qquad \text{if } G = \{i\}$$
$$\neg \exists s'_i \in S_i \; \forall s_{-i} \in S'_{-i} \; M \restriction_G \models s'_i \succ_{s_{-i}} s_i \qquad \text{otherwise,}$$

where by $M \upharpoonright_G \models s_i' \succ_{s_{-i}} s_i$ we mean that $s_i' \succ_{s_{-i}} s_i$ is entailed by those messages in M which G received. Specifically, the **entailment relation**

$$M \upharpoonright_G \models s_i' \succ_{s_{-i}} s_i$$

holds iff there exist messages $(\cdot, G^k, s_i^k \succ_{s_{-i}} s_i^{k+1}) \in M$ for $k \in \{1, \ldots, \ell - 1\}$ such that $G^k \supseteq G$, $s_i^1 = s_i'$ and $s_i^\ell = s_i$.

We now define a generalization of the T_G operator by:

$$T_{G,M}(\mathcal{G}') := (S_1'', \ldots, S_n''),$$

where $\mathcal{G}' = (S_1', \ldots, S_n')$ and for all $i \in N$,

$$S_i'' := \{s_i \in S_i' \mid \phi_{G,M}(s_i, \mathcal{G}')\}.$$

Note that, as before, S_i' remains unchanged for $i \notin G$, since then $\phi_{G,M}(s_i, \mathcal{G}')$ always holds. Indeed, for it to be false, there would have to be some message $(i, G, \cdot) \in M$, which would imply $i \in G$.

Similarly, we now define the **outcome of iterated elimination (of non-ϕ-optimal strategies) with respect to** H, M to be the restriction $\mathcal{G}(H, M)$, where for $i \in N$

$$\mathcal{G}(H, M)_i := T_{\{i\}, M} \left(\bigcap_{A : i \in A \in \overline{H}} T_{A,M}^\infty \right)_i .$$

Here $\overline{H}$ denotes the closure of H under non-empty intersection. That is,

$$\overline{H} = \{A_1 \cap \cdots \cap A_k \mid \{A_1, \ldots, A_k\} \subseteq H\} \setminus \{\emptyset\}.$$

The use of $\overline{H}$ is necessary because certain information may be entailed by messages sent to different hyperarcs. For example, with $(j, A, s_j'' \succ_{s_{-j}} s_j')$, $(j, A', s_j' \succ_{s_{-j}} s_j) \in M$, the combined information that $s_j'' \succ_{s_{-j}} s_j$ is available to the members of $A \cap A'$.

Again, let us "walk through" the definition of $\mathcal{G}(H, M)$. First, a local elimination process is run on each hyperarc of $\overline{H}$, using only information which has been communicated there (which now no longer covers all members' preferences, but only the ones according to the intermediate state M). Then, in the final step, each player combines his insights from all hyperarcs of which he is a member, and eliminates any strategies that he thereby learns not to be optimal.

It is easy to see that in the case where the players have communicated all there is to communicate, i.e., for

$$M_H^{\text{all}} := \{(i, A, s_i' \succ_{s_{-i}} s_i) \mid i \in N, A \in H, \ s_i, s_i' \in S_i \text{ with } s_i' \succ_{s_{-i}} s_i \text{ in } \mathcal{G}\},$$

the intermediate outcome coincides with the previously defined outcome, i.e.,

$$\mathcal{G}(H, M_H^{\text{all}}) = \mathcal{G}(H).$$

This corresponds to the intuition that $\mathcal{G}(H)$ captures the elimination process when all possible communication has taken place. In particular, all entailed information has also been communicated in M_H^{all}, which is why we did not need to consider $\overline{H}$ in 3.1.

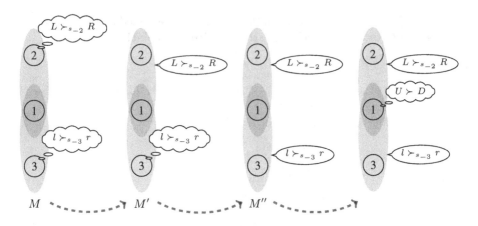

Fig. 2. Illustrating Example 3

Example 3. The process described in this example is illustrated in Fig. 2. Consider again the game $\mathcal{G}$ from Example 1, and the initial state where $M = \emptyset$. We have $T_{A,M}^{\infty} = \mathcal{G}$ for all $A \in \overline{H}$, that is, without communication no strategy can "commonly" be eliminated. However, players 2 and 3 can "privately" eliminate one of their strategies each, since each of them knows his own preferences, so

$$T_{\{1\},M}(\bigcap_{A:1\in A\in\overline{H}} T_{A,M}^{\infty}) = (\{U,D\},\{L,R\},\{l,r\}),$$

$$T_{\{2\},M}(\bigcap_{A:2\in A\in\overline{H}} T_{A,M}^{\infty}) = (\{U,D\},\{L\},\{l,r\}),$$

$$T_{\{3\},M}(\bigcap_{A:3\in A\in\overline{H}} T_{A,M}^{\infty}) = (\{U,D\},\{L,R\},\{l\}),$$

This elimination cannot be iterated upon by other players and the overall outcome is $\mathcal{G}(H,M) = (\{U,D\},\{L\},\{l\})$.

Consider now the intermediate state $M' = \{(2,\{1,2\},L \succ_{s_{-2}} R)|s_{-2} \in S_{-2}\}$, that is, a state in which player 2 has shared with player 1 the information that for any joint strategy of players 1 and 3, he prefers his strategy L over R. Then only the result of player 1 changes:

$$T_{\{1\},M'}(\bigcap_{A:1\in A\in\overline{H}} T_{A,M'}^{\infty}) = (\{U,D\},\{L\},\{l,r\}),$$

while the other results and the overall outcome remain the same. If additionally player 3 communicates all his information in the hyperarc he shares with player 1, that is, if the intermediate state is $M'' = M' \cup \{(3,\{1,3\},l \succ_{s_{-3}} r)|s_{-3} \in S_{-3}\}$, then player 1 can combine all the received information and obtain

$$T_{\{1\},M''}(\bigcap_{A:1\in A\in\overline{H}} T_{A,M''}^{\infty}) = (\{U\},\{L\},\{l\}).$$

This is also the overall outcome $\mathcal{G}(H,M'')$, coinciding with the outcome $\mathcal{G}(H, M_H^{\text{all}})$ where all possible information has been communicated.   □

Let us now illustrate the importance of using entailment in the intermediate optimality notions and $\overline{H}$ (rather than H) in the definition of $\mathcal{G}(H,M)$.

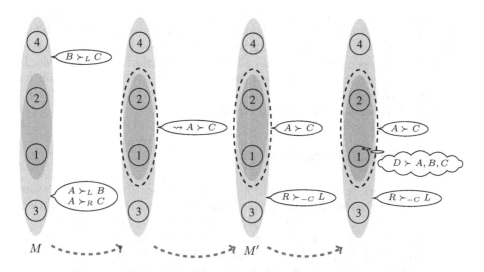

Fig. 3. Illustrating Example 4. Strategies of the dummy players are omitted. $A \succ C$ stands for $A \succ_{s_{-1}} C$, and $\succ_{-C}$ combines $\succ_\alpha$ for $\alpha \in \{A, B, D\}$. Note that in the first step, information is not explicitly communicated but deduced.

Example 4. We look at a game involving four players, but we are only interested in the preferences of two of them. The other two players serve merely to create different hyperarcs. The strategies and payoffs of player 1 and 2 are as follows:

		Pl. 2	
		L	R
	A	3,0	1,1
Pl. 1	B	2,0	1,1
	C	1,1	0,0
	D	0,0	5,1

For players 3 and 4 we assume a "dummy" strategy, denoted respectively by X and Y. Consider the hypergraph $H = \{\{1, 2, 3\}, \{1, 2, 4\}\}$ and the intermediate state

$$M = \{(1, \{1, 2, 3\}, A \succ_{LXY} B),$$
$$(1, \{1, 2, 4\}, B \succ_{LXY} C),$$
$$(1, \{1, 2, 3\}, A \succ_{RXY} C)\}.$$

The fact that player 1, independently of what the remaining players do, strictly prefers A over C is not explicit in these pieces of information, but it is *entailed* by them, since $A \succ_{LXY} B$ and $B \succ_{LXY} C$ imply $A \succ_{LXY} C$. However, this combination of information is only available to the players in $\{1, 2, 3\} \cap \{1, 2, 4\}$.

Player 2 can make use of this fact that C is dominated, and eliminate his own strategy L. If we now look at a state in which player 2 has communicated his relevant preferences, so $M' = M \cup \{(2, \{1, 2, 3\}, R \succ_{\alpha XY} L)|\alpha \in \{A, B, D\}\}$, we notice that player 1 can in turn eliminate A and B, but only by combining information available to the players in $\{1, 2, 3\} \cap \{1, 2, 4\}$. There is no single hyperarc in the original hypergraph which has all the required information available. It thus becomes clear that we need to take into account iterated elimination on intersections of hyperarcs.

The whole process is illustrated in Fig. 3. □

4   Epistemic Foundations

In this section, we provide epistemic foundations for our framework. The aim is to prove that the definition of the outcome $\mathcal{G}(H, M)$ correctly captures what strategies the players can eliminate using all they "know", in a formal sense.

We proceed as follows. First, in Sect. 4.1, we briefly introduce an epistemic model formalizing the players' knowledge. In Sect. 4.2, we give a general epistemic formulation of strict dominance and argue that it correctly captures the notion. Sect. 4.2 also contains the main result of our epistemic analysis, namely that the outcome $\mathcal{G}(H, M)$ indeed yields the outcome stipulated by the epistemic formulation. We rely on the basic framework and results from [3].

We focus on the global version of strict dominance, sd^g, mainly because the presentation is then more concise. However, our results carry over to the local version sd^ℓ due to the equivalence result mentioned in the proof of Theorem 1.

4.1   Epistemic Language and States

Again, we assume a fixed game $\mathcal{G}$ with non-empty set of strategies S_i for each player i, and a hypergraph H representing the interaction structure. Analogously to [3], we use a propositional **epistemic language** with a set At of **atoms** which is divided into disjoint subsets At_i, one for each player i, where $At_i = \{s'_i \succ_{s_{-i}} s_i \mid s_i, s'_i \in S_i, s_{-i} \in S_{-i}\}$.

The set At_i describes all possible strict preferences between pairs of strategies of player i, relative to a joint strategy of the opponents. We consider the usual **connectives** $\wedge$ and $\vee$ (but not the negation $\neg$), and a **common knowledge** operator C_G for any group $G \subseteq N$ of players. As in [3], we write K_i for $C_{\{i\}}$. By $\mathcal{L}^+$ we denote the set of formulas built from the atoms in At using these two connectives and knowledge operators.

A **valuation** V is a subset of At such that for each $s_{-i} \in S_{-i}$, the restriction $V \cap \{\cdot \succ_{s_{-i}} \cdot\}$ is a strict partial order.

Intuitively, a valuation consists of the atoms assumed true. Each specific game $\mathcal{G}$ *induces* exactly one valuation which simply represents its preferences. However, in general we also need to model the fact that players may not have full knowledge of the game. The restriction imposed on the valuations ensures that each of them is induced by some game.

So for example $\{s \succ_a t\}$ is a valuation (given a game with appropriate strategy sets), while $\{s \succ_a t, t \succ_a u\}$ and $\{s \succ_a t, t \succ_a s\}$ are not.

Recall from Sect. 3.2 that a **message** from player i to a hyperarc $A \in H$ has the form $(i, A, s_i' \succ_{s_{-i}} s_i)$, where $i \in A$, $s_i, s_i' \in S_i$, and $s_{-i} \in S_{-i}$. We say that a message $(\cdot, \cdot, p)$ is **truthful** with respect to a valuation V if $p \in V$. A **state**, or **possible world**, is a pair (V, M), where V is a valuation and M is a set of messages that are truthful with respect to V.

This setting is an instance of the framework defined in [3], and the formal **semantics** is as defined there. We repeat here only the intuition that $C_G \varphi$ means that φ is *common knowledge* among G, that is, everybody in G knows φ, everybody knows that everybody knows φ, etc. In particular, $K_i \varphi$ means that player i *knows* φ. We assume that each player i initially knows the true facts in At_i entailed by the initial game $\mathcal{G}$ and that the basic assumptions from Sect. 1.1 are commonly known among the players.

4.2 Correctness Result

We start by giving an epistemic formula describing the global version of iterated elimination of strictly dominated strategies. In contrast to the formulation in Sect. 2, this formula states *player i knows that* a strategy is strictly dominated.

We define, for $i \in N$ and $s_i \in S_i$,

$$dom^1(s_i) := K_i \bigvee_{s_i' \in S_i} \bigwedge_{s_{-i} \in S_{-i}} s_i' \succ_{s_{-i}} s_i,$$

$$dom^{\ell+1}(s_i) := K_i \bigvee_{s_i' \in S_i} \bigwedge_{s_{-i} \in S_{-i}} \left(s_i' \succ_{s_{-i}} s_i \vee \bigvee_{j \in N \setminus \{i\}} dom^\ell(s_j) \right).$$

That is, in the base case, player i knows that s_i is strictly dominated if i knows that there is an alternative strategy s_i' which, for all joint strategies of the other players, is strictly preferred. Furthermore, after iteration $\ell + 1$, i knows that s_i is strictly dominated if i knows that there is an alternative strategy s_i' such that, for all joint strategies s_{-i} of the other players, either s_i' is strictly preferred or some strategy s_j in s_{-i} is already known by player j to be strictly dominated after iteration ℓ.

We restrict our attention to formulas $dom^\ell(s_i)$ with $\ell \in \{1, \dots, \hat{\ell}\}$, where $\hat{\ell} = \sum_{i \in N} |S_i|$. By the semantics of the considered formulas, there is some ℓ within this range such that for all $\ell' \geq \ell$, $dom^{\ell'}$ is equivalent to dom^ℓ. To reflect the fact that this can be seen as the outcome of the iteration, we denote $dom^{\hat{\ell}}$ by dom^∞.

As a first connection with the T_G operator defined in Sect. 3, we have the following epistemic counterpart of Proposition 1. Intuitively, this is due to the fact that if we look at the states in which all communication allowed by a given hypergraph has taken place, then knowledge (of positive formulas) can only grow as that hypergraph grows.

Proposition 2. *If H' extends H, then for all $i \in N$ and $s_i \in S_i$,*

$$(V, M_H^{\mathrm{all}}) \vDash dom^\infty(s_i) \text{ implies } (V, M_{H'}^{\mathrm{all}}) \vDash dom^\infty(s_i),$$

where M^{all} is defined as in Sect. 3.2.

We now proceed to the main result of the paper. We prove that the non-epistemic formulation of iterated elimination of non-sd^g-optimal strategies, as given in Sect. 3, coincides with the epistemic formulation of strict dominance.

Theorem 2. *For any strategic game $\mathcal{G}$, hypergraph H, set of messages M truthful with respect to $\mathcal{G}$, and $i \in N$,*

$$\mathcal{G}(H, M)_i = \{s_i \in S_i \mid (V, M) \not\models dom^\infty(s_i)\},$$

where V is the valuation induced by $\mathcal{G}$.

5   Conclusions

We studied here strategic games in the presence of interaction structures. We assumed that initially the players know only their own preferences, and that they can truthfully communicate information about their own preferences within their parts of the interaction structure. This allowed us to analyze the consequences of locality, formalized by means of an interaction structure, on the outcome of the iterated elimination of strictly dominated strategies. To this end we appropriately adapted the framework introduced in [3] and showed that in any given state of communication this outcome can be described by means of epistemic analysis.

We plan to extend our analysis in a number of ways by:

- Allowing players to send information about the preferences of other players that they learned through interaction. The abstract epistemic framework of [3] includes already this extension,
- Allowing other forms of messages, for example, messages containing information that a strategy has been eliminated, or containing epistemic statements, such as knowing that some strategy of *another player* has been eliminated,
- Considering strategic aspects of communication, even if truthfulness is required (should one send some piece of information or not?)
- Considering formation or evolution of interaction structures, given strategic advantages of certain interaction structures over others.

The last point could connect our research with that on network formation games, see, e.g., [6].

Finally, let us mention that in [3] we already abstracted from the framework considered here and studied a setting in which players send messages that inform a group about some atomic fact that a player knows or has learned. We clarified there, among others, under what conditions common knowledge of the underlying hypergraph matters. The framework there considered could be generalized by allowing players to arrive jointly at some conclusions using their background theories, by means of an interaction through messages sent to groups. From this perspective IESDS could be seen as a metaphor of such a conclusion. Through its focus on the form of allowed messages and background knowledge, this study would differ from the line of research pursued by Fagin et al. [11], where the effects of communication are considered in the framework of distributed systems.

Acknowledgements

We thank Rohit Parikh and Willemien Kets for discussion and helpful suggestions. The first and third authors were supported by a GLoRiClass fellowship funded by the European Commission (Early Stage Research Training Mono-Host Fellowship MEST-CT-2005-020841).

References

[1] Apt, K.R.: The many faces of rationalizability. The B.E. Journal of Theoretical Economics 7(1), Article 18 (2007)

[2] Apt, K.R., Rossi, F., Venable, K.B.: Comparing the notions of optimality in CP-nets, strategic games and soft constraints. Annals of Mathematics and Artificial Intelligence 52(1), 25–54 (2008)

[3] Apt, K.R., Witzel, A., Zvesper, J.A.: Common knowledge in interaction structures. In: Proceedings of the 12th Conference on Theoretical Aspects of Rationality and Knowledge, TARK XII (2009)

[4] Ashlagi, I., Monderer, D., Tennenholtz, M.: Resource selection games with unknown number of players. In: Nakashima, H., Wellman, M.P., Weiss, G., Stone, P. (eds.) Proceedings of the Fifth International Joint Conference on Autonomous Agents and Multiagent Systems, pp. 819–825. ACM Press, New York (2006)

[5] Battigalli, P., Bonanno, G.: Recent results on belief, knowledge and the epistemic foundations of game theory. Research in Economics 53(2), 149–225 (1999)

[6] Bloch, F., Jackson, M.: Definitions of equilibrium in network formation games. International Journal of Game Theory 34(3), 305–318 (2006)

[7] Brandenburger, A., Friedenberg, A., Keisler, H.J.: Fixed points for strong and weak dominance, Working paper (2006),
     http://pages.stern.nyu.edu/abranden/

[8] Chamley, C.P.: Rational herds: Economic models of social learning. Cambridge University Press, Cambridge (2004)

[9] Chen, Y.-C., Van Long, N., Luo, X.: Iterated strict dominance in general games. Games and Economic Behavior 61(2), 299–315 (2007)

[10] Crawford, V.P., Sobel, J.: Strategic information transmission. Econometrica 50(6), 1431–1451 (1982)

[11] Fagin, R., Halpern, J.Y., Vardi, M.Y., Moses, Y.: Reasoning about knowledge. MIT Press, Cambridge (1995)

[12] Farrell, J., Rabin, M.: Cheap talk. The Journal of Economic Perspectives 10(3), 103–118 (1996)

[13] Gerbrandy, J.: Communication strategies in games. Journal of Applied Non-Classical Logics 17(2), 197–211 (2007)

[14] Kearns, M., Littman, M.L., Singh, S.: Graphical models for game theory. In: Breese, J.S., Koller, D. (eds.) Proceedings of the 17th Conference in Uncertainty in Artificial Intelligence, pp. 253–260. Morgan Kaufmann, San Francisco (2001)

[15] Sally, D.: Can I say "bobobo" and mean "There's no such thing as cheap talk"? Journal of Economic Behavior & Organization 57(3), 245–266 (2005)

[16] Tan, T.C.-C., da Costa Werlang, S.R.: The bayesian foundations of solution concepts of games. Journal of Economic Theory 45(2), 370–391 (1988)

The Logic of Knowledge-Based Cooperation in the Social Dilemma

Xiaohong Cui

Department of philosophy, Peking University, China
tsuixiaohong@yahoo.com.cn

Social dilemma is a situation in which individual rationality leads to collective irrationality. That is, individually reasonable behavior leads to a situation in which everyone is worse off than they might have been otherwise [Kol98]. There are four types of social dilemmas: prisoner's dilemmas, assurance games, chicken games, and coordination games. This paper aims to find a possible solution for Social dilemma. In this paper, we focus on a two-person social dilemma: Prisoner's Dilemma and propose a contract which is negotiated by authorities to solve this problem. The contract says that the cooperators can be rewarded and the defectors will be punished. However, the player's collective rationality will ultimately depend on their knowledge about the contract. By introducing the third person: a secretary whose task is to convey the contract to the players and who is completely truthful, we transform the two-player static game into a two-coalition dynamic game, one for the two-player and the other for secretary. In this new game, the secretary's rationality tells her to perform the strategy 'convey', which will bring her better payoff than the strategy 'not convey', since the authorities will punish her seriously if she balks at the task. In addition, the secretary is completely truthful. Thus, once the two players, Bob and Jim, learn the contract from the secretary, they will realize that the cooperation is their best choice as a coalition after knowing that their previous payoffs would have been changed. After introducing the notion of knowledge-based cooperation, we offer an S5, history-based semantics proposed by Parikh and Ramanujam [PR03] to express this notion. These enable us to represent how the contract is transmitted between the two coalitions, and how their knowledge changes, which make the players perform the collective action. The contribution of our work is that we formalize the players' reasoning in this new dynamic game and a semantic and axiomatic system is provided.

References

[Kol98] Kollock, P.: Social Dilemmas: The anatomy of Cooperation. Annual Review of Sociology 24, 183–214 (1998)

[PR03] Parikh, R., Ramanujam, R.: A Knowledge based Semantics of Messages. J. Logic, Language and Information 12, 453–467 (2003)

X. He, J. Horty, and E. Pacuit (Eds.): LORI 2009, LNAI 5834, p. 316, 2009.
© Springer-Verlag Berlin Heidelberg 2009

Getting Together: A Unified Perspective on Modal Logics for Coalitional Interaction

Cédric Dégremont and Lena Kurzen

ILLC, Universiteit van Amsterdam
cedric.uva@gmail.com, lena.kurzen@gmail.com

Cooperation of agents is a major issue in fields such as computer science, economics and philosophy. The conditions under which coalitions are formed occur in various situations involving multiple agents.

Various modal logic (ML) frameworks have been developed for reasoning about coalitional power; an important one is Coalition Logic (**CL**) [7], using modalities of the form $\langle\!\langle C \rangle\!\rangle \, \phi$ saying "coalition C has a joint strategy to ensure that ϕ". **CL** has neighborhood semantics but can be simulated on Kripke models [3]. Another class of cooperation logics explicitly represents the strategies and actions by which groups can achieve something [8].

Another crucial concept for reasoning about interactive situations is that of *preferences*. It also received attention from modal logicians ([5] surveys). Recent works (e.g. [1,2,6]) propose different mixtures of cooperation and preference logics. In such logics, many concepts from *game theory* (GT) and *social choice theory* (SCT) are commonly encountered. Depending on the situations to be modelled, different bundles of notions are important. Ability to express these notions – together with good computational behavior – make a logic appropriate for reasoning about cooperation.

Aim and Methodology. Rather than designing a new logic, we analyze how demanding SCT and GT notions are for MLs in terms of expressivity and complexity. Thus, our work helps making design choices for MLs for cooperation.

We focus on three existing classes of ML models. Then we identify notions inspired by GT and SCT concepts for reasoning about cooperation and preferences. Next, we look at how each notion can be interpreted in the models, and determine the required expressive power for expressing it in ML. This is done both by determining under which operations on models and frames the notions are invariant, and by explicitly defining each notion in some (extended) ML. This way, we obtain upper bounds on the complexity for satisfiability (SAT) and model checking (MC) problems of MLs able express the considered notions.

Three Models of Cooperation. We investigate three different classes of models. In each of them, we represent preferences as TPOs over the statespace.

1. $\wp(\mathtt{N})$-LTS (*labeled transition systems indexed by coalitions* $\wp(\mathtt{N})$)[4].
   The focus is on the interaction of preferences and cooperation; coalitional powers are primitives, directly represented in the accessibility relations.

X. He, J. Horty, and E. Pacuit (Eds.): LORI 2009, LNAI 5834, pp. 317–318, 2009.
© Springer-Verlag Berlin Heidelberg 2009

2. ABC (*action-based coalitional models*). Coalitional power is represented in terms of actions that agents can perform.
3. PBC (*power-based coalitional models*)[3]. The focus is on coalitional power itself, encoding the choices of groups as partitions of the state space.

Some Representative Results

- *"C can guarantee that the next state is one j finds at least as good as the current one."*. Whereas for $\wp(N)$-LTS, this notion can be expressed using the basic language extended with an intersection modality with SAT in PSPACE, for the other models, this is more difficult: we could only express it with formulas having alternation of $\downarrow x$ and boxes, which leads to SAT in Π_1^0.
- *Nash-stability* vs. *strong Nash-stability*. In $\wp(N)$-LTS, strong Nash-stability is easy to express (in a logic in PSPACE), but for Nash-stability seems to be only expressible in logics with undecidable (Π_1^0) SAT. For the other two models, we saw an opposite effect. Nash-stability can be expressed in logics in EXPTIME whereas the strong version seems to require logics in Π_1^0.
- *"Only coalitions containing a majority of the agents have non-trivial power."* This global property involving the inability of groups to achieve anything is difficult to axiomatize in ABC, which we could only do in logics with EXPTIME-hard SAT. For the other models, it can be done in PSPACE.

Conclusion. Our results show that in general global notions are not very demanding and mostly expressible in basic ML. Local notions however are more demanding and many are not invariant under bounded morprhisms. Our analysis shows that when designing logics for coalitional power the choice of primitives is not only conceptually important but also has an impact on complexity required to express certain notions: e.g. whether *weak* or *strong* efficiency notions are "dangerous" w.r.t. complexity, heavily depends on the choice of models.

References

1. Ågotnes, T., Dunne, P.E., van der Hoek, W., Wooldridge, M.: Logics for coalitional games. In: LORI 2007, Beijing, China (2007)
2. Ågotnes, T., Dunne, P.E., van der Hoek, W., Wooldridge, M.: Reasoning about coalitional games. Artificial Intelligence 173(1), 45–79 (2009)
3. Broersen, J., Herzig, A., Troquard, N.: Normal Coalition Logic and its conformant extension. In: TARK 2007, pp. 91–101 (2007)
4. Dégremont, C., Kurzen, L.: Modal logics for preferences and cooperation: Expressivity and complexity. In: ILLC PP-2008-39, University of Amsterdam (2008)
5. Girard, P.: Modal Logic for Preference Change. PhD thesis, Stanford (2008)
6. Kurzen, L.: Reasoning about cooperation, actions and preferences. Synthese 169(2), 223–240 (2009)
7. Pauly, M.: A modal logic for coalitional power in games. Journal of Logic and Computation 12(1), 149–166 (2002)
8. Walther, D., van der Hoek, W., Wooldridge, M.: Alternating-time temporal logic with explicit strategies. In: TARK 2007, pp. 269–278 (2007)

Oppositional Logic

Guoping Du[1,2], Hongguang Wang[1], and Jie Shen[1]

[1] Modern Logic and Application Institute,
Nanjing University, Nanjing, China
[2] Department of Computer,
Nanjing University of Aeronautics and Astronautics, Nanjing, China
dgpnju@163.com

Abstract. In intuitionistic logic system, constructive negation operator complies with the law of contradiction but not the law of excluded middle in intuitionistic logic system.In da Costa's paraconsistent logic system , paraconsistent negation operator complies with the law of excluded middle but not the law of contradiction. Putting aside classical negation operator, both intuitionistic logic and da Costa's paraconsistent logic establish logic systems by directly introducing new negation operators basing on the positive proposition logic. This paper attempts to make constructive negation operator and paraconsistent negation operator satisfying the conditions mentioned above in classical logical system.

Oppositional logic is an extended system of classical propositional logic. It can be obtained from the classical propositional logic by adding an unary connective * and introducing the definitions of two unary connectives $\triangle$ and $\triangledown$. In oppositional logic system, there are four kinds of negation: the classical negation $\neg$ complying with both law of contradiction and law of excluded middle, the constructive negation $\triangledown$ complying with law of contradiction but not law of excluded middle, the paraconsistent negation $\triangle$ complying with law of excluded middle but not law of contradiction, as well as the dialectical negation * complying with neither law of contradiction nor law of excluded middle.

This paper gives the proof of the soundness and completenesstheorem of oppositional logic. It also gives the following conclusions:

[1] Oppositonal logic can be a kind of tools for paraconsistent theory and intuitionistic theory; the famous Duns Scotus law does not hold according to the paraconsistent negation and the dialectical negation;

[2] In oppositonal logic, according to the unary connective $\neg$, * , $\triangledown$ and $\triangle$, A is in contradictory opposition with $\neg A$; A is in subaltern opposition with *A; A is in contrary opposition with $\triangledown A$; A is in subcontrary opposition with $\triangle A$. In this sense, we call the logical system mentioned above oppositional logic.

Keywords: negation, oppositional logic, paraconsistent, intuitionism.

References

1. Gabbay, D.M., Guenthner, F.: Handbook of Philosophical Logic. Kluwer Academic Publishers, Boston (2001)
2. da Costa, N.C.A., Krause, D., Bueno, O.: Paraconsistent Logics and Paraconsistency: Technical and Philosophical Developments. CLE e-prints, 4(3), Section Logic (2004), http://www.cle.unicamp.br/e-prints/vol_4,n_3,2004.html
3. Hamilton, A.G.: Logic for Mathematicians. Cambridge University Press, London (1998)
4. Du, G.: The Basis of Classical Logic and Non-classical Logic. 经典逻辑与非经典逻辑基础. Higher Education Press, Beijing (2006)

X. He, J. Horty, and E. Pacuit (Eds.): LORI 2009, LNAI 5834, p. 319, 2009.

Deliberate Contrary-to-Law Action

Qing Jia

School of Philosophy, Renmin University of China, Beijing, China 100872
v100jq@yahoo.com.cn

Abstract. Deliberate contrary-to-law action is a kind of action that caused by the inconsistent of law and personal intention. This kind of action is not encouraged, however, it offen happens in our society as a kind of irrational actions. In this paper I will use the theory of agents and choices to show that deliberate contrary-to-law action is not a kind of irrational actions. It is a pragmatic problem which can be interpreted by perspectival act utilitarianism. But perspectival act utilitarianism is based on decision theory especialy the theory of normative behavior. I think the theory of descirptive behavior is more suitable to interpret deliberate contrary-to-law action so I will alter some disciplines in John Hortys perspectival act utilitarianism to show why people do deliberate contrary-to-law action. The modified perspectival act utilitarianism can also be a kind of interpretation about irrational actions.

Keywords: deliberate contrary-to-law action, irrational action, perspectival act utilitarianism.

References

1. Aqvist, L.: Combinations of tense and deontic modality: On the Rt approach to temporal logic with historical necessity and conditional obligation. Journal of Applied Logic 3, 421–460 (2005)
2. Belnap, N., Perloff, M., Xu, M.: Facing the Future: Agents and Choices in Our Indeterminist World. Oxford University Press, Oxford (2001)
3. Davidson, D.: Essays on Actions and Events. Clarendon Press (1980)
4. Gabbay, D.M., Guenthner, F.: Handbook of Philosophical Logic, 2nd edn., vol. 8. Kluwer Academic Publishers, Dordrecht (2002)
5. Goble, L.: A logic for deontic dilemmas. Journal of Applied Logic 3, 461–483 (2005)
6. Hansen, J.: Conflicting imperatives and dyadic deontic logic. Journal of Applied Logic 3, 484–511 (2005)
7. Horty, J.: Moral Dilemmas and Nomonotonic Logic. Journal of Philosophical Logic 23, 35–65 (1994)
8. Horty, J.: Agency and deontic logic. Oxford University Press, Oxford (2001)
9. Horty, J.: Reasoning with Moral Conflicts. Nous 37, 557–605 (2003)
10. Horty, J.: Perspectival Act Utilitarianism (manuscript, 2009)

X. He, J. Horty, and E. Pacuit (Eds.): LORI 2009, LNAI 5834, p. 320, 2009.

Mono-Agent Dynamics

Sebastian Sequoiah-Grayson

[1] Postdoctoral Research Fellow, Formal Epistemology Project, Centre for Logic and Analytical Philosophy, *University of Leuven* - Belgium
[2] Senior Research Associate, IEG - Computing Laboratory, *University of Oxford*
[3] Research Associate, GPI - *University of Hertfordshire*

We model the information flow between different states of a single agent as that agent reasons deductively. K–axiom–based epistemic closure for explicit knowledge is rejected for even the most trivial cases of inferential reasoning on account of the fact that the closure axiom does not extend beyond a raw consequence relation.

The resource management of the database of agent states for the deductive reasoning fragment in question is covered by the logic corresponding to the *non–associative Lambek Calculus with permutation, bottom, and identity*: $\mathbf{NLP_{0,1}}$.

$A, B \ldots$ are types of propositional formula $\phi, \psi, \ldots$ such that $\phi : A$ is read as *formula ϕ is of type A*. The language of our logic is given as follows:

$$A \ ::= \ \phi \ | \ A \ | \ \mathbf{0} \ | \ \mathbf{1} \ | \ A \otimes B \ | \ A \multimap B \ | \ A^{\perp} \tag{1}$$

We have an information frame $\mathbf{F} \ \langle S, \sqsubseteq, \bullet \rangle$ with weak (one place) commutation and Cut. S is a set of incomplete, or partial information states $x, y, \ldots$.[1] The binary relation $\sqsubseteq$ is a partial order on S of informational development/inclusion. $\bullet$ is the (non-associative, but weakly commutative) binary composition operator on information states. $\otimes$ is merge/fusion, and $\multimap$ is implication. $^{\perp}$ is interactive negation on account of its being defined in terms of $\multimap$ and $\mathbf{0}$: $A^{\perp} := A \multimap \mathbf{0}$. $\mathbf{0}$ is bottom, and $\mathbf{1}$ is unit/identity such that: $\mathbf{1} \otimes A = A = A \otimes \mathbf{1}$.

A model $\mathbf{M} := \langle \mathbf{F}, \Vdash \rangle$ is an ordered pair $\mathbf{F} \ \langle S, \sqsubseteq, \bullet \rangle$ and $\Vdash$ such that $\Vdash$ is an evaluation relation that holds between members of S and formulas constructed out of our binary connectives $\otimes$, and $\multimap$, and constant $\mathbf{0}$. Where A is a propositional formula, and $x, y, z \in \mathbf{F}$, $\Vdash$ obeys the heredity or monotonicity condition:[2]

$$\text{For all } A, \text{if } x \Vdash A \text{ and } x \sqsubseteq y, \text{then } y \Vdash A, \tag{2}$$

And also obeys the following conditions for each of our connectives and constant $\mathbf{0}$:

$$x \Vdash A \otimes B \ \textit{iff} \text{ for some } y, z, \in \mathbf{F} \text{ s.t. } y \bullet z \sqsubseteq x, y \Vdash A \text{ and } z \Vdash B. \tag{3}$$

$$x \Vdash A \multimap B \ \textit{iff} \text{ for all } y, z \in \mathbf{F} \text{ s.t. } x \bullet y \sqsubseteq z, \text{if } y \Vdash A \text{ then } z \Vdash B. \tag{4}$$

---

[1] In doxastic cases, we would allow for inconsistency also. But since knowledge is factive if anything is, we disallow this property here.

[2] We would drop this condition for certain doxastic scenarios where non–monotonicity is a distinctive property.

X. He, J. Horty, and E. Pacuit (Eds.): LORI 2009, LNAI 5834, pp. 321–322, 2009.
© Springer-Verlag Berlin Heidelberg 2009

$x \Vdash \mathbf{0}$ for no $x \in \mathbf{F}$.   (5)

$x \Vdash A^{\perp}[A \multimap \mathbf{0}]$ *iff* for all $y, z \in \mathbf{F}$ s.t. $x \bullet y \sqsubseteq z$, if $y \Vdash A$ then $z \Vdash \mathbf{0}$.   (6)

Now set the following:

$$\phi \Rightarrow \psi : A, \sigma \Rightarrow \phi : B, \sigma : C, \phi : D, \psi : E \quad (7)$$

Allowing multiple types, then given that $\phi \Rightarrow \psi : A$ and $\sigma : C$, it is also the case that $\phi \Rightarrow \psi : C^{\perp}$. Hence $\phi \Rightarrow \psi : C \multimap \mathbf{0}$. For any state $x \Vdash C \multimap \mathbf{0}$, we know that it is that case that if we combine this information with any other state y s.t. $y \Vdash A$, then the result will be a state z s.t. $z \Vdash \mathbf{0}$ via (6). However, we know that $\mathbf{0}$ is not supported via any state via (5). Hence associativity fails.

Take $A \otimes (C \otimes B) \vdash E$, we have the following corresponding step–wise information state combination:

$x \Vdash A \otimes (C \otimes B)$ *iff for some* $w, y, z \in \mathbf{F}$ s.t. $w \bullet (y \bullet z) \sqsubseteq x$,

$w \Vdash A, y \Vdash C$, and $z \Vdash B$.   (8)

The information states $x, y, \ldots \in S$ may be naturally interpreted as states of α as α reasons deductively. In this case, the information state combination $w \bullet (y \bullet z)[\sqsubseteq x]$ specifies the step–wise reasoning procedure that α must engage in in order to be truthfully said to know the result of the merged propositions, namely ψ. Since, $y \Vdash C$, and $z \Vdash B$ via (8), and $\sigma : C$ and $\sigma \Rightarrow \phi : B$ via (7), $y \bullet z \sqsubseteq v$, where $v \Vdash D$, and $\phi : D$ via (7). Since $w \Vdash A$ via (8) and $\phi \Rightarrow \psi : A$ via (7), $w \bullet v \sqsubseteq x$, where $x \Vdash E$ and $\psi : E$ via (7).   □

We can transform interaction structures into iterated conditional information *processing structures*. From $A \otimes (C \otimes B)$, we get:

$$\mathbf{1} \vdash B \multimap (C \multimap (A \multimap E)) \quad (9)$$

Similarly to interaction structures, the processing structures/function types are individuated by information states. With respect to (9), and via (4), we have the following:

$x \Vdash B \multimap (C \multimap (A \multimap E))$ *iff for all* $s, t, v, w, y, z \in \mathbf{F}$ s.t. $((x \bullet y) \bullet v) \bullet t \sqsubseteq s$,

if $z \Vdash C \multimap (A \multimap E)$, and $y \Vdash B$, and $w \Vdash A \multimap E$, and $v \Vdash C$, and $t \Vdash A$,

then $s \Vdash E$.   (10)

Since $x \bullet y \sqsubseteq z, z \bullet v \sqsubseteq w$, and $w \bullet t \sqsubseteq s$.   □

We interpret processing structures as *instructions*, and their corresponding interaction structures as the result of carrying out or executing the corresponding instruction, i.e., as *executions*.

By following the instructions laid out in the processing structure, α can extract the very interaction structure who's "activation" will cause her to know explicitly *that* ψ. This fact has a straightforward interpretation in terms of the data–base structure, or grammar, of the "cognitive langauge" of deductive reasoning.

Modal Expressivity and Definability over Sets

Jing Shi

Institute for Modern Logic,
Central University of Finance and Economics, Beijing, China, 100081

The link between modal logic and non-well-founded sets has been shown by P. Aczel [1988], and systematically by J. Barwise and L. Moss [1996]. A. Baltag [1998] also proved some important theorems about characterizing sets by modal sentences. The aim of this paper is to explore the relationship between modal logic and sets more deeply in the expressive power of modal languages and modal definability over sets. Let's consider both basic and infinitary modal languages.

For each set a, the support of a (notation: $support(a)$) is the set of propositional letters in the transitive closure of a. A set a is pure, if $support(a) = \varnothing$. Define $V_{afa}[\Phi] := \{a : a \text{ is a set and } support(a) \subseteq \Phi\}$. The satisfaction relation $a \models \phi$ is defined recursively as follows: 1) $a \models p$ iff $p \in a$ for all propositional letter $p \in \Phi$; 2) $a \models \Diamond\phi$ iff there exists a set $b \in a$ with $b \models \phi$; 3) $a \models \bigvee \Sigma$ iff $a \models \sigma$ for some $\sigma \in \Sigma$. A formula ϕ is valid, if $a \models \phi$ for all sets $a \in V_{afa}[\Phi]$. A subset $W \subseteq V_{afa}[\Phi]$ is transitive on sets, if for any set $b \in a \in W$, $b \in W$. Write $W \models \phi$, if $a \models \phi$ for all $a \in W$.

Next, let's define some non-standard operations on set which are used to prove some preservation results. Given a set $a \in V_{afa}[\Phi]$, define the *transitive closure on sets* of $\{a\}$ (written: $STC(\{a\})$) as the minimal set satisfying the following two conditions: $a \in STC(\{a\})$; if a set $b \in c \in a$ then $b \in STC(\{a\})$. Define satisfaction relation $STC(\{a\}), x \models \phi$ like the satisfaction relation between set and formula such that $STC(\{a\}), x \models \Diamond\phi$ iff there exists $y \in x$ with $STC(\{a\}), y \models \phi$. **(i)** Given a family of sets $\{a_i\}_{i \in I}$, the *disjoint union* $\bigcup_{i \in I} STC(\{a_i\})$ of all transitive closures on sets $STC(\{a_i\})$ is defined as $\bigcup_{i \in I}(\{i\} \times STC(\{a_i\}))$. **(ii)** For all sets a, given any non-empty subset $X \subseteq STC(\{a\})$, the *subset* $Y \subseteq STC(\{a\})$ *generated from* X is defined as the minimal set satisfying conditions: $X \subseteq Y$; and if a set $a \in b \in X$ then $a \in Y$. **(iii)** Given any two sets a and b, a function $f : STC(\{a\}) \to STC(\{b\})$ is a *p-morphism*, if the following conditions hold: f is surjective; if $f(x) = y$, then $x \cap \Phi = y \cap \Phi$; if $u \in x \in STC(\{a\})$ then $f(u) \in f(x) \in STC(\{b\})$; if $v \in f(x) \in STC(\{b\})$ then there exists $u \in STC(\{a\})$ with $f(u) = v$ and $u \in x$. **(iv)** Let $b \in STC(\{a\})$. The *unraveling* set $unr(STC(\{a\}), b)$ of $STC(\{a\})$ from b is defined as the minimal set of all finite sequences of sets in $STC(\{a\})$ satisfying the following three conditions: the sequence $(b) \in unr(STC(\{a\}), b)$; if $(x_1, \ldots, x_n) \in unr(STC(\{a\}), b)$ and $x_{n+1} \in x_n$ then $(x_1, \ldots, x_{n+1}) \in unr(STC(\{a\}), b)$; p is true at $(x_1, \ldots, x_n)$ iff $p \in x_n$ for all propositional letters p. **(v)** A *bisimulation* relation Z on sets is a non-empty binary relation between sets with the following properties: if aZb then (1) for all sets $c \in a$ there exists $d \in b$ with cZd, (2) for all sets $d \in b$ there exists $c \in a$ with cZd, and (3) $a \cap \Phi = b \cap \Phi$. It is also easy to check that the above non-standard operations are special cases of bisimulation.

X. He, J. Horty, and E. Pacuit (Eds.): LORI 2009, LNAI 5834, pp. 323–324, 2009.
© Springer-Verlag Berlin Heidelberg 2009

Theorem 1. *Let ϕ be any (infinitary) modal formula, a and $\{a_i\}_{i \in I}$ sets. Then the following hold: 1) for every $x \in STC(\{a_i\})$, $\bigcup_{i \in I} STC(\{a_i\}), x \models \phi$ iff $STC(\{a_i\}), x \models \phi$; 2) given any subset $X \subseteq STC(\{a\})$, let Y be the subset of $STC(\{a\})$ generated from X. Then for any $y \in Y$, $Y, y \models \phi$ iff $STC(\{a\}), y \models \phi$; 3) let f be a p-morphism from $STC(\{a\})$ to $STC(\{b\})$. Then for any $x \in STC(\{a\})$, $STC(\{a\}), x \models \phi$ iff $STC(\{b\}), f(x) \models \phi$; 4) for any $b \in STC(\{a\})$, let $c \in STC(\{a\})$ and x be a finite sequence of sets in $STC(\{a\})$ which ends with the set c. Then $unr(STC(\{a\}), b), x \models \phi$ iff $STC(\{a\}), c \models \phi$.*

J. Barwise and L. Moss [1996] obtained one result with respect to the modal formula $T := \Box p \to p$. A set a is called *reflexive*, if $a \in a$. The class of all hereditarily reflexive sets is the largest class $HRefl$ satisfying the following condition: if $a \in HRefl$ then $a \in a$ and for all sets $b \in a$, $b \in HRefl$. The class of all hereditarily non-empty sets is the largest class HNe satisfying the following condition: if $a \in HNe$ then there exists $b \in a$ and for all sets $b \in a$, $b \in HNe$. Define the set of infinitary modal formulas Θ_T as follows: let $\Theta_0 := \{\Box \psi \to \psi : \psi$ is an infinitary modal formula$\}$, and $\Theta_{n+1} := \Theta_n \cup \{\Box \phi : \phi \in \Theta_n\}$. Define $\Theta_T = \bigcap_{n \in \omega} \Theta_n$. Then set Θ_T defines $HRefl$. The following theorem is a family of similar results. Consider the class HNe and the following: i) let $Tran$ be the class of sets satisfying the following condition: if a set $b \in a \in Tran$ then $b \in Tran$; (ii) Let Sym be the largest class of sets satisfying the following condition: if a set $b \in a \in Sym$ then $a \in b \in Sym$; (iii) define the class Euc as the largest class of sets satisfying the following condition: if a set $b \in a \in Euc$ and $c \in a$ then $b \in c \in Euc$; (iv) let WF be the class of all well-founded sets in $V_{afa}[\Phi]$.

Theorem 2. *The classes of sets HNe, $Tran$, Sym, Euc, and WF are definable in the infinitary modal language by sets of formulas defined like Θ_T through formulas $\Diamond \top$, $\Box p \to \Box \Box p$, $p \to \Box \Diamond p$, $\Diamond p \to \Box \Diamond p$, and $\Box(\Box p \to p) \to \Box p$ respectively.*

Although the idea of combining modal logic and non-well-founded set theory seems to simple, results are fruitful in this direction. One more interesting direction not contained in this paper is to introduce coalgebra into this area.

References

1. Aczel, P.: Non-Well-Founded Sets. Stanford CSLI publications (1988)
2. Baltag, A.: STS: A Structural Theory of Sets. Ph.D. dissertation, Indiana University (1998)
3. Blackburn, P., de Rijke, M., de Venema, Y.: Modal Logic. Cambridge University Press, Cambridge (2001)
4. Segerberg, K.: An Essay in Classical Modal Logic. Filosofiska Studier 13, University of Uppsala (1971)
5. Jech, T.: Set Theory. Springer, Heidelberg (2003)
6. Barwise, J., Moss, L.: Vicious Circles: On the Mathematics of Non-Well-Founded Phenomena. CSLI publications, Stanford (1996)

Dynamic Logics for Explicit and Implicit Information

Fernando R. Velázquez-Quesada

Institute for Logic, Language and Computation, Universiteit van Amsterdam.
P.O. Box 94242, 1090 GE Amsterdam, The Netherlands
F.R.VelazquezQuesada@uva.nl

1 Introduction

Classical *Epistemic Logic* (*EL*) is a compact and powerful framework for representing an agent's information. In its dynamic versions (*Dynamic Epistemic Logic*), it also describes the information flow driven by observation and communication. Nevertheless, it makes a strong idealization: the agent's information is closed under logical consequence, making truth-preserving inference uninformative. This criticism extends to its dynamics versions: acts of observation and communication provides the agent not only with the new information but also with all logical consequences of it. Thus, dynamic epistemic logics lack of an account of the step-by-step information flow driven by agent's inferences, a concern that arises not only in epistemic contexts, but also in doxastic areas. The extended version of the present abstract [1] combines ideas from the earlier literature proposing a unified framework to address these problems.

2 The Modal-Access Framework

In *EL*, an agent is informed about φ iff φ is true in all the worlds she considers possible. The definition assumes that the agent can access all true formulas of every world she considers possible, but that does not need to be the case.

Our semantic model, $M = \langle W, R, V, Y, Z \rangle$, extends a Kripke model with two functions. The first, Y, returns the set of propositional formulas the agent can access at each possible world; the second, Z, returns the set of rules (a rule is a pair (Γ, γ) with Γ its finite set of premises and γ its conclusion, all of them prop. formulas) she can apply at each world. We ask for every $\gamma \in Y(w)$ to be true at w, and for the translation TR of each rule[1] $\rho \in Z(w)$ to be true at w.[2]

Our language extends that of *EL* with two kind of formulas: $A\gamma$ is true at world w iff $\gamma \in Y(w)$, and $L\rho$ is true at w iff $\rho \in Z(w)$. The basic system K plus $A\gamma \rightarrow \gamma$ and $L\rho \rightarrow \text{TR}(\rho)$ provides a sound and complete axiomatization.

---

[1] TR is defined as an implication with the conjunction of the premises as antecedent and the conclusion as consequent.

[2] In general these properties are not preserved by model operations; that is the reason to restrict formulas and rules in Y and Z to only propositional formulas.

X. He, J. Horty, and E. Pacuit (Eds.): LORI 2009, LNAI 5834, pp. 325–326, 2009.
© Springer-Verlag Berlin Heidelberg 2009

Defining Implicit and Explicit Information. Other works have proposed similar models for representing the information of non-omniscient agents, the main difference being the definition of explicit information.

For implicit information, we say that the agent is *implicitly* informed about φ iff $\Box\varphi$ is the case. For explicit information, different from previous approaches, we use a definition with access to formulas in the scope of the modal box: the agent is *explicitly* informed about γ (γ a propositional formula) iff $\Box A\,\gamma$ holds. We can talk about implicit information about any φ in the language, but given our restriction for access sets, explicit information is limited to the propositional case. Similar definitions apply for the case of rules ($\Box\mathrm{TR}(\rho)$ and $\Box L\,\rho$, respectively).

3 Knowledge

For the notion of knowledge, we interpret R as an *indistinguishability* relation, asking for it to be reflexive, transitive and symmetric. The notions of implicit and explicit *knowledge* are given by $\Box\varphi$ and $\Box A\,\gamma$, just as before.

We define operations over the model, representing the effect of *rule-based inference* and *observation*. The first one adds the conclusion of a rule ρ to the formulas the agent can access at w whenever she can access both the rule and its premises at w. The second one has two variants: an *implicit* observation of γ simply removes those worlds where γ does not hold; an *explicit* one also adds γ to the access sets of the remaining worlds. We add to the language modalities to express the effect of these operations. Sound and complete axiom systems (based in reduction axioms) as well as properties of the operations are presented.

4 Belief

We define belief as what is true in the most plausible worlds. This is formalized by interpreting R as a *plausibility* relation and by asking for it to be reflexive, transitive and connected. Because of connectedness, the notions of implicit and explicit *belief* can be defined as $\Diamond\Box\varphi$ and $\Diamond\Box A\,\gamma$, respectively.

We define analogous operations for implicit and explicit dynamics, this time for the notion of beliefs. The *inference* operation adds the conclusion of the applied rule to the sets of those worlds where the agent can access the rule and its premises. The *upgrade* operation is defined on its implicit version: an upgrade with φ makes all φ-worlds more plausible than all $\neg\varphi$-worlds, keeping the original order within the two zones. New modalities for these new operations are added to the language, and again, sound and complete axiom systems as well as properties of the operations are presented.

Reference

1. Velázquez-Quesada, F.R.: Dynamic logics for explicit and implicit information (2009), http://staff.science.uva.nl/~fvelazqu/docs/InfBeliefs-06-25.pdf

Existence of Satisfied Alternative and the Occurring of Morph-Dictator

Zhiyuan Wang

Nanjing University, Nanjing Jiangsu 210093, P.R. China
Wangzhiyuan.nju@gmail.com

In social life of human being, individuals or collective group will be always confronted with choices. The occurring of choice implies that rational action agent (individual, collective group or social group in wide sense) must make a satisfying decision based on the alternatives set whose cardinal number is at least 2. Choice depends on preferences (Fishburn (1979)), so the nature of choice can be deemed to preference whether for individual or collective group. Preference is the ordering of alternatives given by rational agent according to his own will based on the sensibility and proneness. Preference can be crisp and fuzzy also.

The group's choice represents the reduction of the individuals' preferences to a single collective preference, i.e. to a group preference (Hwang and Lin (1987)), so group decision-making is basically an extension of the individual decision-making activity (Cheng (2004)).

Whether for individual or collective group, when facing choice agent has in mind a transcendental preferential choice complex which, as a premise but not a criterion in the theory of Feldman and Serrano (2006), says that people always choose one alternative which is preferred to or is indifferent with all the other alternatives, further more, one alternative may become satisfied alternative if it is not inferior to others. That is, rational agent will always chooses the "best" or satisfied alternative from his point of view. It is called complex of preferential choice (CPC), popularly, we also call this complex rational choice principle or rational choice presumption.

Let $C(X)$ be choice function over alternatives set X, the formal description of CPC is, $C_i(X) = x \Leftrightarrow \{x \in X \mid xR_i y, \forall y \in X\}$.

In section 2, we proposed and proved theorem existence of satisfied alternative (TESA) which says that if preferential relations satisfy completeness and IIA, then choice function is nonempty. Following TESA, we give a feasible method to attain satisfied alternative, which can be programmed through a computer to attain satisfied alternative. TESA can works well even if the set of alternatives is infinite, and the method is applicable both to individual choice and group choice.

Section 3 shows the applications of TESA to special cases in choice and constructed general models based on applications. The applications show that rational agent can achieves satisfied alternative and make well-pleasing choice because of the existence of satisfied alternative or TESA, and that TESA is an important theorem when human being confronted with choices especially the specific cases, such as Condorcet (1785) voting and choice of Buridan's Ass, M_{CC}, M_{ECC},

X. He, J. Horty, and E. Pacuit (Eds.): LORI 2009, LNAI 5834, pp. 327–328, 2009.

M_{BC}, M_{EBC} and M_{HC}, on the one hand, cases of exigency and emergency on the other hand, occurs.

However, each decision maker may have unique motivations or goals and may approach the decision process from a different angle (Evangelos (2000), Marakas (2003)), the application of TESA may lead to the occurring of Morph-dictator. For example, we can see that both choice making of Condorcet Voting Paradox and Buridan's Ass are finally consistent with preference of Morph-Dictator, in details, M_{CC}, M_{ECC}, M_{BC}, M_{EBC} and M_{HC} all implies the possibility of the occurring of Morph-Dictator in certain circumstances. When alternatives satisfied M_{CC}, M_{ECC}, M_{BC}, M_{EBC}, M_{HC}, agent can legally decides which alternative be the satisfied alternative according to his own preference, viz. Agent's preference would automatically and legally becomes social preference or collective preference if he want to realizes his dictatorship.

Morph-dictator is a concept not derived from the literature. For Morph-dictator, (1) choice function is non-emptiness for Morph-dictator as well as for traditional dictator, and, (2) Morph-Dictator is Rational, and, (3) there are conditions and non-conditions to attain satisfied alternative, and (4) choice is legitimate for Morph-dictator in a sense. We think that Morph-Dictator is rational and legal agent who, based on TESA, can automatically turn his preference into social preference through precisely logic reasoning from premises of common knowledge. Morph-Dictator is also called rational dictator or legal dictator.

Nevertheless, in addition to gain satisfied alternative before action when agent facing common choice situations, we think that TESA is a quite feasible way to ravel out Condorcet Paradox and its homothetic choice problem especially when decision must be made in a given time or when an immediate decision has to be reached.

References

[Arr]   Arrow, K.J.: Social choice and individual values. Wiley, New York (1963)
[Che]   Cheng, C.B.: Group Opinion Aggregation Based on a Grading Process: A Method for Constructing Triangular Fuzzy Numbers. Computers and Mathematics with Applications 48, 1619–1632 (2004)
[Eps]   Epstein, L.G.: The Unimportance of the Intransitivity of Seperable Preferences. International Economic Review 28, 315–322 (1978)
[Eva]   Evangelos, T.: Multi-Criteria Decision Making Methods: A Comparative Study. Kluwer Academic Publishers, Dordrecht (2000)
[Fes]   Feldman, A.M., Serrano, R.: Welfare economics and social choice theory. Springer, Heidelberg (2006)
[Fi1]   Fishburn, P.C.: Intransitive Indifference in Preference Theory: A Survey. Operations Research 18, 207–228 (1970)
[Fi2]   Fishburn, P.C.: Utility theory for decision making. Robert E. Krieger Publishing Company, Robert (1979)
[Hwl]   Hwang, C.L., Lin, M.L.: Group decision making under multiple criteria: methods and applications. Springer, Berlin (1987)
[Mar]   Marakas, G.H.: Decision Support Systems in the 21st Century. Pearson Education, Inc., New Jersey (2003)

Author Index